Graduate Texts in Mathematics 214

Graduate Texts in Mathematics

(continued after index)

Jürgen Jost

Partial Differential Equations

Second Edition

 Springer

Jürgen Jost
Max Planck Institute for Mathematics
  in the Sciences
04103 Leipzig
Germany
jjost@mis.mpg.de

Mathematics Subject Classification (2000): 35-01, 35Jxx, 35Kxx, 35Axx, 35Bxx

ISBN 978-1-4419-2380-6

e-ISBN-13: 978-0-387-49319-0

Printed on acid-free paper.

springer.com

Preface

This textbook is intended for students who wish to obtain an introduction to the theory of partial differential equations (PDEs, for short), in particular, those of elliptic type. Thus, it does not offer a comprehensive overview of the whole field of PDEs, but tries to lead the reader to the most important methods and central results in the case of elliptic PDEs. The guiding question is how one can find a solution of such a PDE. Such a solution will, of course, depend on given constraints and, in turn, if the constraints are of the appropriate type, be uniquely determined by them. We shall pursue a number of strategies for finding a solution of a PDE; they can be informally characterized as follows:

(0) *Write down an* **explicit formula** *for the solution in terms of the given data (constraints).*

This may seem like the best and most natural approach, but this is possible only in rather particular and special cases. Also, such a formula may be rather complicated, so that it is not very helpful for detecting qualitative properties of a solution. Therefore, mathematical analysis has developed other, more powerful, approaches.

(1) *Solve a sequence of auxiliary problems that* **approximate** *the given one, and show that their solutions converge to a solution of that original problem.*

Differential equations are posed in spaces of functions, and those spaces are of infinite dimension. The strength of this strategy lies in carefully choosing finite-dimensional approximating problems that can be solved explicitly or numerically and that still share important crucial features with the original problem. Those features will allow us to control their solutions and to show their convergence.

(2) *Start anywhere, with the required constraints satisfied, and let things* **flow** *toward a solution.*

This is the diffusion method. It depends on characterizing a solution of the PDE under consideration as an asymptotic equilibrium state for a diffusion process. That diffusion process itself follows a PDE, with an additional independent variable. Thus, we are solving a PDE that is more complicated than the original one. The advantage lies in the fact that we can simply start anywhere and let the PDE control the evolution.

(3) *Solve an* **optimization** *problem, and identify an optimal state as a solution of the PDE.*

This is a powerful method for a large class of elliptic PDEs, namely, for those that characterize the optima of variational problems. In fact, in applications in physics, engineering, or economics, most PDEs arise from such optimization problems. The method depends on two principles. First, one can demonstrate the existence of an optimal state for a variational problem under rather general conditions. Second, the optimality of a state is a powerful property that entails many detailed features: If the state is not very good at every point, it could be improved and therefore could not be optimal.

(4) **Connect** *what you want to know to what you know already.*

This is the continuity method. The idea is that, if you can connect your given problem continuously with another, simpler, problem that you can already solve, then you can also solve the former. Of course, the continuation of solutions requires careful control.

The various existence schemes will lead us to another, more technical, but equally important, question, namely, the one about the regularity of solutions of PDEs. If one writes down a differential equation for some function, then one might be inclined to assume explicitly or implicitly that a solution satisfies appropriate differentiability properties so that the equation is meaningful. The problem, however, with many of the existence schemes described above is that they often only yield a solution in some function space that is so large that it also contains nonsmooth and perhaps even noncontinuous functions. The notion of a solution thus has to be interpreted in some generalized sense. It is the task of regularity theory to show that the equation in question forces a generalized solution to be smooth after all, thus closing the circle. This will be the second guiding problem of the present book.

The existence and the regularity questions are often closely intertwined. Regularity is often demonstrated by deriving explicit estimates in terms of the given constraints that any solution has to satisfy, and these estimates in turn can be used for compactness arguments in existence schemes. Such estimates can also often be used to show the uniqueness of solutions, and of course, the problem of uniqueness is also fundamental in the theory of PDEs.

After this informal discussion, let us now describe the contents of this book in more specific detail.

Our starting point is the Laplace equation, whose solutions are the harmonic functions. The field of elliptic PDEs is then naturally explored as a generalization of the Laplace equation, and we emphasize various aspects on the way. We shall develop a multitude of different approaches, which in turn will also shed new light on our initial Laplace equation. One of the important approaches is the heat equation method, where solutions of elliptic PDEs are obtained as asymptotic equilibria of parabolic PDEs. In this sense, one chapter treats the heat equation, so that the present textbook definitely is not confined to elliptic equations only. We shall also treat the wave equation as the prototype of a hyperbolic PDE and discuss its relation to the Laplace and heat equations. In the context of the heat equation, another chapter develops the theory of semigroups and explains the connection with Brownian motion.

Other methods for obtaining the existence of solutions of elliptic PDEs, like the difference method, which is important for the numerical construction of solutions; the Perron method; and the alternating method of H.A. Schwarz; are based on the maximum principle. We shall present several versions of the maximum principle that are also relevant for applications to nonlinear PDEs.

In any case, it is an important guiding principle of this textbook to develop methods that are also useful for the study of nonlinear equations, as those present the research perspective of the future. Most of the PDEs occurring in applications in the sciences, economics, and engineering are of nonlinear types. One should keep in mind, however, that, because of the multitude of occurring equations and resulting phenomena, there cannot exist a unified theory of nonlinear (elliptic) PDEs, in contrast to the linear case. Thus, there are also no universally applicable methods, and we aim instead at doing justice to this multitude of phenomena by developing very diverse methods.

Thus, after the maximum principle and the heat equation, we shall encounter variational methods, whose idea is represented by the so-called Dirichlet principle. For that purpose, we shall also develop the theory of Sobolev spaces, including fundamental embedding theorems of Sobolev, Morrey, and John–Nirenberg. With the help of such results, one can show the smoothness of the so-called weak solutions obtained by the variational approach. We also treat the regularity theory of the so-called strong solutions, as well as Schauder's regularity theory for solutions in Hölder spaces. In this context, we also explain the continuity method that connects an equation that one wishes to study in a continuous manner with one that one understands already and deduces solvability of the former from solvability of the latter with the help of a priori estimates.

The final chapter develops the Moser iteration technique, which turned out to be fundamental in the theory of elliptic PDEs. With that technique one can extend many properties that are classically known for harmonic functions (Harnack inequality, local regularity, maximum principle) to solutions of a large class of general elliptic PDEs. The results of Moser will also allow us to prove the fundamental regularity theorem of de Giorgi and Nash for minimizers of variational problems.

At the end of each chapter, we briefly summarize the main results, occasionally suppressing the precise assumptions for the sake of saliency of the statements. I believe that this helps in guiding the reader through an area of mathematics that does not allow a unified structural approach, but rather derives its fascination from the multitude and diversity of approaches and methods, and consequently encounters the danger of getting lost in the technical details.

Some words about the logical dependence between the various chapters: Most chapters are composed in such a manner that only the first sections are necessary for studying subsequent chapters. The first—rather elementary—chapter, however, is basic for understanding almost all remaining chapters. Section 2.1 is useful, although not indispensable, for Chapter 3. Sections 4.1 and 4.2 are important for Chapters 6 and 7. Sections 8.1 to 8.4 are fundamental for Chapters 9 and 12, and Section 9.1 will be employed in Chapters 10 and 12. With those exceptions, the various chapters can be read independently. Thus, it is also possible to vary the order in which the chapters are studied. For example, it would make sense to read Chapter 8 directly after Chapter 1, in order to see the variational aspects of the Laplace equation (in particular, Section 8.1) and also the transformation formula for this equation with respect to changes of the independent variables. In this way one is naturally led to a larger class of elliptic equations. In any case, it is usually not very efficient to read a mathematical textbook linearly, and the reader should rather try first to grasp the central statements.

The present book can be utilized for a one-year course on PDEs, and if time does not allow all the material to be covered, one could omit certain sections and chapters, for example, Section 3.3 and the first part of Section 3.4 and Chapter 10. Of course, the lecturer may also decide to omit Chapter 12 if he or she wishes to keep the treatment at a more elementary level.

This book is based on a one-year course that I taught at the Ruhr University Bochum, with the support of Knut Smoczyk. Lutz Habermann carefully checked the manuscript and offered many valuable corrections and suggestions. The LaTeX work is due to Micaela Krieger and Antje Vandenberg.

The present book is a somewhat expanded translation of the original German version. I have also used this opportunity to correct some misprints in that version. I am grateful to Alexander Mielke, Andrej Nitsche, and Friedrich Tomi for pointing out that Lemma 4.2.3, and to C.G. Simader and Matthias Stark that the proof of Corollary 8.2.1 were incorrect in the German version.

Leipzig, Germany   Jürgen Jost

Preface to the 2nd Edition

For this new edition, I have written a new chapter on reaction-diffusion equations and systems. Such equations or systems combine a linear elliptic or parabolic differential operator, of the type extensively studied in this book, with a non-linear reaction term. The result are phenomena that can be obtained by neither of the two processes – linear diffusion or non-linear reaction as in ordinary differential equations or systems – in isolation. The patterns resulting from this interplay of local non-linear self-interactions and global diffusion in space, such as travelling waves or Turing patterns, have been proposed as models for many biological and chemical structures and processes. Therefore, such reaction-diffusion systems are very popular in mathematical biology and other fields concerned with non-linear pattern formation. In mathematical terms, their success stems from the fact that, through a combination of the PDE techniques developed in this book and some dynamical systems methods, a penetrating and often rather complete mathematical analysis can be achieved. – This new chapter is inserted after Chapter 4 that deals with linear parabolic equations, since this is the area of PDEs that is basic for studying reaction-diffusion equations. While the new chapter thus finds its most natural place there, occasionally, we also need to invoke some results from subsequent chapters, in particular from §9.5 about eigenvalues of the Laplace operator. Still, we find it preferable to discuss reaction-diffusion equations and systems at this earlier place so that we can emphasize the parabolic diffusion phenomena. This chapter also provides us with the opportunity of a glimpse at systems of PDEs as opposed to single equations. That is, we study scalar functions each of which satisfies a PDE and which are coupled through non-linear interaction terms. Of course, the field of systems of PDEs is richer than this, and more difficult couplings are possible and important, but this seems to be the point to which we can reasonably get in an introductory textbook.

I have also rewritten §11.1 (§10.1 in the previous edition, but due to the insertion of the new chapter, subsequent chapter numberings are shifted in the present edition) on the Hölder regularity of solutions of the Poisson equation. The previous proof had a problem. While that problem could have been resolved, I preferred to write a new proof based on scaling relations that is perhaps more insightful than the previous one.

The new edition also contains numerous other additions, about Neumann boundary value problems, Poincaré inequalities, expansions,..., as well as some minor (mostly typographical) corrections. I thank some careful readers for relevant comments.

Leipzig, Aug.2006   Jürgen Jost

Contents

Introduction:
What Are Partial Differential Equations?

As a first answer to the question, What are partial differential equations, we would like to give a definition:

Definition 1: *A partial differential equation (PDE) is an equation involving derivatives of an unknown function* $u\colon \Omega \to \mathbb{R}$, *where Ω is an open subset of $\mathbb{R}^d$, $d \geq 2$ (or, more generally, of a differentiable manifold of dimension $d \geq 2$).*

Often, one also considers systems of partial differential equations for vector-valued functions $u\colon \Omega \to \mathbb{R}^N$, or for mappings with values in a differentiable manifold.

The preceding definition, however, is misleading, since in the theory of PDEs one does not study arbitrary equations but concentrates instead on those equations that naturally occur in various applications (physics and other sciences, engineering, economics) or in other mathematical contexts.

Thus, as a second answer to the question posed in the title, we would like to describe some typical examples of PDEs. We shall need a little bit of notation: A partial derivative will be denoted by a subscript,

$$u_{x^i} := \frac{\partial u}{\partial x^i} \quad \text{for } i = 1, \ldots, d.$$

In case $d = 2$, we write x, y in place of x^1, x^2. Otherwise, x is the vector $x = (x^1, \ldots, x^d)$.

Examples: (1) The Laplace equation

$$\Delta u := \sum_{i=1}^{d} u_{x^i x^i} = 0 \quad (\Delta \text{ is called the Laplace operator}),$$

or, more generally, the Poisson equation

$$\Delta u = f \quad \text{for a given function} \quad f : \Omega \to \mathbb{R}.$$

For example, the real and imaginary parts u and v of a holomorphic function $u\colon \Omega \to \mathbb{C}$ ($\Omega \subset \mathbb{C}$ open) satisfy the Laplace equation. This easily follows from the Cauchy–Riemann equations:

$$u_x = v_y, \quad \text{with} \quad z = x + iy$$
$$u_y = -v_x,$$

implies

$$u_{xx} + u_{yy} = 0 = v_{xx} + v_{yy}.$$

The Cauchy–Riemann equations themselves represent a system of PDEs. The Laplace equation also models many equilibrium states in physics, and the Poisson equation is important in electrostatics.

(2) The heat equation:

Here, one coordinate t is distinguished as the "time" coordinate, while the remaining coordinates $x^1, \ldots, x^d$ represent spatial variables. We consider

$$u : \Omega \times \mathbb{R}^+ \to \mathbb{R}, \quad \Omega \text{ open in } \mathbb{R}^d, \quad \mathbb{R}^+ := \{t \in \mathbb{R} : t > 0\},$$

and pose the equation

$$u_t = \Delta u, \quad \text{where again } \Delta u := \sum_{i=1}^{d} u_{x^i x^i}.$$

The heat equation models heat and other diffusion processes.

(3) The wave equation:

With the same notation as in (2), here we have the equation

$$u_{tt} = \Delta u.$$

It models wave and oscillation phenomena.

(4) The Korteweg–de Vries equation

$$u_t - 6uu_x + u_{xxx} = 0$$

(notation as in (2), but with only one spatial coordinate x) models the propagation of waves in shallow waters.

(5) The Monge–Ampère equation

$$u_{xx}u_{yy} - u_{xy}^2 = f,$$

or in higher dimensions

$$\det \left(u_{x^i x^j} \right)_{i,j=1,\ldots,d} = f,$$

with a given function f, is used for finding surfaces (or hypersurfaces) with prescribed curvature.

(6) The minimal surface equation

$$\left(1 + u_y^2\right) u_{xx} - 2u_x u_y u_{xy} + \left(1 + u_x^2\right) u_{yy} = 0$$

describes an important class of surfaces in $\mathbb{R}^3$.

(7) The Maxwell equations for the electric field strength $E = (E_1, E_2, E_3)$ and the magnetic field strength $B = (B_1, B_2, B_3)$ as functions of (t, x^1, x^2, x^3):

$$\operatorname{div} B = 0 \qquad \text{(magnetostatic law)},$$
$$B_t + \operatorname{curl} E = 0 \qquad \text{(magnetodynamic law)},$$
$$\operatorname{div} E = 4\pi\varrho \qquad \text{(electrostatic law, } \varrho = \text{charge density)},$$
$$E_t - \operatorname{curl} E = -4\pi j \qquad \text{(electrodynamic law, } j = \text{current density)},$$

where div and curl are the standard differential operators from vector analysis with respect to the variables $(x^1, x^2, x^3) \in \mathbb{R}^3$.

(8) The Navier–Stokes equations for the velocity $v(x,t)$ and the pressure $p(x,t)$ of an incompressible fluid of density ϱ and viscosity η:

$$\varrho v_t^j + \varrho \sum_{i=1}^{3} v^i v_{x^i}^j - \eta \Delta v^j = -p_{x^j} \quad \text{for } j = 1, 2, 3,$$

$$\operatorname{div} v = 0$$

$(d = 3, v = (v^1, v^2, v^3))$.

(9) The Einstein field equations of the theory of general relativity for the curvature of the metric (g_{ij}) of space-time:

$$R_{ij} - \frac{1}{2} g_{ij} R = \kappa T_{ij} \quad \text{for } i, j = 0, 1, 2, 3 \quad \begin{array}{l} \text{(the index 0 stands for the} \\ \text{time coordinate } t = x^0). \end{array}$$

Here, κ is a constant, T_{ij} is the energy-momentum tensor (considered as given), while

$$R_{ij} := \sum_{k=0}^{3} \left(\frac{\partial}{\partial x^k} \Gamma_{ij}^k - \frac{\partial}{\partial x^j} \Gamma_{ik}^k + \sum_{l=0}^{3} \left(\Gamma_{lk}^k \Gamma_{ij}^l - \Gamma_{lj}^k \Gamma_{ik}^l \right) \right)$$

(Ricci curvature)

with

$$\Gamma_{ij}^k := \frac{1}{2} \sum_{l=0}^{3} g^{kl} \left(\frac{\partial}{\partial x^i} g_{jl} + \frac{\partial}{\partial x^j} g_{il} - \frac{\partial}{\partial x^l} g_{ij} \right)$$

and

$$(g^{ij}) := (g_{ij})^{-1} \text{ (inverse matrix)}$$

and

$$R := \sum_{i,j=0}^{3} g^{ij} R_{ij} \text{ (scalar curvature)}.$$

Thus R and R_{ij} are formed from first and second derivatives of the unknown metric (g_{ij}).

(10) The Schrödinger equation

$$i\hbar u_t = -\frac{\hbar^2}{2m} \Delta u + V(x, u)$$

(m = mass, V = given potential, $u \colon \Omega \to \mathbb{C}$) from quantum mechanics is formally similar to the heat equation, in particular in the case $V = 0$. The factor $i\,(= \sqrt{-1})$, however, leads to crucial differences.

(11) The plate equation

$$\Delta \Delta u = 0$$

even contains 4th derivatives of the unknown function.

We have now seen many rather different-looking PDEs, and it may seem hopeless to try to develop a theory that can treat all these diverse equations. This impression is essentially correct, and in order to proceed, we want to look for criteria for classifying PDEs. Here are some possibilities:

(I) Algebraically, i.e., according to the algebraic structure of the equation:
  (a) Linear equations, containing the unknown function and its derivatives only linearly. Examples (1), (2), (3), (7), (11), as well as (10) in the case where V is a linear function of u.
  An important subclass is that of the linear equations with constant coefficients. The examples just mentioned are of this type; (10), however, only if $V(x, u) = v_0 \cdot u$ with constant v_0. An example of a linear equation with nonconstant coefficients is

$$\sum_{i,j=1}^{d} \frac{\partial}{\partial x^i}\left(a^{ij}(x) u_{x^j}\right) + \sum_{i=1}^{d} \frac{\partial}{\partial x^i}\left(b^i(x) u\right) + c(x) u = 0$$

  with nonconstant functions a^{ij}, b^i, c.
  (b) Nonlinear equations.
  Important subclasses:
  – Quasilinear equations, containing the highest-occurring derivatives of u linearly. This class contains all our examples with the exception of (5).

– Semilinear equations, i.e., quasilinear equations in which the term with the highest-occurring derivatives of u does not depend on u or its lower-order derivatives. Example (6) is a quasilinear equation that is not semilinear.

Naturally, linear equations are simpler than nonlinear ones. We shall therefore first study some linear equations.

(II) According to the order of the highest-occurring derivatives:
The Cauchy–Riemann equations and (7) are of first order; (1), (2), (3), (5), (6), (8), (9), (10) are of second order; (4) is of third order; and (11) is of fourth order. Equations of higher order rarely occur, and most important PDEs are second-order PDEs. Consequently, in this textbook we shall almost exclusively study second-order PDEs.

(III) In particular, for second-order equations the following partial classifications turns out to be useful:
Let

$$F\left(x, u, u_{x^i}, u_{x^i x^j}\right) = 0$$

be a second-order PDE. We introduce dummy variables and study the function

$$F\left(x, u, p_i, p_{ij}\right).$$

The equation is called *elliptic* in Ω at $u(x)$ if the matrix

$$F_{p_{ij}}\left(x, u(x), u_{x^i}(x), u_{x^i x^j}(x)\right)_{i,j=1,\ldots,d}$$

is positive definite for all $x \in \Omega$. (If this matrix should happen to be negative definite, the equation becomes elliptic by replacing F by $-F$.) Note that this may depend on the function u. For example, if $f(x) > 0$ in (5), the equation is elliptic for any solution u with $u_{xx} > 0$. (For verifying ellipticity, one should write in place of (5)

$$u_{xx} u_{yy} - u_{xy} u_{yx} - f = 0,$$

which is equivalent to (5) for a twice continuously differentiable u.) Examples (1) and (6) are always elliptic.

The equation is called *hyperbolic* if the above matrix has precisely one negative and $(d-1)$ positive eigenvalues (or conversely, depending on a choice of sign). Example (3) is hyperbolic, and so is (5), if $f(x) < 0$, for a solution u with $u_{xx} > 0$. Example (9) is hyperbolic, too, because the metric (g_{ij}) is required to have signature $(-, +, +, +)$. Finally, an equation that can be written as

$$u_t = F(t, x, u, u_{x^i}, u_{x^i x^j})$$

with elliptic F is called *parabolic*. Note, however, that there is no longer a free sign here, since a negative definite $(F_{p_{ij}})$ is not allowed. Example

(2) is parabolic. Obviously, this classification does not cover all possible cases, but it turns out that other types are of minor importance only. Elliptic, hyperbolic, and parabolic equations require rather different theories, with the parabolic case being somewhat intermediate between the elliptic and hyperbolic ones, however.

(IV) According to solvability:

We consider a second-order PDE

$$F\left(x, u, u_{x^i}, u_{x^i x^j}\right) = 0 \text{ for } u : \Omega \to \mathbb{R},$$

and we wish to impose additional conditions upon the solution u, typically prescribing the values of u or of certain first derivatives of u on the boundary $\partial\Omega$ or part of it.

Ideally, such a boundary value problem satisfies the three conditions of Hadamard for a well-posed problem:

- Existence of a solution u for given boundary values;
- Uniqueness of this solution;
- Stability, meaning continuous dependence on the boundary values.

The third requirement is important, because in applications, the boundary data are obtained through measurements and thus are given only up to certain error margins, and small measurement errors should not change the solution drastically.

The existence requirement can be made more precise in various senses: The strongest one would be to ask that the solution be obtained by an explicit formula in terms of the boundary values. This is possible only in rather special cases, however, and thus one is usually content if one is able to deduce the existence of a solution by some abstract reasoning, for example by deriving a contradiction from the assumption of nonexistence. For such an existence procedure, often nonconstructive techniques are employed, and thus an existence theorem does not necessarily provide a rule for constructing or at least approximating some solution.

Thus, one might refine the existence requirement by demanding a constructive method with which one can compute an approximation that is as accurate as desired. This is particularly important for the numerical approximation of solutions. However, it turns out that it is often easier to treat the two problems separately, i.e., first deducing an abstract existence theorem and then utilizing the insights obtained in doing so for a constructive and numerically stable approximation scheme. Even if the numerical scheme is not rigorously founded, one might be able to use one's knowledge about the existence or nonexistence of a solution for a heuristic estimate of the reliability of numerical results.

Exercise: Find five more examples of important PDEs in the literature.

1. The Laplace Equation as the Prototype of an Elliptic Partial Differential Equation of Second Order

1.1 Harmonic Functions. Representation Formula for the Solution of the Dirichlet Problem on the Ball (Existence Techniques 0)

In this section Ω is a bounded domain in $\mathbb{R}^d$ for which the divergence theorem holds; this means that for any vector field V of class $C^1(\Omega) \cap C^0(\bar{\Omega})$,

$$\int_\Omega \operatorname{div} V(x)dx = \int_{\partial\Omega} V(z) \cdot \nu(z)do(z), \tag{1.1.1}$$

where the dot $\cdot$ denotes the Euclidean product of vectors in $\mathbb{R}^d$, ν is the exterior normal of $\partial\Omega$, and $do(z)$ is the volume element of $\partial\Omega$. Let us recall the definition of the divergence of a vector field $V = (V^1, \ldots, V^d) : \Omega \to \mathbb{R}^d$:

$$\operatorname{div} V(x) := \sum_{i=1}^d \frac{\partial V^i}{\partial x^i}(x).$$

In order that (1.1.1) hold, it is, for example, sufficient that $\partial\Omega$ be of class C^1.

Lemma 1.1.1: *Let* $u, v \in C^2(\bar{\Omega})$. *Then we have Green's 1st formula*

$$\int_\Omega v(x)\Delta u(x)dx + \int_\Omega \nabla u(x) \cdot \nabla v(x)dx = \int_{\partial\Omega} v(z)\frac{\partial u}{\partial\nu}(z)do(z) \tag{1.1.2}$$

(here, ∇u is the gradient of u), and Green's 2nd formula

$$\int_\Omega \{v(x)\Delta u(x) - u(x)\Delta v(x)\} dx = \int_{\partial\Omega} \left\{ v(z)\frac{\partial u}{\partial\nu}(z) - u(z)\frac{\partial v}{\partial\nu}(z) \right\} do(z). \tag{1.1.3}$$

Proof: With $V(x) = v(x)\nabla u(x)$, (1.1.2) follows from (1.1.1). Interchanging u and v in (1.1.2) and subtracting the resulting formula from (1.1.2) yields (1.1.3). $\square$

In the sequel we shall employ the following notation:

$$B(x,r) := \{y \in \mathbb{R}^d : |x - y| \leq r\} \qquad \text{(closed ball)}$$

and

$$\mathring{B}(x,r) := \{y \in \mathbb{R}^d : |x - y| < r\} \qquad \text{(open ball)}$$

for $r > 0$, $x \in \mathbb{R}^d$.

Definition 1.1.1: *A function $u \in C^2(\Omega)$ is called harmonic (in Ω) if*

$$\Delta u = 0 \quad \text{in } \Omega.$$

In Definition 1.1.1, Ω may be an arbitrary open subset of $\mathbb{R}^d$. We begin with the following simple observation:

Lemma 1.1.2: *The harmonic functions in Ω form a vector space.*

Proof: This follows because Δ is a *linear* differential operator.   □

Examples of harmonic functions:

(1) In $\mathbb{R}^d$, all constant functions and, more generally, all affine linear functions are harmonic.
(2) There also exist harmonic polynomials of higher order, e.g.,

$$u(x) = \left(x^1\right)^2 - \left(x^2\right)^2$$

for $x = \left(x^1, \ldots, x^d\right) \in \mathbb{R}^d$.
(3) For $x, y \in \mathbb{R}^d$ with $x \neq y$, we put

$$\Gamma(x,y) := \Gamma(|x - y|) := \begin{cases} \frac{1}{2\pi} \log |x - y| & \text{for } d = 2, \\ \frac{1}{d(2-d)\omega_d} |x - y|^{2-d} & \text{for } d > 2, \end{cases} \qquad (1.1.4)$$

where ω_d is the volume of the d-dimensional unit ball $B(0,1) \subset \mathbb{R}^d$. We have

$$\frac{\partial}{\partial x^i} \Gamma(x,y) = \frac{1}{d\omega_d} \left(x^i - y^i\right) |x - y|^{-d},$$

$$\frac{\partial^2}{\partial x^i \partial x^j} \Gamma(x,y) = \frac{1}{d\omega_d} \left\{|x - y|^2 \delta_{ij} - d \left(x^i - y^i\right) \left(x^j - y^j\right)\right\} |x - y|^{-d-2}.$$

Thus, as a function of x, Γ is harmonic in $\mathbb{R}^d \setminus \{y\}$. Since Γ is symmetric in x and y, it is then also harmonic as a function of y in $\mathbb{R}^d \setminus \{x\}$. The reason for the choice of the constants employed in (1.1.4) will become apparent after (1.1.8) below.

Definition 1.1.2: *Γ from (1.1.4) is called the fundamental solution of the Laplace equation.*

What is the reason for this particular solution Γ of the Laplace equation in $\mathbb{R}^d \setminus \{y\}$? The answer comes from the rotational symmetry of the Laplace operator. The equation

$$\Delta u = 0$$

is invariant under rotations about an arbitrary center y. (If $A \in O(d)$ (orthogonal group) and $y \in \mathbb{R}^d$, then for a harmonic $u(x)$, $u(A(x - y) + y)$ is likewise harmonic.) Because of this invariance of the operator, one then also searches for invariant solutions, i.e., solutions of the form

$$u(x) = \varphi(r) \quad \text{with } r = |x - y|.$$

The Laplace equation then is transformed into the following equation for y as a function of r, with $'$ denoting a derivative with respect to r,

$$\varphi''(r) + \frac{d-1}{r}\varphi'(r) = 0.$$

Solutions have to satisfy

$$\varphi'(r) = cr^{1-d}$$

with constant c. Fixing this constant plus one further additive constant leads to the fundamental solution $\Gamma(r)$.

Theorem 1.1.1 (Green representation formula): *If $u \in C^2(\bar{\Omega})$, we have for $y \in \Omega$,*

$$u(y) = \int_{\partial\Omega} \left\{ u(x)\frac{\partial\Gamma}{\partial\nu_x}(x,y) - \Gamma(x,y)\frac{\partial u}{\partial\nu}(x) \right\} do(x) + \int_{\Omega} \Gamma(x,y)\Delta u(x)dx \tag{1.1.5}$$

(here, the symbol $\frac{\partial}{\partial\nu_x}$ indicates that the derivative is to be taken in the direction of the exterior normal with respect to the variable x).

Proof: For sufficiently small $\varepsilon > 0$,

$$B(y, \varepsilon) \subset \Omega,$$

since Ω is open. We apply (1.1.3) for $v(x) = \Gamma(x, y)$ and $\Omega \setminus B(y, \varepsilon)$ (in place of Ω). Since Γ is harmonic in $\Omega \setminus \{y\}$, we obtain

$$\int_{\Omega\setminus B(y,\varepsilon)} \Gamma(x,y)\Delta u(x)dx = \int_{\partial\Omega} \left\{ \Gamma(x,y)\frac{\partial u}{\partial\nu}(x) - u(x)\frac{\partial\Gamma(x,y)}{\partial\nu_x} \right\} do(x)$$

$$+ \int_{\partial B(y,\varepsilon)} \left\{ \Gamma(x,y)\frac{\partial u}{\partial\nu}(x) - u(x)\frac{\partial\Gamma(x,y)}{\partial\nu_x} \right\} do(x). \tag{1.1.6}$$

In the second boundary integral, ν denotes the exterior normal of $\Omega \setminus B(y, \varepsilon)$, hence the interior normal of $B(y, \varepsilon)$.

We now wish to evaluate the limits of the individual integrals in this formula for $\varepsilon \to 0$. Since $u \in C^2(\bar{\Omega})$, Δu is bounded. Since Γ is integrable, the left-hand side of (1.1.6) thus tends to

$$\int_\Omega \Gamma(x, y) \Delta u(x) dx.$$

On $\partial B(y, \varepsilon)$, we have $\Gamma(x, y) = \Gamma(\varepsilon)$. Thus, for $\varepsilon \to 0$,

$$\left| \int_{\partial B(y,\varepsilon)} \Gamma(x, y) \frac{\partial u}{\partial \nu}(x) do(x) \right| \leq d\omega_d \varepsilon^{d-1} \Gamma(\varepsilon) \sup_{B(y,\varepsilon)} |\nabla u| \to 0.$$

Furthermore,

$$-\int_{\partial B(y,\varepsilon)} u(x) \frac{\partial \Gamma(x, y)}{\partial \nu_x} do(x) = \frac{\partial}{\partial \varepsilon} \Gamma(\varepsilon) \int_{\partial B(y,\varepsilon)} u(x) do(x)$$

$$\text{(since } \nu \text{ is the interior normal of } B(y, \varepsilon))$$

$$= \frac{1}{d\omega_d \varepsilon^{d-1}} \int_{\partial B(y,\varepsilon)} u(x) do(x) \to u(y).$$

Altogether, we get (1.1.5).   □

Remark: Applying the Green representation formula for a so-called test function $\varphi \in C_0^\infty(\Omega)$,[1] we obtain

$$\varphi(y) = \int_\Omega \Gamma(x, y) \Delta\varphi(x) dx. \tag{1.1.7}$$

This can be written symbolically as

$$\Delta_x \Gamma(x, y) = \delta_y, \tag{1.1.8}$$

where Δ_x is the Laplace operator with respect to x, and δ_y is the Dirac delta distribution, meaning that for $\varphi \in C_0^\infty(\Omega)$,

$$\delta_y[\varphi] := \varphi(y).$$

In the same manner, $\Delta\Gamma(\cdot, y)$ is defined as a distribution, i.e.,

$$\Delta\Gamma(\cdot, y)[\varphi] := \int_\Omega \Gamma(x, y) \Delta\varphi(x) dx.$$

Equation (1.1.8) explains the terminology "fundamental solution" for Γ, as well as the choice of constant in its definition.

---

[1] $C_0^\infty(\Omega) := \{f \in C^\infty(\Omega), \operatorname{supp}(f) := \overline{\{x : f(x) \neq 0\}} \text{ is a compact subset of } \Omega\}.$

Remark: By definition, a distribution is a linear functional ℓ on C_0^∞ that is continuous in the following sense:
Suppose that $(\varphi_n)_{n\in\mathbb{N}} \subset C_0^\infty(\Omega)$ satisfies $\varphi_n = 0$ on $\Omega \setminus K$ for all n and some fixed compact $K \subset \Omega$ as well as $\lim_{n\to\infty} D^\alpha \varphi_n(x) = 0$ uniformly in x for all partial derivatives D^α (of arbitrary order). Then

$$\lim_{n\to\infty} \ell[\varphi_n] = 0$$

must hold.

We may draw the following consequence from the Green representation formula: If one knows Δu, then u is completely determined by its values and those of its normal derivative on $\partial\Omega$. In particular, a harmonic function on Ω can be reconstructed from its boundary data. One may then ask conversely whether one can construct a harmonic function for arbitrary given values on $\partial\Omega$ for the function and its normal derivative. Even ignoring the issue that one might have to impose certain regularity conditions like continuity on such data, we shall find that this is not possible in general, but that one can prescribe essentially only one of these two data. In any case, the divergence theorem (1.1.1) for $V(x) = \nabla u(x)$ implies that because of $\Delta = \operatorname{div} \operatorname{grad}$, a harmonic u has to satisfy

$$\int_{\partial\Omega} \frac{\partial u}{\partial \nu} do(x) = \int_\Omega \Delta u(x) dx = 0, \tag{1.1.9}$$

so that the normal derivative cannot be prescribed completely arbitrarily.

Definition 1.1.3: *A function $G(x,y)$, defined for $x, y \in \bar{\Omega}$, $x \neq y$, is called a Green function for Ω if*

(1) $G(x,y) = 0$ for $x \in \partial\Omega$;
(2) $h(x,y) := G(x,y) - \Gamma(x,y)$ is harmonic in $x \in \Omega$ (thus in particular also at the point $x = y$).

We now assume that a Green function $G(x,y)$ for Ω exists (which indeed is true for all Ω under consideration here), and put $v(x) = h(x,y)$ in (1.1.3) and add the result to (1.1.5), obtaining

$$u(y) = \int_{\partial\Omega} u(x) \frac{\partial G(x,y)}{\partial \nu_x} do(x) + \int_\Omega G(x,y) \Delta u(x) dx. \tag{1.1.10}$$

Equation (1.1.10) in particular implies that a harmonic u is already determined by its boundary values $u_{|\partial\Omega}$.

This construction now raises the converse question: If we are given functions $\varphi : \partial\Omega \to \mathbb{R}$, $f : \Omega \to \mathbb{R}$, can we obtain a solution of the Dirichlet problem for the Poisson equation

$$\begin{aligned} \Delta u(x) &= f(x) \quad \text{for } x \in \Omega, \\ u(x) &= \varphi(x) \quad \text{for } x \in \partial\Omega, \end{aligned} \tag{1.1.11}$$

by the representation formula

$$u(y) = \int_{\partial\Omega} \varphi(x)\frac{\partial G(x,y)}{\partial\nu_x}do(x) + \int_{\Omega} f(x)G(x,y)dx? \qquad (1.1.12)$$

After all, if u is a solution, it does satisfy this formula by (1.1.10).

Essentially, the answer is yes; to make it really work, however, we need to impose some conditions on φ and f. A natural condition should be the requirement that they be continuous. For φ, this condition turns out to be sufficient, provided that the boundary of Ω satisfies some mild regularity requirements. If Ω is a ball, we shall verify this in Theorem 1.1.2 for the case $f = 0$, i.e., the Dirichlet problem for harmonic functions. For f, the situation is slightly more subtle. It turns out that even if f is continuous, the function u defined by (1.1.12) need not be twice differentiable, and so one has to exercise some care in assigning a meaning to the equation $\Delta u = f$. We shall return to this issue in Sections 10.1 and 11.1 below. In particular, we shall show that if we require a little more about f, namely, that it be Hölder continuous, then the function u given by (1.1.12) is twice continuously differentiable and satisfies

$$\Delta u = f.$$

Analogously, if $H(x,y)$ for $x,y \in \bar{\Omega}$, $x \neq y$ is defined with[2]

$$\frac{\partial}{\partial\nu_x}H(x,y) = \frac{-1}{\|\partial\Omega\|} \quad \text{for } x \in \partial\Omega$$

and a harmonic difference $H(x,y) - \Gamma(x,y)$ as before, we obtain

$$u(y) = \frac{1}{\|\partial\Omega\|}\int_{\partial\Omega} u(x)do(x) - \int_{\partial\Omega} H(x,y)\frac{\partial u}{\partial\nu}(x)do(x)$$
$$+ \int_{\Omega} H(x,y)\Delta u(x)dx. \quad (1.1.13)$$

If now u_1 and u_2 are two harmonic functions with

$$\frac{\partial u_1}{\partial\nu} = \frac{\partial u_2}{\partial\nu} \quad \text{on } \partial\Omega,$$

applying (1.1.13) to the difference $u = u_1 - u_2$ yields

$$u_1(y) - u_2(y) = \frac{1}{\|\partial\Omega\|}\int_{\partial\Omega} (u_1(x) - u_2(x))\,do(x). \qquad (1.1.14)$$

Since the right-hand side of (1.1.14) is independent of y, $u_1 - u_2$ must be constant in Ω. In other words, a solution of the Neumann boundary value problem

---

[2] Here, $\|\partial\Omega\|$ denotes the measure of the boundary $\partial\Omega$ of Ω; it is given as $\int_{\partial\Omega} do(x)$.

$$\Delta u(x) = 0 \qquad \text{for } x \in \Omega,$$
$$\frac{\partial u}{\partial \nu} = g(x) \quad \text{for } x \in \partial\Omega \tag{1.1.15}$$

is determined only up to a constant, and, conversely, by (1.1.9), a necessary condition for the existence of a solution is

$$\int_{\partial\Omega} g(x) do(x) = 0. \tag{1.1.16}$$

Boundary conditions tend to make the theory of PDEs difficult. Actually, in many contexts, the Neumann condition is more natural and easier to handle than the Dirichlet condition, even though we mainly study Dirichlet boundary conditions in this book as those occur more frequently. – There is in fact another, even easier, boundary condition, which actually is not a boundary condition at all, the so-called periodic boundary condition. This means the following. We consider a domain of the form $\Omega = (0, L_1) \times \cdots \times (0, L_d) \subset \mathbb{R}^d$ and require for $u : \bar\Omega \to \mathbb{R}$ that

$$u(x_1, \ldots, x_{i-1}, L_i, x_{i+1}, \ldots, x_d) = u(x_1, \ldots, x_{i-1}, 0, x_{i+1}, \ldots, x_d) \quad (1.1.17)$$

for all $x = (x_1, \ldots, x_d) \in \Omega$, $i = 1, \ldots, d$. This means that u can be periodically extended from Ω to all of $\mathbb{R}^d$. A reader familiar with basic geometric concepts will view such a u as a function on the torus obtained by identifying opposite sides in Ω. More generally, one may then consider solutions of PDEs on compact manifolds.

Anyway, we now turn to the Dirichlet problem on a ball. As a preparation, we compute the Green function G for such a ball $B(0, R)$. For $y \in \mathbb{R}^d$, we put

$$\bar{y} := \begin{cases} \frac{R^2}{|y|^2} y & \text{for } y \neq 0, \\ \infty & \text{for } y = 0. \end{cases}$$

($\bar{y}$ is the point obtained from y by reflection across $\partial B(0, R)$.) We then put

$$G(x, y) := \begin{cases} \Gamma(|x - y|) - \Gamma\left(\frac{|y|}{R} |x - \bar{y}|\right) & \text{for } y \neq 0, \\ \Gamma(|x|) - \Gamma(R) & \text{for } y = 0. \end{cases} \tag{1.1.18}$$

For $x \neq y$, $G(x, y)$ is harmonic in x, since for $y \in \mathring{B}(0, R)$, the point $\bar{y}$ lies in the exterior of $B(0, R)$. The function $G(x, y)$ has only one singularity in $B(0, R)$, namely at $x = y$, and this singularity is the same as that of $\Gamma(x, y)$. The formula

$$G(x, y) = \Gamma\left(\left(|x|^2 + |y|^2 - 2x \cdot y\right)^{1/2}\right) - \Gamma\left(\left(\frac{|x|^2 |y|^2}{R^2} + R^2 - 2x \cdot y\right)^{1/2}\right)$$

$$\tag{1.1.19}$$

then shows that for $x \in \partial B(0, R)$, i.e., $|x| = R$, we have indeed

$$G(x, y) = 0.$$

Therefore, the function $G(x, y)$ defined by (1.1.18) is the Green function of $B(0, R)$.

Equation (1.1.19) also implies the symmetry

$$G(x, y) = G(y, x). \tag{1.1.20}$$

Furthermore, since $\Gamma(|x-y|)$ is monotonic in $|x-y|$, we conclude from (1.1.19) that

$$G(x, y) \leq 0 \quad \text{for } x, y \in B(0, R). \tag{1.1.21}$$

Since for $x \in \partial B(0, R)$,

$$|x|^2 + |y|^2 - 2x \cdot y = \frac{|x|^2 |y|^2}{R^2} + R^2 - 2x \cdot y,$$

(1.1.19) furthermore implies for $x \in \partial B(0, R)$ that

$$\frac{\partial}{\partial \nu_x} G(x, y) = \frac{\partial}{\partial |x|} G(x, y) = \frac{1}{d\omega_d} \frac{|x|}{|x - y|^d} - \frac{1}{d\omega_d} \frac{|x|}{|x - y|^d} \frac{|y|^2}{R^2}$$
$$= \frac{R^2 - |y|^2}{d\omega_d R} \frac{1}{|x - y|^d}.$$

Inserting this result into (1.1.10), we obtain a representation formula for a harmonic $u \in C^2(B(0, R))$ in terms of its boundary values on $\partial B(0, R)$:

$$u(y) = \frac{R^2 - |y|^2}{d\omega_d R} \int_{\partial B(0,R)} \frac{u(x)}{|x - y|^d} do(x). \tag{1.1.22}$$

The regularity condition here can be weakened; in fact, we have the following theorem:

Theorem 1.1.2 (Poisson representation formula; solution of the Dirichlet problem on the ball): *Let* $\varphi : \partial B(0, R) \to \mathbb{R}$ *be continuous. Then* u, *defined by*

$$u(y) := \begin{cases} \frac{R^2 - |y|^2}{d\omega_d R} \int_{\partial B(0,R)} \frac{\varphi(x)}{|x-y|^d} do(x) & \text{for } y \in \mathring{B}(0, R), \\ \varphi(y) & \text{for } y \in \partial B(0, R), \end{cases} \tag{1.1.23}$$

is harmonic in the open ball $\mathring{B}(0, R)$ *and continuous in the closed ball* $B(0, R)$.

Proof: Since G is harmonic in y, so is the kernel of the Poisson representation formula

$$K(x,y) := \frac{\partial G}{\partial \nu_x}(x,y) = \frac{R^2 - |y|^2}{d\omega_d R}\,|x-y|^{-d}.$$

Thus u is harmonic as well.

It remains only to show continuity of u on $\partial B(0,R)$. We first insert the harmonic function $u \equiv 1$ in (1.1.22), yielding

$$\int_{\partial B(0,R)} K(x,y)do(x) = 1 \quad \text{for all } y \in \mathring{B}(0,R). \tag{1.1.24}$$

We now consider $y_0 \in \partial B(0,R)$. Since y is continuous, for every $\varepsilon > 0$ there exists $\delta > 0$ with

$$|\varphi(y) - \varphi(y_0)| < \frac{\varepsilon}{2} \quad \text{for } |y - y_0| < 2\delta. \tag{1.1.25}$$

With

$$\mu := \sup_{y \in \partial B(0,R)} |\varphi(y)|,$$

by (1.1.23), (1.1.24) we have for $|y - y_0| < \delta$ that

$$\left| u(y) - u(y_0) \right| = \left| \int_{\partial B(0,R)} K(x,y)\,(\varphi(x) - \varphi(y_0))\,do(x) \right|$$

$$\leq \int_{|x-y_0| \leq 2\delta} K(x,y)\,|\varphi(x) - \varphi(y_0)|\,do(x)$$

$$+ \int_{|x-y_0| > 2\delta} K(x,y)\,|\varphi(x) - \varphi(y_0)|\,do(x)$$

$$\leq \frac{\varepsilon}{2} + 2\mu\left(R^2 - |y|^2\right)R^{d-1}. \tag{1.1.26}$$

(For estimating the second integral, note that because of $|y - y_0| < \delta$, for $|x - y_0| > 2\delta$ also $|x - y| \geq \delta$.) Since $|y_0| = R$, for sufficiently small $|y - y_0|$ then also the second term on the right-hand side of (1.1.26) becomes smaller than $\varepsilon/2$, and we see that u is continuous at y_0. □

Corollary 1.1.1: *For $\varphi \in C^0(\partial B(0,R))$, there exists a unique solution $u \in C^2(\mathring{B}(0,R)) \cap C^0(B(0,R))$ of the Dirichlet problem*

$$\Delta u(x) = 0 \qquad \text{for } x \in \mathring{B}(0,R),$$
$$u(x) = \varphi(x) \quad \text{for } x \in \partial B(0,R).$$

Proof: Theorem 1.1.2 shows the existence. Uniqueness follows from (1.1.10); however, in (1.1.10) we have assumed $u \in C^2(B(0,R))$, while more generally, here we consider continuous boundary values. This difficulty is easily overcome: Since u is harmonic in $\mathring{B}(0,R)$, it is of class C^2 in $\mathring{B}(0,R)$, for example by Corollary 1.1.2 below. Consequently, for $|y| < r < R$, applying (1.1.22) with r in place of R, we get

$$u(y) = \frac{r^2 - |y|^2}{d\omega_d r} \int_{\partial B(0,r)} \frac{u(x)}{|x-y|^d} do(x),$$

and since u is continuous in $B(0,R)$, we may let r tend to R in order to get the representation formula in its full generality. □

Corollary 1.1.2: *Any harmonic function $u : \Omega \to \mathbb{R}$ is real analytic in Ω.*

Proof: Let $z \in \Omega$ and choose R such that $B(z,R) \subset \Omega$. Then by (1.1.22), for $y \in \mathring{B}(z,R)$,

$$u(y) = \frac{R^2 - |y-z|^2}{d\omega_d R} \int_{\partial B(z,R)} \frac{u(x)}{|x-y|^d} do(x),$$

which is a real analytic function of $y \in \mathring{B}(z,R)$. □

1.2 Mean Value Properties of Harmonic Functions. Subharmonic Functions. The Maximum Principle

Theorem 1.2.1 (Mean value formulae): *A continuous $u : \Omega \to \mathbb{R}$ is harmonic if and only if for any ball $B(x_0, r) \subset \Omega$,*

$$u(x_0) = S(u, x_0, r) := \frac{1}{d\omega_d r^{d-1}} \int_{\partial B(x_0, r)} u(x) do(x) \quad \text{(spherical mean)},$$
$$\text{(1.2.1)}$$

or equivalently, if for any such ball

$$u(x_0) = K(u, x_0, r) := \frac{1}{\omega_d r^d} \int_{B(x_0, r)} u(x) dx \quad \text{(ball mean)}. \quad \text{(1.2.2)}$$

Proof: "$\Rightarrow$":
Let u be harmonic. Then (1.2.1) follows from Poisson's formula (1.1.22) (since we have written (1.1.22) only for the ball $B(0,R)$, take the harmonic function $v(x) := u(x + x_0)$ and apply the formula at the point $x = 0$). Alternatively,

we may prove (1.2.1) from the following observation:
Let $u \in C^2(\mathring{B}(y,r))$, $0 < \varrho < r$. Then by (1.1.1)

$$\int_{B(y,\varrho)} \Delta u(x)dx = \int_{\partial B(y,\varrho)} \frac{\partial u}{\partial \nu}(x)do(x)$$

$$= \int_{\partial B(0,1)} \frac{\partial u}{\partial \varrho}(y + \varrho\omega)\varrho^{d-1}d\omega$$

in polar coordinates $\omega = \dfrac{x-y}{\varrho}$

$$= \varrho^{d-1}\frac{\partial}{\partial \varrho}\int_{\partial B(0,1)} u(y + \varrho\omega)d\omega$$

$$= \varrho^{d-1}\frac{\partial}{\partial \varrho}\left(\varrho^{1-d}\int_{\partial B(y,\varrho)} u(x)do(x)\right)$$

$$= d\omega_d\varrho^{d-1}\frac{\partial}{\partial \varrho}S(u,y,\varrho). \tag{1.2.3}$$

If u is harmonic, this yields $\frac{\partial}{\partial \varrho}S(u,y,\varrho) = 0$, and so $S(u,y,\varrho)$ is constant in ρ. Because of

$$u(y) = \lim_{\varrho \to 0} S(u,y,\varrho), \tag{1.2.4}$$

for a continuous u this implies the spherical mean value property. Because of

$$K(u,x_0,r) = \frac{d}{r^d}\int_0^r S(u,x_0,\varrho)\varrho^{d-1}d\varrho \tag{1.2.5}$$

we also get (1.2.2) if (1.2.1) holds for all radii ϱ with $B(x_0,\varrho) \subset \Omega$.
"$\Leftarrow$":

We have just seen that the spherical mean value property implies the ball mean value property. The converse also holds:
If $K(u,x_0,r)$ is constant as a function of r, i.e., by (1.2.5)

$$0 = \frac{\partial}{\partial r}K(u,x_0,r) = \frac{d}{r}S(u,x_0,r) - \frac{d}{r}K(u,x_0,r),$$

then $S(u,x_0,r)$ is likewise constant in r, and by (1.2.4) it thus always has to equal $u(x_0)$.

Suppose now (1.2.1) for $B(x_0,r) \subset \Omega$. We want to show first that u then has to be smooth. For this purpose, we use the following general construction:
Put

$$\varrho(t) := \begin{cases} c_d \exp\left(\frac{1}{t^2-1}\right) & \text{if } 0 \leq t < 1, \\ 0 & \text{otherwise}, \end{cases}$$

where the constant c_d is chosen such that

$$\int_{\mathbb{R}^d} \varrho(|x|)dx = 1.$$

The reader should note that $\varrho(|x|)$ is infinitely differentiable with respect to x. For $f \in L^1(\Omega)$, $B(y,r) \subset \Omega$, $\overline{B(y,r)} \subset \Omega$ we consider the so-called mollification

$$f_r(y) := \frac{1}{r^d} \int_\Omega \varrho\left(\frac{|y - x|}{r}\right) f(x)dx. \tag{1.2.6}$$

Then f_r is infinitely differentiable with respect to y.

If now (1.2.1) holds, we have

$$u_r(y) = \frac{1}{r^d} \int_0^r \int_{\partial B(y,s)} \varrho\left(\frac{s}{r}\right) u(x)do(x)ds$$

$$= \frac{1}{r^d} \int_0^r \varrho\left(\frac{s}{r}\right) d\omega_d s^{d-1} S(u,y,s)ds$$

$$= u(y) \int_0^1 \varrho(\sigma)d\omega_d \sigma^{d-1} d\sigma$$

$$= u(y) \int_{B(0,1)} \varrho(|x|) dx$$

$$= u(y).$$

Thus a function satisfying the mean value property also satisfies

$$u_r(x) = u(x), \quad \text{provided that } B(x,r) \subset \Omega.$$

Thus, with u_r also u is infinitely differentiable. We may thus again consider (1.2.3), i.e.,

$$\int_{B(y,\varrho)} \Delta u(x)dx = d\omega_d \varrho^{d-1} \frac{\partial}{\partial \varrho} S(u,y,\varrho). \tag{1.2.7}$$

If (1.2.7) holds, then $S(u,x_0,\varrho)$ is constant in ϱ, and therefore, the right-hand side of (1.2.7) vanishes for all y and ϱ with $B(y,\varrho) \subset \Omega$. Thus, also

$$\Delta u(y) = 0$$

for all $y \in \Omega$, and u is harmonic.   □

Instead of requiring that u be continuous, it suffices to require that u be measurable and locally integrable in Ω. The preceding theorem and its proof then remain valid since in the second part we have not used the continuity of u.

With this observation, we easily obtain the following corollary:

Corollary 1.2.1 (Weyl's lemma): *Let $u : \Omega \to \mathbb{R}$ be measurable and lo-cally integrable in Ω. Suppose that for all $\varphi \in C_0^\infty(\Omega)$,*

$$\int_\Omega u(x)\Delta\varphi(x)dx = 0.$$

Then u is harmonic and, in particular, smooth.

Proof: We again consider the mollifications

$$u_r(x) = \frac{1}{r^d} \int_\Omega \varrho\left(\frac{|y-x|}{r}\right) u(y) dy.$$

For $\varphi \in C_0^\infty$ and $r < \text{dist}(\text{supp}(\varphi), \partial\Omega)$, we obtain

$$\int_\Omega u_r(x)\Delta\varphi(x)dx = \int_\Omega \frac{1}{r^d} \int_\Omega \varrho\left(\frac{|y-x|}{r}\right) u(y) dy \Delta\varphi(x)dx$$

$$= \int_\Omega u(y)\Delta\varphi_r(y)dy$$

exchanging the integrals and observing that $(\Delta\varphi)_r = \Delta(\varphi_r)$, so that the Laplace operator commutes with the mollification

$$= 0,$$

since by our assumption for r also $\varphi_r \in C_0^\infty(\Omega)$.

Since u_r is smooth, this also implies

$$\int_\Omega \Delta u_r(x)\varphi(x)dx = 0 \quad \text{for all } \varphi \in C_0^\infty(\Omega_r),$$

with $\Omega_r := \{x \in \Omega : \text{dist}(x, \partial\Omega) > r\}$.

Hence,

$$\Delta u_r = 0 \quad \text{in } \Omega_r.$$

Thus, u_r is harmonic in Ω_r.

We consider $R > 0$ and $0 < r \leq \frac{1}{2}R$. Then u_r satisfies the mean value property on any ball with center in Ω_r and radius $\leq \frac{1}{2}R$. Since

$$\int_{\Omega_r} |u_r(y)| \, dy \leq \int_{\Omega_r} \frac{1}{r^d} \int_\Omega \varrho\left(\frac{|x-y|}{r}\right) |u(x)| \, dx \, dy$$

$$\leq \int_\Omega |u(x)| \, dx$$

obtained by exchanging the integrals and using $\int_{\mathbb{R}^d} \frac{1}{r^d}\varrho\left(\frac{|x-y|}{r}\right) dy = 1$, the u_r have uniformly bounded norms in $L^1(\Omega)$, if $u \in L^1(\Omega)$. If u is only locally integrable, the preceding reasoning has to be applied locally in Ω, in order to get the local uniform integrability of the u_r. Since this is easily done, we assume for simplicity $u \in L^1(\Omega)$.

Since the u_r satisfy the mean value property on balls of radius $\frac{1}{2}R$, this implies that they are also uniformly bounded (keeping R fixed and letting r tend to 0). Furthermore, because of

$$|u_r(x_1) - u_r(x_2)| \leq \frac{1}{\omega_d} \left(\frac{2}{R}\right)^d \int_{\substack{B(x_1,R/2)\backslash B(x_2,R/2) \\ \cup B(x_2,R/2)\backslash B(x_1,R/2)}} |u_r(x)| \, dx$$

$$\leq \frac{1}{\omega_d} \left(\frac{2}{R}\right)^d \sup|u_r| \, 2\mathrm{Vol}\,(B(x_1,R/2) \backslash B(x_2,R/2)),$$

the u_r are also equicontinuous. Thus, by the Arzela–Ascoli theorem, for $r \to 0$, a subsequence of the u_r converges uniformly towards some continuous function v. We must have $u = v$, because u is (locally) in $L^1(\Omega)$, and so for almost all $x \in \Omega$, $u(x)$ is the limit of $u_r(x)$ for $r \to 0$ (cf. Lemma A.3). Thus, u is continuous, and since all the u_r satisfy the mean value property, so does u. Theorem 1.2.1 now implies the claim.   □

Definition 1.2.1: *Let $v : \Omega \to [-\infty, \infty)$ be upper semicontinuous, but not identically $-\infty$. Such a v is called* subharmonic *if for every subdomain $\Omega' \subset\subset \Omega$ and every harmonic function $u : \Omega' \to \mathbb{R}$ (we assume $u \in C^0(\bar{\Omega}')$) with*

$$v \leq u \quad on \ \partial\Omega'$$

we have

$$v \leq u \quad on \ \Omega'.$$

A function $w : \Omega \to (-\infty, \infty]$, lower semicontinuous, $w \not\equiv \infty$, is called superharmonic *if $-w$ is subharmonic.*

Theorem 1.2.2: *A function $v : \Omega \to [-\infty, \infty)$ (upper semicontinuous, $\not\equiv -\infty$) is subharmonic if and only if for every ball $B(x_0, r) \subset \Omega$,*

$$v(x_0) \leq S(v, x_0, r), \tag{1.2.8}$$

or, equivalently, if for every such ball

$$v(x_0) \leq K(v, x_0, r). \tag{1.2.9}$$

Proof: "⇒"
Since v is upper semicontinuous, there exists a monotonically decreasing sequence $(v_n)_{n \in \mathbb{N}}$ of continuous functions with $v = \lim_{n \in \mathbb{N}} v_n$. By Theorem 1.1.2, for every u, there exists a harmonic

$$u_n : B(x_0, r) \to \mathbb{R}$$

with

$$u_n|_{\partial B(x_0,r)} = v_n|_{\partial B(x_0,r)} \quad \left(\geq v|_{\partial B(x_0,r)}\right);$$

hence, in particular,

$$S(u_n, x_0, r) = S(v_n, x_0, r).$$

Since v is subharmonic and u_n is harmonic, we obtain

$$v(x_0) \leq u_n(x_0) = S(u_n, x_0, r) = S(v_n, x_0, r).$$

Now $n \to \infty$ yields (1.2.8). The mean value inequality for balls follows from that for spheres (cf. (1.2.5)). For the converse direction, we employ the following lemma:

Lemma 1.2.1: *Suppose v satisfies the mean value inequality (1.2.8) or (1.2.9) for all $B(x_0,r) \subset \Omega$. Then v also satisfies the maximum principle, meaning that if there exists some $x_0 \in \Omega$ with*

$$v(x_0) = \sup_{x \in \Omega} v(x),$$

then v is constant. In particular, if Ω is bounded and $v \in C^0(\bar{\Omega})$, then

$$v(x) \leq \max_{y \in \partial \Omega} v(y) \quad \text{for all } x \in \Omega.$$

Remark: We shall soon see that the assumption of Lemma 1.2.1 is equivalent to v being subharmonic, and therefore, the lemma will hold for subharmonic functions.

Proof: Assume

$$v(x_0) = \sup_{x \in \Omega} v(x) =: M.$$

Thus,

$$\Omega^M := \{y \in \Omega : v(y) = M\} \neq \emptyset.$$

Let $y \in \Omega^M$, $B(y,r) \subset \Omega$. Since (1.2.8) implies (1.2.9) (cf. (1.2.5)), we may apply (1.2.9) in any case to obtain

$$0 = v(y) - M \leq \frac{1}{\omega_d r^d} \int_{B(y,r)} (v(x) - M)dx. \tag{1.2.10}$$

Since M is the supremum of v, always $v(x) \leq M$, and we obtain $v(x) = M$ for all $x \in B(y,r)$. Thus Ω^M contains together with y all balls $B(y,r) \subset \Omega$, and it thus has to coincide with Ω, since Ω is assumed to be connected. Thus $u(x) = M$ for all $x \in \Omega$. $\qquad\square$

We may now easily conclude the proof of Theorem 1.2.2:
Let u be as in Definition 1.2.1. Then $v - u$ likewise satisfies the mean value inequality, hence the maximum principle, and so

$$v \leq u \quad \text{in } \Omega',$$

if $v \leq u$ on $\partial\Omega'$.   □

Corollary 1.2.2: *A function v of class $C^2(\Omega)$ is subharmonic precisely if*

$$\Delta v \geq 0 \quad in \quad \Omega.$$

Proof: "⇒":
Let $B(y, r) \subset \Omega$, $0 < \varrho < r$. Then by (1.2.3)

$$0 \leq \int_{B(y,\varrho)} \Delta v(x) dx = d\omega_d \varrho^{d-1} \frac{\partial}{\partial \varrho} S(v, y, \varrho).$$

Integrating this inequality yields, for $0 < \varrho < r$,

$$S(v, y, \varrho) \leq S(v, y, r),$$

and since the left-hand side tends to $v(y)$ for $\varrho \to 0$, we obtain

$$v(y) \leq S(v, y, r).$$

By Theorem 1.2.2, v then is subharmonic.
"⇒": Assume $\Delta v(y) < 0$. Since $v \in C^2(\Omega)$, we could then find a ball $B(y, r) \subset \Omega$ with $\Delta v < 0$ on $B(y, r)$. Applying the first part of the proof to $-v$ would yield

$$v(y) > S(v, y, r),$$

and v could not be subharmonic.   □

Examples of subharmonic functions:

(1) Let $d \geq 2$. We compute

$$\Delta |x|^\alpha = (d\alpha + \alpha(\alpha - 2)) |x|^{\alpha - 2}.$$

Thus $|x|^\alpha$ is subharmonic for $\alpha \geq 2 - d$. (This is not unexpected because $|x|^{2-d}$ is harmonic.)

(2) Let $u : \Omega \to \mathbb{R}$ be harmonic and positive, $\beta \geq 1$. Then

$$\Delta u^\beta = \sum_{i=1}^{d} \left(\beta u^{\beta-1} u_{x^i x^i} + \beta(\beta - 1) u^{\beta-2} u_{x^i} u_{x^i} \right)$$

$$= \sum_{i=1}^{d} \beta(\beta - 1) u^{\beta-2} u_{x^i} u_{x^i},$$

since u is harmonic. Since u is assumed to be positive and $\beta \geq 1$, this implies that u^β is subharmonic.

(3) Let $u : \Omega \to \mathbb{R}$ again be harmonic and positive. Then

$$\Delta \log u = \sum_{i=1}^{d} \left(\frac{u_{x^i x^i}}{u} - \frac{u_{x^i} u_{x^i}}{u^2} \right) = -\sum_{i=1}^{d} \frac{u_{x^i} u_{x^i}}{u^2},$$

since u is harmonic. Thus, $\log u$ is superharmonic, and $-\log u$ then is subharmonic.

(4) The preceding examples can be generalized as follows:
Let $u : \Omega \to \mathbb{R}$ be harmonic, $f : u(\Omega) \to \mathbb{R}$ convex. Then $f \circ u$ is subharmonic. To see this, we first assume $f \in C^2$. Then

$$\Delta f(u(x)) = \sum_{i=1}^{d} \left(f'(u(x)) u_{x^i x^i} + f''(u(x)) u_{x^i} u_{x^i} \right)$$

$$= \sum_{i=1}^{d} f''(u(x)) \left(u_{x^i} \right)^2 \quad \text{(since } u \text{ is harmonic)}$$

$$\geq 0,$$

since for a convex C^2-function $f'' \geq 0$. If the convex function f is not of class C^2, there exists a sequence $(f_n)_{n \in \mathbb{N}}$ of convex C^2-functions converging to f locally uniformly. By the preceding, $f_n \circ u$ is subharmonic, and hence satisfies the mean value inequality. Since $f_n \circ u$ converges to $f \circ u$ locally uniformly, $f \circ u$ satisfies the mean value inequality as well and so is subharmonic by Theorem 1.2.2.

We now return to studying harmonic functions. If u is harmonic, u and $-u$ both are subharmonic, and we obtain from Lemma 1.2.1 the following result:

Corollary 1.2.3 (Strong maximum principle): *Let u be harmonic in Ω. If there exists $x_0 \in \Omega$ with*

$$u(x_0) = \sup_{x \in \Omega} u(x) \quad or \quad u(x_0) = \inf_{x \in \Omega} u(x),$$

then u is constant in Ω.

A weaker version of Corollary 1.2.3 is the following:

Corollary 1.2.4 (Weak maximum principle): *Let Ω be bounded and $u \in C^0(\bar{\Omega})$ harmonic. Then for all $x \in \Omega$,*

$$\min_{y \in \partial \Omega} u(y) \leq u(x) \leq \max_{y \in \partial \Omega} u(y).$$

Proof: Otherwise, u would achieve its supremum or infimum in some interior point of Ω. Then u would be constant by Corollary 1.2.3, and the claim would also hold true. $\qquad \square$

Corollary 1.2.5 (Uniqueness of solutions of the Poisson equation):
Let $f \in C^0(\Omega)$, Ω bounded, $u_1, u_2 \in C^0(\bar{\Omega}) \cap C^2(\Omega)$ solutions of the Poisson equation

$$\Delta u_i(x) = f(x) \quad \text{for } x \in \Omega \quad (i = 1, 2).$$

If $u_1(z) \leq u_2(z)$ for all $z \in \partial\Omega$, then also

$$u_1(x) \leq u_2(x) \quad \text{for all } x \in \Omega.$$

In particular, if

$$u_1|_{\partial\Omega} = u_2|_{\partial\Omega},$$

then

$$u_1 = u_2.$$

Proof: We apply the maximum principle to the harmonic function $u_1 - u_2$.
□

In particular, for $f = 0$, we once again obtain the uniqueness of harmonic functions with given boundary values.

Remark: The reverse implication in Theorem 1.2.1 can also be seen as follows: We observe that the maximum principle needs only the mean value inequalities. Thus, the uniqueness of Corollary 1.2.5 holds for functions that satisfy the mean value formulae. On the other hand, by Theorem 1.1.2, for continuous boundary values there exists a harmonic extension on the ball, and this harmonic extension also satisfies the mean value formulae by the first implication of Theorem 1.2.1. By uniqueness, therefore, any continuous function satisfying the mean value property must be harmonic on every ball in its domain of definition Ω, hence on all of Ω.

As an application of the weak maximum principle we shall show the removability of isolated singularities of harmonic functions:

Corollary 1.2.6: *Let $x_0 \in \Omega \subset \mathbb{R}^d (d \geq 2)$, $u : \Omega \setminus \{x_0\} \to \mathbb{R}$ harmonic and bounded. Then u can be extended as a harmonic function on all of Ω; i.e., there exists a harmonic function*

$$\tilde{u} : \Omega \to \mathbb{R}$$

that coincides with u on $\Omega \setminus \{x_0\}$.

Proof: By a simple transformation, we may assume $x_0 = 0$ and that Ω contains the ball $B(0,2)$. By Theorem 1.1.2, we may then solve the following Dirichlet problem:

$$\Delta \tilde{u} = 0 \quad \text{in } \mathring{B}(0,1),$$
$$\tilde{u} = u \quad \text{on } \partial B(0,1).$$

We consider the following Green function on $B(0,1)$ for $y = 0$:

$$G(x) = \begin{cases} \frac{1}{2\pi} \log |x| & \text{for } d = 2, \\ \frac{1}{d(2-d)\omega_d}(|x|^{2-d} - 1) & \text{for } d \geq 3. \end{cases}$$

For $\varepsilon > 0$, we put

$$u_\varepsilon(x) := \tilde{u}(x) - \varepsilon G(x) \quad (0 < |x| \leq 1).$$

First of all,

$$u_\varepsilon(x) = \tilde{u}(x) = u(x) \quad \text{for } |x| = 1. \tag{1.2.11}$$

Since on the one hand, u as a smooth function possesses a bounded derivative along $|x| = 1$, and on the other hand (with $r = |x|$), $\frac{\partial}{\partial r} G(x) > 0$, we obtain, for sufficiently large ε,

$$u_\varepsilon(x) > u(x) \quad \text{for } 0 < |x| < 1.$$

But we also have

$$\lim_{x \to 0} u_\varepsilon(x) = \infty \quad \text{for } \varepsilon > 0.$$

Since u is bounded, consequently, for every $\varepsilon > 0$ there exists $r(\varepsilon) > 0$ with

$$u_\varepsilon(x) > u(x) \quad \text{for } |x| < r(\varepsilon). \tag{1.2.12}$$

From these arguments, we may find a smallest $\varepsilon_0 \geq 0$ with

$$u_{\varepsilon_0}(x) \geq u(x) \quad \text{for } |x| \leq 1.$$

We now wish to show that $\varepsilon_0 = 0$.

Assume $\varepsilon_0 > 0$. By (1.2.11), (1.2.12), we could then find z_0, $r(\frac{\varepsilon_0}{2}) < |z_0| < 1$, with

$$u_{\frac{\varepsilon_0}{2}}(z_0) < u(z_0).$$

This would imply

$$\min_{x \in \mathring{B}(0,1) \setminus B(0, r(\frac{\varepsilon_0}{2}))} \left(u_{\frac{\varepsilon_0}{2}}(x) - u(x) \right) < 0,$$

while by (1.2.11), (1.2.12)

$$\min_{y \in \partial B(0,1) \cup \partial B(0,r(\frac{\varepsilon_0}{2}))} \left(u_{\frac{\varepsilon_0}{2}}(y) - u(y) \right) = 0.$$

This contradicts Corollary 1.2.4, because $u_{\frac{\varepsilon_0}{2}} - u$ is harmonic in the annular region considered here. Thus, we must have $\varepsilon_0 = 0$, and we conclude that

$$u \le u_0 = \tilde{u} \quad \text{in } B(0,1) \setminus \{0\}.$$

In the same way, we obtain the opposite inequality

$$u \ge \tilde{u} \quad \text{in } B(0,1) \setminus \{0\}.$$

Thus, u coincides with $\tilde{u}$ in $B(0,1) \setminus \{0\}$. Since $\tilde{u}$ is harmonic in all of $B(0,1)$, we have found the desired extension. □

From Corollary 1.2.6 we see that not every Dirichlet problem for a harmonic function is solvable. For example, there is no solution of

$$\Delta u(x) = 0 \quad \text{in } \mathring{B}(0,1) \setminus \{0\},$$
$$u(x) = 0 \quad \text{for } |x| = 1,$$
$$u(0) = 1.$$

Namely, by Corollary 1.2.6 any solution u could be extended to a harmonic function on the entire ball $\mathring{B}(0,1)$, but such a harmonic function would have to vanish identically by Corollary 1.2.4, since its boundary values on $\partial B(0,1)$ vanish, and so it could not assume the prescribed value 1 at $x = 0$.

Another consequence of the maximum principle for subharmonic functions is a gradient estimate for solutions of the Poisson equation:

Corollary 1.2.7: *Suppose that in Ω,*

$$\Delta u(x) = f(x)$$

with a bounded function f. Let $x_0 \in \Omega$ and $R := \text{dist}(x_0, \partial \Omega)$. Then

$$|u_{x^i}(x_0)| \le \frac{d}{R} \sup_{\partial B(x_0,R)} |u| + \frac{R}{2} \sup_{B(x_0,R)} |f| \quad \text{for } i = 1, \ldots, d. \qquad (1.2.13)$$

Proof: We consider the case $i = 1$. For abbreviation, put

$$\mu := \sup_{\partial B(x_0,R)} |u|, \quad M := \sup_{B(x_0,R)} |f|.$$

Without loss of generality, suppose again $x_0 = 0$. The auxiliary function

$$v(x) := \frac{\mu}{R^2} |x|^2 + x^1 \left(R - x^1\right) \left(\frac{d\mu}{R^2} + \frac{M}{2}\right)$$

satisfies, in $B(0, R)$,

$$\Delta v(x) = -M,$$
$$v\left(0, x^2, \ldots, x^d\right) \geq 0 \quad \text{for all } x^2, \ldots, x^d,$$
$$v(x) \geq \mu \quad \text{for } |x| = R, \; x^1 \geq 0.$$

We now consider

$$\bar{u}(x) := \frac{1}{2} \left(u\left(x^1, \ldots, x^d\right) - u\left(-x^1, x^2, \ldots, x^d\right)\right).$$

In $B(0, R)$, we have

$$|\Delta \bar{u}(x)| \leq M,$$
$$\bar{u}(0, x^2, \ldots, x^d) = 0 \quad \text{for all } x^2, \ldots, x^d,$$
$$|\bar{u}(x)| \leq \mu \quad \text{for all } |x| = R.$$

We consider the half-ball $B^+ := \{|x| \leq R, \; x^1 > 0\}$. The preceding inequalities imply

$$\Delta(v \pm \bar{u}) \leq 0 \quad \text{in } \mathring{B}^+,$$
$$v \pm \bar{u} \geq 0 \quad \text{on } \partial B^+.$$

The maximum principle (Lemma 1.2.1) yields

$$|\bar{u}| \leq v \quad \text{in } B^+.$$

We conclude that

$$|u_{x^1}(0)| = \lim_{\substack{x^1 \to 0 \\ x^1 > 0}} \left|\frac{\bar{u}(x^1, 0, \ldots, 0)}{x^1}\right| \leq \lim_{\substack{x^1 \to 0 \\ x^1 > 0}} \frac{v(x^1, 0, \ldots, 0)}{x^1} = \frac{d\mu}{R} + \frac{R}{2} M,$$

i.e., (1.2.13). $\square$

Other consequences of the mean value formulae are the following:

Corollary 1.2.8 (Liouville theorem): *Let* $u : \mathbb{R}^d \to \mathbb{R}$ *be harmonic and bounded. Then* u *is constant.*

Proof: For $x_1, x_2 \in \mathbb{R}^d$, by (1.2.2) for all $r > 0$,

$$u(x_1) - u(x_2) = \frac{1}{\omega_d r^d} \left(\int_{B(x_1, r)} u(x) dx - \int_{B(x_2, r)} u(x) dx\right)$$

$$= \frac{1}{\omega_d r^d} \left(\int_{B(x_1, r) \backslash B(x_2, r)} u(x) dx - \int_{B(x_2, r) \backslash B(x_1, r)} u(x) dx\right).$$

$$(1.2.14)$$

By assumption

$$|u(x)| \leq M,$$

and for $r \to \infty$,

$$\frac{1}{\omega_d r^d} \text{Vol}\left(B(x_1, r) \setminus B(x_2, r)\right) \to 0.$$

This implies that the right-hand side of (1.2.14) converges to 0 for $r \to \infty$. Therefore, we must have

$$u(x_1) = u(x_2).$$

Since x_1 and x_2 are arbitrary, u has to be constant. □

Another *proof* of Corollary 1.2.8 follows from Corollary 1.2.7:
By Corollary 1.2.7, for all $x_0 \in \mathbb{R}^d$, $R > 0$, $i = 1, \ldots, d$,

$$|u_{x^i}(x_0)| \leq \frac{d}{R} \sup_{\mathbb{R}^d} |u|.$$

Since u is bounded by assumption, the right-hand side tends to 0 for $R \to \infty$, and it follows that u is constant. This proof also works under the weaker assumption

$$\lim_{R \to \infty} \frac{1}{R} \sup_{B(x_0, R)} |u| = 0.$$

This assumption is sharp, since affine linear functions are harmonic functions on $\mathbb{R}^d$ that are not constant.

Corollary 1.2.9 (Harnack inequality): *Let $u : \Omega \to \mathbb{R}$ be harmonic and nonnegative. Then for every subdomain $\Omega' \subset\subset \Omega$ there exists a constant $c = c(d, \Omega, \Omega')$ with*

$$\sup_{\Omega'} u \leq c \inf_{\Omega'} u. \tag{1.2.15}$$

Proof: We first consider the special case $\Omega' = \mathring{B}(x_0, r)$, assuming $B(x_0, 4r) \subset \Omega$. Let $y_1, y_2 \in B(x_0, r)$. By (1.2.2),

$$u(y_1) = \frac{1}{\omega_d r^d} \int_{B(y_1,r)} u(y)dy$$

$$\leq \frac{1}{\omega_d r^d} \int_{B(x_0,2r)} u(y)dy,$$

since $u \geq 0$ and $B(y_1,r) \subset B(x_0,2r)$

$$= \frac{3^d}{\omega_d (3r)^d} \int_{B(x_0,2r)} u(y)dy$$

$$\leq \frac{3^d}{\omega_d (3r)^d} \int_{B(y_2,3r)} u(y)dy,$$

since $u \geq 0$ and $B(x_0,2r) \subset B(y_2,3r)$

$$= 3^d u(y_2),$$

and in particular,

$$\sup_{B(x_0,r)} u \leq 3^d \inf_{B(x_0,r)} u,$$

which is the claim in this special case.

For an arbitrary subdomain $\Omega' \subset\subset \Omega$, we choose $r > 0$ with

$$r < \frac{1}{4} \operatorname{dist}(\Omega', \partial\Omega).$$

Since Ω' is bounded and connected, there exists $m \in \mathbb{N}$ such that any two points $y_1, y_2 \in \Omega'$ can be connected in Ω' by a curve that can be covered by at most m balls of radius r with centers in Ω'. Composing the preceding inequalities for all these balls, we get

$$u(y_1) \leq 3^{md} u(y_2).$$

Thus, we have verified the claim for $c = 3^{md}$. □

The Harnack inequality implies the following result:

Corollary 1.2.10 (Harnack convergence theorem): *Let $u_n : \Omega \to \mathbb{R}$ be a monotonically increasing sequence of harmonic functions. If there exists $y \in \Omega$ for which the sequence $(u_n(y))_{n \in \mathbb{N}}$ is bounded, then u_n converges on any subdomain $\Omega' \subset\subset \Omega$ uniformly towards a harmonic function.*

Proof: The monotonicity and boundedness imply that $u_n(y)$ converges for $n \to \infty$. For $\varepsilon > 0$, there thus exists $N \in \mathbb{N}$ such that for $n \geq m \geq N$,

$$0 \leq u_n(y) - u_m(y) < \varepsilon.$$

Then $u_n - u_m$ is a nonnegative harmonic function (by monotonicity), and by Corollary 1.2.9,

$$\sup_{\Omega'}(u_n - u_m) \leq c\varepsilon, \quad (\text{wlog } y \in \Omega'),$$

where c depends on d, Ω, and Ω'. Thus $(u_n)_{n\in\mathbb{N}}$ converges uniformly in all of Ω'. The uniform limit of harmonic functions has to satisfy the mean value formulae as well, and it is hence harmonic itself by Theorem 1.2.1. $\square$

Summary

In this chapter we encountered some basic properties of harmonic functions, i.e., of solutions of the Laplace equation

$$\Delta u = 0 \quad \text{in } \Omega,$$

and also of solutions of the Poisson equation

$$\Delta u = f \quad \text{in } \Omega$$

with given f.

We found the unique solution of the Dirichlet problem on the ball (Theorem 1.1.2), and we saw that solutions are smooth (Corollary 1.1.2) and even satisfy explicit estimates (Corollary 1.2.7) and in particular the maximum principle (Corollary 1.2.3, Corollary 1.2.4), which actually already holds for subharmonic functions (Lemma 1.2.1). All these results are typical and characteristic for solutions of elliptic PDEs. The methods presented in this chapter, however, mostly do not readily generalize, since they have used heavily the rotational symmetry of the Laplace operator. In subsequent chapters we thus need to develop different and more general methods in order to show analogues of these results for larger classes of elliptic PDEs.

Exercises

1.1 Determine the Green function of the half-space
$$\{x = (x^1, \ldots, x^d) \in \mathbb{R}^d : x^1 > 0\}.$$

1.2 On the unit ball $B(0,1) \subset \mathbb{R}^d$, determine a function $H(x,y)$, defined for $x \neq y$, with
  (i) $\frac{\partial}{\partial \nu_x} H(x,y) = 1$ for $x \in \partial B(0,1)$;
  (ii) $H(x,y) - \Gamma(x,y)$ is a harmonic function of $x \in B(0,1)$. (Here, $\Gamma(x,y)$ is a fundamental solution.)

1.3 Use the result of Exercise 1.2 to study the Neumann problem for the Laplace equation on the unit ball $B(0,1) \subset \mathbb{R}^d$:
  Let $g : \partial B(0,1) \to \mathbb{R}$ with $\int_{\partial B(0,1)} g(y)\, do(y) = 0$ be given. We wish to find a solution of
$$\Delta u(x) = 0 \qquad \text{for } x \in \mathring{B}(0,1),$$
$$\frac{\partial u}{\partial \nu}(x) = g(x) \quad \text{for } x \in \partial B(0,1).$$

1.4 Let $u : B(0, R) \to \mathbb{R}$ be harmonic and nonnegative. Prove the following version of the Harnack inequality:

$$\frac{R^{d-2}(R - |x|)}{(R + |x|)^{d-1}} u(0) \le u(x) \le \frac{R^{d-2}(R + |x|)}{(R - |x|)^{d-1}} u(0)$$

for all $x \in B(0, R)$.

1.5 Let $u : \mathbb{R}^d \to \mathbb{R}$ be harmonic and nonnegative. Show that u is constant. (Hint: Use the result of Exercise 1.4.)

1.6 Let u be harmonic with periodic boundary conditions. Use the maximum principle to show that u is constant.

1.7 Let $\Omega \subset \mathbb{R}^3 \setminus \{0\}$, $u : \Omega \to \mathbb{R}$ harmonic. Show that

$$v(x^1, x^2, x^3) := \frac{1}{|x|} u \left(\frac{x^1}{|x|^2}, \frac{x^2}{|x|^2}, \frac{x^3}{|x|^2} \right)$$

is harmonic in the region $\Omega' := \left\{ x \in \mathbb{R}^3 : \left(\frac{x^1}{|x|^2}, \frac{x^2}{|x|^2}, \frac{x^3}{|x|^2} \right) \in \Omega \right\}$.

 – Is there a deeper reason for this?
 – Is there an analogous result for arbitrary dimension d?

1.8 Let Ω be the unbounded region $\{x \in \mathbb{R}^d : |x| > 1\}$. Let $u \in C^2(\Omega) \cap C^0(\bar{\Omega})$ satisfy $\Delta u = 0$ in Ω. Furthermore, assume

$$\lim_{|x| \to \infty} u(x) = 0.$$

Show that

$$\sup_{\Omega} |u| = \max_{\partial \Omega} |u|.$$

1.9 **(Schwarz reflection principle):**
Let $\Omega^+ \subset \{x^d > 0\}$,

$$\Sigma := \partial \Omega^+ \cap \{x^d = 0\} \neq \emptyset.$$

Let u be harmonic in Ω^+, continuous on $\Omega^+ \cup \Sigma$, and suppose $u = 0$ on Σ. We put

$$\bar{u}(x^1, \ldots, x^d) := \begin{cases} u(x^1, \ldots, x^d) & \text{for } x^d \ge 0, \\ -u(x^1, \ldots, -x^d) & \text{for } x^d < 0. \end{cases}$$

Show that $\bar{u}$ is harmonic in $\Omega^+ \cup \Sigma \cup \Omega^-$, where $\Omega^- := \{x \in \mathbb{R}^d : (x^1, \ldots, -x^d) \in \Omega^+\}$.

1.10 Let $\Omega \subset \mathbb{R}^d$ be a bounded domain for which the divergence theorem holds. Assume $u \in C^2(\bar{\Omega})$, $u = 0$ on $\partial \Omega$. Show that for every $\varepsilon > 0$,

$$2 \int_{\Omega} |\nabla u(x)|^2 \, dx \le \varepsilon \int_{\Omega} (\Delta u(x))^2 \, dx + \frac{1}{\varepsilon} \int_{\Omega} u^2(x) \, dx.$$

2. The Maximum Principle

Throughout this chapter, Ω is a bounded domain in $\mathbb{R}^d$. All functions u are assumed to be of class $C^2(\Omega)$.

2.1 The Maximum Principle of E. Hopf

We wish to study linear elliptic differential operators of the form

$$Lu(x) = \sum_{i,j=1}^{d} a^{ij}(x) u_{x^i x^j}(x) + \sum_{i=1}^{d} b^i(x) u_{x^i}(x) + c(x) u(x),$$

where we impose the following conditions on the coefficients:

(i) Symmetry: $a^{ij}(x) = a^{ji}(x)$ for all i, j and $x \in \Omega$ (this is no serious restriction).

(ii) Ellipticity: There exists a constant $\lambda > 0$ with

$$\lambda |\xi|^2 \leq \sum_{i,j=1}^{d} a^{ij}(x) \xi^i \xi^j \quad \text{for all } x \in \Omega, \xi \in \mathbb{R}^d$$

(this is the key condition).

In particular, the matrix $(a^{ij}(x))_{i,j=1,\ldots,d}$ is positive definite for all x, and the smallest eigenvalue is greater than or equal to λ.

(iii) Boundedness of the coefficients: There exists a constant K with

$$\left| a^{ij}(x) \right|, \left| b^i(x) \right|, |c(x)| \leq K \quad \text{for all } i, j \text{ and } x \in \Omega.$$

Obviously, the Laplace operator satisfies all three conditions. The aim of the present chapter is to prove maximum principles for solutions of $Lu = 0$. It turns out that for that purpose, we need to impose an additional condition on the sign of $c(x)$, since otherwise no maximum principle can hold, as the following simple example demonstrates: The Dirichlet problem

$$u''(x) + u(x) = 0 \quad \text{on } (0, \pi),$$
$$u(0) = 0 = u(\pi),$$

has the solutions

$$u(x) = \alpha \sin x$$

for arbitrary u, and depending on the sign of α, these solutions assume a strict interior maximum or minimum at $x = \pi/2$. The Dirichlet problem

$$u''(x) - u(x) = 0,$$
$$u(0) = 0 = u(\pi),$$

however, has 0 as its only solution.

As a start, let us present a proof of the weak maximum principle for subharmonic functions (Lemma 1.2.1) that does not depend on the mean value formulae:

Lemma 2.1.1: *Let $u \in C^2(\Omega) \cap C^0(\bar{\Omega})$, $\Delta u \geq 0$ in Ω. Then*

$$\sup_{\Omega} u = \max_{\partial\Omega} u. \tag{2.1.1}$$

(Since u is continuous and Ω is bounded, and the closure $\bar{\Omega}$ thus is compact, the supremum of u on Ω coincides with the maximum of u on $\bar{\Omega}$.)

Proof: We first consider the case where we even have

$$\Delta u > 0 \quad \text{in } \Omega.$$

Then u cannot assume an interior maximum at some $x_0 \in \Omega$, since at such a maximum, we would have

$$u_{x^i x^i}(x_0) \leq 0 \quad \text{for } i = 1, \ldots, d,$$

and thus also

$$\Delta u(x_0) \leq 0.$$

We now come to the general case $\Delta u \geq 0$ and consider the auxiliary function

$$v(x) = e^{x^1},$$

which satisfies

$$\Delta v = v > 0.$$

For each $\varepsilon > 0$, then

$$\Delta(u + \varepsilon v) > 0 \quad \text{in } \Omega,$$

and from the case studied in the beginning, we deduce

$$\sup_{\Omega}(u + \varepsilon v) = \max_{\partial\Omega}(u + \varepsilon v).$$

Then

$$\sup_{\Omega} u + \varepsilon \inf_{\Omega} v \le \max_{\partial\Omega} u + \varepsilon \max_{\partial\Omega} v,$$

and since this holds for every $\varepsilon > 0$, we obtain (2.1.1). $\square$

Theorem 2.1.1: *Assume $c(x) \equiv 0$, and let u satisfy in Ω*

$$Lu \ge 0,$$

i.e.,

$$\sum_{i,j=1}^{d} a^{ij}(x)u_{x^i x^j} + \sum_{i=1}^{d} b^i(x)u_{x^i} \ge 0. \tag{2.1.2}$$

Then also

$$\sup_{x \in \Omega} u(x) = \max_{x \in \partial\Omega} u(x). \tag{2.1.3}$$

In the case $Lu \le 0$, a corresponding result holds for the infimum.

Proof: As in the proof of Lemma 2.1.1, we first consider the case

$$Lu > 0.$$

Since at an interior maximum x_0 of u, we must have

$$u_{x^i}(x_0) = 0 \quad \text{for } i = 1, \dots, d,$$

and

$$(u_{x^i x^j}(x_0))_{i,j=1,\dots,d} \quad \text{negative semidefinite,}$$

and thus by the ellipticity condition also

$$Lu(x_0) = \sum_{i,j=1}^{d} a^{ij}(x_0)u_{x^i x^j}(x_0) \le 0,$$

such an interior maximum cannot occur.

Returning to the general case $Lu \ge 0$, we now consider the auxiliary function

$$v(x) = e^{\alpha x^1}$$

for $\alpha > 0$. Then

$$Lv(x) = \left(\alpha^2 a^{11}(x) + \alpha b^1(x)\right) v(x).$$

Since Ω and the coefficients b^i are bounded and the coefficients satisfy $a^{ii}(x) \geq \lambda$, we have for sufficiently large α,

$$Lv > 0,$$

and applying what we have proved already to $u + \varepsilon v$

$$(L(u + \varepsilon v) > 0),$$

the claim follows as in the proof of Lemma 2.1.1. The case $Lu \leq 0$ can be reduced to the previous one by considering $-u$. $\qquad\square$

Corollary 2.1.1: *Let L be as in Theorem 2.1.1, and let $f \in C^0(\Omega)$, $\varphi \in C^0(\partial\Omega)$ be given. Then the Dirichlet problem*

$$\begin{aligned}
Lu(x) &= f(x) \quad \text{for } x \in \Omega, \qquad\qquad (2.1.4)\\
u(x) &= \varphi(x) \quad \text{for } x \in \partial\Omega,
\end{aligned}$$

admits at most one solution.

Proof: The difference $v(x) = u_1(x) - u_2(x)$ of two solutions satisfies

$$\begin{aligned}
Lv(x) &= 0 \quad \text{in } \Omega,\\
v(x) &= 0 \quad \text{on } \partial\Omega,
\end{aligned}$$

and by Theorem 2.1.1 it then has to vanish identically on Ω. $\qquad\square$

Theorem 2.1.1 supposes $c(x) \equiv 0$. This assumption can be weakened as follows:

Corollary 2.1.2: *Suppose $c(x) \leq 0$ in Ω. Let $u \in C^2(\Omega) \cap C^0(\bar{\Omega})$ satisfy*

$$Lu \geq 0 \quad \text{in } \Omega.$$

With $u^+(x) := \max(u(x), 0)$, we then have

$$\sup_{\Omega} u^+ \leq \max_{\partial\Omega} u^+. \qquad\qquad (2.1.5)$$

Proof: Let $\Omega^+ := \{x \in \Omega : u(x) > 0\}$. Because of $c \leq 0$, we have in Ω^+,

$$\sum_{i,j=1}^{d} a^{ij}(x) u_{x^i x^j} + \sum_{i=1}^{d} b^i(x) u_{x^i} \geq 0,$$

and hence by Theorem 2.1.1,

$$\sup_{\Omega^+} u \leq \max_{\partial\Omega^+} u. \qquad\qquad (2.1.6)$$

We have

$$u = 0 \quad \text{on } \partial\Omega^+ \cap \Omega \quad \text{(by continuity of } u\text{)},$$

$$\max_{\partial\Omega^+ \cap \partial\Omega} u \leq \max_{\partial\Omega} u,$$

and hence, since $\partial\Omega^+ = (\partial\Omega^+ \cap \Omega) \cup (\partial\Omega^+ \cap \partial\Omega)$,

$$\max_{\partial\Omega^+} u \leq \max_{\partial\Omega} u^+. \tag{2.1.7}$$

Since also

$$\sup_{\Omega} u^+ = \sup_{\Omega^+} u, \tag{2.1.8}$$

(2.1.5) follows from (2.1.6), (2.1.7).   □

We now come to the strong maximum principle of E. Hopf:

Theorem 2.1.2: *Suppose $c(x) \equiv 0$, and let u satisfy in Ω,*

$$Lu \geq 0. \tag{2.1.9}$$

If u assumes its maximum in the interior of Ω, it has to be constant. More generally, if $c(x) \leq 0$, u has to be constant if it assumes a nonnegative interior maximum.

For the proof, we need the boundary point lemma of E. Hopf:

Lemma 2.1.2: *Suppose $c(x) \leq 0$ and*

$$Lu \geq 0 \quad \text{in } \Omega' \subset \mathbb{R}^d,$$

and let $x_0 \in \partial\Omega'$. Moreover, assume

(i) u is continuous at x_0;
(ii) $u(x_0) \geq 0$ if $c(x) \not\equiv 0$;
(iii) $u(x_0) > u(x)$ for all $x \in \Omega'$;
(iv) there exists a ball $\mathring{B}(y, R) \subset \Omega'$ with $x_0 \in \partial B(y, R)$.

We then have, with $r := |x - y|$,

$$\frac{\partial u}{\partial r}(x_0) > 0,$$

provided that this derivative (in the direction of the exterior normal of Ω') exists.

Proof: We may assume

$$\partial B(y, R) \cap \partial \Omega' = \{x_0\}.$$

For $0 < \rho < R$, on the annular region $\mathring{B}(y, R) \setminus B(y, \rho)$ we consider the auxiliary function

$$v(x) := e^{-\gamma |x-y|^2} - e^{-\gamma R^2}.$$

We have

$$
\begin{aligned}
Lv(x) = \Bigg\{ & 4\gamma^2 \sum_{i,j=1}^{d} a^{ij}(x) \left(x^i - y^i\right) \left(x^j - y^j\right) \\
& - 2\gamma \sum_{i=1}^{d} a^{ii}(x) + b^i(x) \left(x^i - y^i\right) \Bigg\} e^{-\gamma |x-y|^2} \\
& + c(x) \left(e^{-\gamma |x-y|^2} - e^{-\gamma R^2}\right).
\end{aligned}
$$

For sufficiently large γ, because of the assumed boundedness of the coefficients of L and the ellipticity condition, we have

$$Lv \geq 0 \quad \text{in } \mathring{B}(y, R) \setminus B(y, \rho). \tag{2.1.10}$$

By (iii) and (iv),

$$u(x) - u(x_0) < 0 \quad \text{for } x \in \mathring{B}(y, R).$$

Therefore, we may find $\varepsilon > 0$ with

$$u(x) - u(x_0) + \varepsilon v(x) \leq 0 \quad \text{for } x \in \partial B(y, \rho). \tag{2.1.11}$$

Since $v = 0$ on $\partial B(y, R)$, (2.1.11) continues to hold on $\partial B(y, R)$. On the other hand,

$$L\left(u(x) - u(x_0) + \varepsilon v(x)\right) \geq -c(x)u(x_0) \geq 0 \tag{2.1.12}$$

by (2.1.10) and (ii) and because of $c(x) \leq 0$. Thus, we may apply Corollary 2.1.2 on $\mathring{B}(y, R) \setminus B(y, \rho)$ and obtain

$$u(x) - u(x_0) + \varepsilon v(x) \leq 0 \quad \text{for } x \in \mathring{B}(y, R) \setminus B(y, \rho).$$

Provided that the derivative exists, it follows that

$$\frac{\partial}{\partial r} \left(u(x) - u(x_0) + \varepsilon v(x)\right) \geq 0 \text{ at } x = x_0,$$

and hence for $x = x_0$,

$$\frac{\partial}{\partial r} u(x) \geq -\varepsilon \frac{\partial v(x)}{\partial r} = \varepsilon \left(2\gamma R e^{-\gamma R^2}\right) > 0.$$

$\square$

Proof of Theorem 2.1.2: We assume by contradiction that u is not constant, but has a maximum m (≥ 0 in case $c \not\equiv 0$) in Ω. We then have

$$\Omega' := \{x \in \Omega : u(x) < m\} \neq \emptyset$$

and

$$\partial\Omega' \cap \Omega \neq \emptyset.$$

We choose some $y \in \Omega'$ that is closer to $\partial\Omega'$ than to $\partial\Omega$. Let $\mathring{B}(y, R)$ be the largest ball with center y that is contained in Ω'. We then get

$$u(x_0) = m \quad \text{for some } x_0 \in \partial B(y, R),$$

and

$$u(x) < u(x_0) \quad \text{for } x \in \Omega'.$$

By Lemma 2.1.2,

$$Du(x_0) \neq 0,$$

which, however, is not possible at an interior maximum point. This contradiction demonstrates the claim.   □

2.2 The Maximum Principle of Alexandrov and Bakelman

In this section, we consider differential operators of the same type as in the previous one, but for technical simplicity, we assume that the coefficients $c(x)$ and $b^i(x)$ vanish. While similar results as those presented here continue to hold for vanishing $b^i(x)$ and nonpositive $c(x)$, here we wish only to present the key ideas in a situation that is as simple as possible.

Theorem 2.2.1: *Suppose that $u \in C^2(\Omega) \cap C^0(\bar{\Omega})$ satisfies*

$$Lu(x) := \sum_{i,j=1}^{d} a^{ij}(x)u_{x^i x^j} \geq f(x), \tag{2.2.1}$$

where the matrix $(a^{ij}(x))$ is positive definite and symmetric for each $x \in \Omega$. Moreover, let

$$\int_\Omega \frac{|f(x)|^d}{\det(a^{ij}(x))} dx < \infty. \tag{2.2.2}$$

We then have

$$\sup_\Omega u \leq \max_{\partial\Omega} u + \frac{\operatorname{diam}(\Omega)}{d\omega_d^{1/d}} \left(\int_\Omega \frac{|f(x)|^d}{\det(a^{ij}(x))} dx \right)^{1/d}. \tag{2.2.3}$$

In contrast to those estimates that are based on the Hopf maximum principle (cf., e.g., Theorem 2.3.2 below), here we have only an integral norm of f on the right-hand side, i.e., a norm that is weaker than the supremum norm. In this sense, the maximum principle of Alexandrov and Bakelman is stronger than that of Hopf.

For the *proof of Theorem 2.2.1*, we shall need some geometric constructions. For $v \in C^0(\Omega)$, we define the upper contact set

$$T^+(v) := \{y \in \Omega : \exists p \in \mathbb{R}^d \quad \forall x \in \Omega : v(x) \leq v(y) + p \cdot (x - y)\}. \quad (2.2.4)$$

The dot "$\cdot$" here denotes the Euclidean scalar product of $\mathbb{R}^d$. The p that occurs in this definition in general will depend on y; that is, $p = p(y)$. The set $T^+(v)$ is that subset of Ω in which the graph of v lies below a hyperplane in $\mathbb{R}^{d+1}$ that touches the graph of v at $(y, v(y))$. If v is differentiable at $y \in T^+(v)$, then necessarily $p(y) = Dv(y)$. Finally, v is concave precisely if $T^+(v) = \Omega$.

Lemma 2.2.1: *For $v \in C^2(\Omega)$, the Hessian*

$$\left(v_{x^i x^j}\right)_{i,j=1,\ldots,d}$$

is negative definite on $T^+(v)$.

Proof: For $y \in T^+(v)$, we consider the function

$$w(x) := v(x) - v(y) - p(y) \cdot (x - y).$$

Then $w(x) \leq 0$ on Ω, since $y \in T^+(v)$ and $w(y) = 0$. Thus, w has a maximum at y, implying that $(w_{x^i x^j}(y))$ is negative semidefinite. Since $v_{x^i x^j} = w_{x^i x^j}$ for all i, j, the claim follows.   □

If v is not differentiable at $y \in T^+(v)$, then $p = p(y)$ need not be unique, but there may exist several p's satisfying the condition in (2.2.4). We assign to $y \in T^+(v)$ the set of all those p's, i.e., consider the set-valued map

$$\tau_v(y) := \{p \in \mathbb{R}^d : \forall x \in \Omega : v(x) \leq v(y) + p \cdot (x - y)\}.$$

For $y \notin T^+(v)$, we put $\tau_v(y) := \emptyset$.

Example 2.2.1: $\Omega = \mathring{B}(0,1)$, $\beta > 0$,

$$v(x) = \beta(1 - |x|).$$

The graph of v thus is a cone with a vertex of height β at 0 and having the unit sphere as its base. We have $T^+(v) = \mathring{B}(0,1)$,

$$\tau_v(y) = \begin{cases} B(0,\beta) & \text{for } y = 0, \\ \left\{-\beta \frac{y}{|y|}\right\} & \text{for } y \neq 0. \end{cases}$$

For the cone with vertex of height β at x_0 and base $\partial B(x_o, R)$,

$$v(x) = \beta \left(1 - \frac{|x - x_0|}{R} \right)$$

and $\Omega = \mathring{B}(x_0, R)$, and analogously,

$$\tau_v \left(\mathring{B}(x_0, R) \right) = \tau_v(x_0) = B(0, \beta/R). \tag{2.2.5}$$

We now consider the image of Ω under τ_v,

$$\tau_v(\Omega) = \bigcup_{y \in \Omega} \tau_v(y) \subset \mathbb{R}^d.$$

We will let $\mathcal{L}_d$ denote d-dimensional Lebesgue measure. Then we have the following lemma:

Lemma 2.2.2: *Let $v \in C^2(\Omega) \cap C^0(\bar{\Omega})$. Then*

$$\mathcal{L}_d \left(\tau_v(\Omega) \right) \leq \int_{T^+(v)} \left| \det \left(v_{x^i x^j}(x) \right) \right| dx. \tag{2.2.6}$$

Proof: First of all,

$$\tau_v(\Omega) = \tau_v(T^+(v)) = Dv(T^+(v)), \tag{2.2.7}$$

since v is differentiable. By Lemma 2.2.1, the Jacobian matrix of $Dv : \Omega \to \mathbb{R}^d$, namely $(v_{x^i x^j})$, is negative semidefinite on $T^1(v)$. Thus $Dv - \varepsilon\,\mathrm{Id}$ has maximal rank for $\varepsilon > 0$. From the transformation formula for multiple integrals, we then get

$$\mathcal{L}_d \left((Dv - \varepsilon\,\mathrm{Id}) \left(T^+(v) \right) \right) \leq \int_{T^+(v)} \left| \det \left(v_{x^i x^j}(x) - \varepsilon \delta_{ij} \right)_{i,j=1,\ldots,d} \right| dx. \tag{2.2.8}$$

Letting ε tend to 0, the claim follows because of (2.2.7). $\qquad \square$

We are now able to *prove Theorem 2.2.1:* We may assume

$$u \leq 0 \quad \text{on } \partial\Omega$$

by replacing u by $u - \max_{\partial\Omega} u$ if necessary.

Now let $x_0 \in \Omega$, $u(x_0) > 0$. We consider the function κ_{x_0} on $B(x_0, \delta)$ with $\delta = \mathrm{diam}(\Omega)$ whose graph is the cone with vertex of height $u(x_0)$ at x_0 and base $\partial B(x_0, \delta)$. From the definition of the diameter $\delta = \mathrm{diam}\,\Omega$,

$$\Omega \subset B(x_0, \delta).$$

Since we assume $u \leq 0$ on $\partial\Omega$, for each hyperplane that is tangent to this cone there exists some parallel hyperplane that is tangent to the graph of u. (In order to see this, we simply move such a hyperplane parallel to its original position from above towards the graph of u until it first becomes tangent to it. Since the graph of u is at least of height $u(x_0)$, i.e., of the height of the cone, and since $u \leq 0$ on $\partial\Omega$ and $\partial\Omega \subset B(x_0, \delta)$, such a first tangency cannot occur at a boundary point of Ω, but only at an interior point x_1. Thus, the corresponding hyperplane is contained in $\tau_v(x_1)$.) This means that

$$\tau_{\kappa_{x_0}}(\Omega) \subset \tau_u(\Omega). \tag{2.2.9}$$

By (2.2.5),

$$\tau_{\kappa_{x_0}}(\Omega) = B\left(0, u(x_0)/\delta\right). \tag{2.2.10}$$

Relations (2.2.6), (2.2.9), (2.2.10) imply

$$\mathcal{L}_d\left(B\left(0, u(x_0)/\delta\right)\right) \leq \int_{T^+(u)} \left|\det\left(u_{x^i x^j}(x)\right)\right| dx,$$

and hence

$$
\begin{aligned}
u(x_0) &\leq \frac{\delta}{\omega_d^{1/d}} \left(\int_{T^+(u)} \left|\det\left(u_{x^i x^j}(x)\right)\right| dx\right)^{1/d} \\
&= \frac{\delta}{\omega_d^{1/d}} \left(\int_{T^+(u)} (-1)^d \det\left(u_{x^i x^j}(x)\right) dx\right)^{1/d}
\end{aligned}
\tag{2.2.11}
$$

by Lemma 2.2.1. Without assuming $u \leq 0$ on $\partial\Omega$, we get an additional term $\max_{\partial\Omega} u$ on the right-hand side of (2.2.11). Since the formula holds for all $x_0 \in \Omega$, we have the following result:

Lemma 2.2.3: For $u \in C^2(\Omega) \cap C^0(\bar{\Omega})$,

$$\sup_\Omega u \leq \max_{\partial\Omega} u + \frac{\operatorname{diam}(\Omega)}{\omega_d^{1/d}} \left(\int_{T^+(u)} (-1)^d \det\left(u_{x^i x^j}(x)\right) dx\right)^{1/d}. \tag{2.2.12}$$

$\square$

In order to deduce Theorem 2.2.1 from this result, we need the following elementary lemma:

Lemma 2.2.4: On $T^+(u)$,

$$(-1)^d \det\left(u_{x^i x^j}(x)\right) \leq \frac{1}{\det\left(a^{ij}(x)\right)} \left(-\frac{1}{d} \sum_{i,j=1}^{d} a^{ij}(x) u_{x^i x^j}(x)\right)^d. \tag{2.2.13}$$

Proof: It is well known that for symmetric, positive definite matrices A, B,

$$\det A \det B \leq \left(\frac{1}{d} \text{ trace } AB\right)^d,$$

which is readily verified by diagonalizing one of the matrices, which is possible if that matrix is symmetric.

Inserting $A = (-u_{x^i x^j})$, $B = (a^{ij})$ (which is possible by Lemma 2.2.1 and the ellipticity assumption), we obtain (2.2.13). $\square$

Inequalities (2.2.12), (2.2.13) imply

$$\sup_\Omega u \leq \max_{\partial\Omega} u + \frac{\text{diam}(\Omega)}{d\omega_d^{1/d}} \left(\int_{T^+(u)} \frac{\left(-\sum_{i,j=1}^d a^{ij}(x)u_{x^i x^j}(x)\right)^d}{\det\left(a^{ij}(x)\right)} dx\right)^{1/d}.$$

$$(2.2.14)$$

In turn (2.2.14) directly implies Theorem 2.2.1, since by assumption, $-\sum a^{ij} u_{x^i x^j} \leq -f$, and the left-hand side of this inequality is nonnegative on $T^+(u)$ by Lemma 2.2.1. $\square$

We wish to apply Theorem 2.2.1 to some nonlinear equation, namely, the two-dimensional Monge–Ampère equation.

Thus, let Ω be open in $\mathbb{R}^2 = \{(x^1, x^2)\}$, and let $u \in C^2(\Omega)$ satisfy

$$u_{x^1 x^1}(x)u_{x^2 x^2}(x) - u_{x^1 x^2}^2(x) = f(x) \quad \text{in } \Omega, \qquad (2.2.15)$$

with given f. In order that (2.2.15) be elliptic:

(i) the Hessian of u must be positive definite, and hence also
(ii) $f(x) > 0$ in Ω.

Condition (i) means that u is a convex function. Thus, u cannot assume a maximum in the interior of Ω, but a minimum is possible. In order to control the minimum, we observe that if u is a solution of (2.2.15), then so is $(-u)$. However, equation (2.2.15) is no longer elliptic at $(-u)$, since the Hessian of $(-u)$ is negative, and not positive, so that Theorem 2.2.1 cannot be applied directly. We observe, however, that Lemma 2.2.3 does not need an ellipticity assumption, and obtain the following corollary:

Corollary 2.2.1: *Under the assumptions (i), (ii), a solution u of the Monge–Ampère equation (2.2.15) satisfies*

$$\inf_\Omega u \geq \min_{\partial\Omega} u - \frac{\text{diam}(\Omega)}{2\sqrt{\pi}} \left(\int_\Omega f(x)dx\right)^{\frac{1}{2}}.$$

$\square$

The crucial point here is that the nonlinear Monge–Ampère equation for a solution u can be formally written as a linear differential equation. Namely, with

$$a^{11}(x) = \frac{1}{2}u_{x^2 x^2}(x), \qquad a^{12}(x) = a^{21}(x) = \frac{1}{2}u_{x^1 x^2}(x),$$

$$a^{22}(x) = \frac{1}{2}u_{x^1 x^1}(x)$$

(2.2.15) becomes

$$\sum_{i,j=1}^{2} a^{ij}u_{x^i x^j}(x) = f(x),$$

and is thus of the type considered. Consequently, in order to deduce properties of a solution u, we have only to check whether the required conditions for the coefficients $a^{ij}(x)$ hold under our assumptions about u. It may happen, however, that these conditions are satisfied for some, but not for all, solutions u. For example, under the assumptions (i), (ii), (2.2.15) was no longer elliptic at the solution $(-u)$.

2.3 Maximum Principles for Nonlinear Differential Equations

We now consider a general differential equation of the form

$$F[u] = F(x, u, Du, D^2 u) = 0, \tag{2.3.1}$$

with $F : S := \Omega \times \mathbb{R} \times \mathbb{R}^d \times S(d, \mathbb{R}) \to \mathbb{R}$, where $S(d, \mathbb{R})$ is the space of symmetric, real-valued, $d \times d$ matrices. Elements of S are written as (x, z, p, r); here $p = (p_1, \ldots, p_d) \in \mathbb{R}^d$, $r = (r_{ij})_{i,j=1,\ldots,d} \in S(d, \mathbb{R})$. We assume that F is differentiable with respect to the r_{ij}.

Definition 2.3.1: *The differential equation (2.3.1) is called elliptic at $u \in C^2(\Omega)$ if*

$$\left(\frac{\partial F}{\partial r_{ij}} \left(x, u(x), Du(x), D^2 u(x)\right) \right)_{i,j=1,\ldots,d} \qquad \text{is positive definite.} \tag{2.3.2}$$

For example, the Monge–Ampère equation (2.2.15) is elliptic in this sense if the conditions (i), (ii) at the end of Section 2.2 hold.

It is not completely clear what the appropriate generalization of the maximum principle from linear to nonlinear equations is, because in the linear case, we always have to make assumptions on the lower-order terms. One interpretation that suggests a possible generalization is to consider the maximum principle as a statement comparing a solution with a constant that under different conditions was a solution of $Lu \leq 0$. Because of the linear structure, this immediately led to a comparison theorem for arbitrary solutions u_1, u_2 of $Lu = 0$. For this reason, in the nonlinear case we also start with a comparison theorem:

Theorem 2.3.1: *Let $u_0, u_1 \in C^2(\Omega) \cap C^0(\bar{\Omega})$, and suppose*

(i) $F \in C^1(S)$,
(ii) F is elliptic at all functions $tu_1 + (1-t)u_0$, $0 \leq t \leq 1$,
(iii) for each fixed (x, p, r), F is monotonically decreasing in z.

If

$$u_1 \leq u_0 \quad \text{on } \partial\Omega$$

and

$$F[u_1] \geq F[u_0] \quad \text{in } \Omega,$$

then either

$$u_1 < u_0 \quad \text{in } \Omega$$

or

$$u_0 \equiv u_1 \quad \text{in } \Omega.$$

Proof: We put

$$v := u_1 - u_0,$$
$$u_t := tu_1 + (1-t)u_0 \quad \text{for } 0 \leq t \leq 1,$$
$$a^{ij}(x) := \int_0^1 \frac{\partial F}{\partial r_{ij}}\left(x, u_t(x), Du_t(x), D^2 u_t(x)\right) dt,$$
$$b^i(x) := \int_0^1 \frac{\partial F}{\partial p_i}\left(x, u_t(x), Du_t(x), D^2 u_t(x)\right) dt,$$
$$c(x) := \int_0^1 \frac{\partial F}{\partial z}\left(x, u_t(x), Du_t(x), D^2 u_t(x)\right) dt$$

(note that we are integrating a total derivative with respect to t, namely, $\frac{d}{dt} F(x, u_t(x), Du_t(x), D^2 u_t(x))$, and consequently, we can convert the integral into boundary terms, leading to the correct representation of Lv below; cf. (2.3.3)),

$$Lv := \sum_{i,j=1}^d a^{ij}(x)v_{x^i x^j}(x) + \sum_{i=1}^d b^i(x)v_{x^i}(x) + c(x)v(x).$$

Then

$$Lv = F[u_1] - F[u_0] \geq 0 \quad \text{in } \Omega. \tag{2.3.3}$$

The equation L is elliptic because of (ii), and by (iii), $c(x) \leq 0$. Thus, we may apply Theorem 2.1.2 for v and obtain the conclusions of the theorem. $\square$

The theorem holds in particular for solutions of $F[u] = 0$. The key point in the proof of Theorem 2.3.1 then is that since the solutions u_0 and u_1 of the *nonlinear* equation $F[u] = 0$ are already given, we may interpret quantities that depend on u_0 and u_1 and their derivatives as coefficients of a linear differential equation for the difference.

We also would like to formulate the following uniqueness result for the Dirichlet problem for $F[u] = f$ with given f:

Corollary 2.3.1: *Under the assumptions of Theorem 2.3.1, suppose $u_0 = u_1$ on $\partial\Omega$, and*

$$F[u_0] = F[u_1] \quad in \ \Omega.$$

Then $u_0 = u_1$ in Ω.   □

As an example, we consider the *minimal surface equation*: Let $\Omega \subset \mathbb{R}^2 = \{(x, y)\}$. The minimal surface equation then is the quasilinear equation

$$\left(1 + u_y^2\right) u_{xx} - 2u_x u_y u_{xy} + \left(1 + u_x^2\right) u_{yy} = 0. \tag{2.3.4}$$

Theorem 2.3.1 implies the following corollary:

Corollary 2.3.2: *Let $u_0, u_1 \in C^2(\Omega)$ be solutions of the minimal surface equation. If the difference $u_0 - u_1$ assumes a maximum or minimum at an interior point of Ω, we have*

$$u_0 - u_1 \equiv const \quad in \ \Omega.$$

□

We now come to the following maximum principle:

Theorem 2.3.2: *Let $u \in C^2(\Omega) \cap C^0(\bar{\Omega})$, and let $F \in C^2(S)$. Suppose that for some $\lambda > 0$, the ellipticity condition*

$$\lambda |\xi|^2 \le \sum_{i,j=1}^{d} \frac{\partial F}{\partial r_{ij}}(x, z, p, r)\xi^i \xi^j \tag{2.3.5}$$

holds for all $\xi \in \mathbb{R}^d$, $(x, z, p, r) \in S$. Moreover, assume that there exist constants μ_1, μ_2 such that for all (x, z, p),

$$\frac{F(x, z, p, 0)\,\text{sign}(z)}{\lambda} \le \mu_1 |p| + \frac{\mu_2}{\lambda}. \tag{2.3.6}$$

If

$$F[u] = 0 \quad in \ \Omega,$$

then

$$\sup_{\Omega} |u| \le \max_{\partial\Omega} |u| + c\frac{\mu_2}{\lambda}, \tag{2.3.7}$$

where the constant c depends on μ_1 and the diameter $\text{diam}(\Omega)$.

Here, one should think of (2.3.6) as an analogue of the sign condition $c(x) \leq 0$ and the bound for the $b^i(x)$ as well as a bound of the right-hand side f of the equation $Lu = f$.

Proof: We shall follow a similar strategy as in the proof of Theorem 2.3.1 and shall reduce the result to the maximum principle from Section 2.1 for linear equations. Here v is an auxiliary function to be determined, and $w := u - v$. We consider the operator

$$Lw := \sum_{i,j=1}^{d} a^{ij}(x) w_{x^i x^j} + \sum_{i=1}^{d} b^i(x) w_{x^i}$$

with

$$a^{ij}(x) := \int_0^1 \frac{\partial F}{\partial r_{ij}} \left(x, u(x), Du(x), tD^2u(x) \right) dt, \qquad (2.3.8)$$

while the coefficients $b^i(x)$ are defined through the following equation:

$$\sum_{i=1}^{d} b^i(x) w_{x^i} = \sum_{i,j=1}^{d} \int_0^1 \left(\frac{\partial F}{\partial r_{ij}} \left(x, u(x), Du(x), tD^2u(x) \right) \right.$$
$$\left. - \frac{\partial F}{\partial r_{ij}} \left(x, u(x), Dv(x), tD^2u(x) \right) \right) dt \cdot v_{x^i x^j}$$
$$+ F\left(x, u(x), Du(x), 0 \right) - F\left(x, u(x), Dv(x), 0 \right). \qquad (2.3.9)$$

(That this is indeed possible follows from the mean value theorem and the assumption $F \in C^2$. It actually suffices to assume that F is twice continuously differentiable with respect to the variables r only.) Then L satisfies the assumptions of Theorem 2.1.1. Now

$$Lw = L(u - v)$$

$$= \sum_{i,j=1}^{d} \left(\int_0^1 \frac{\partial F}{\partial r_{ij}} \left(x, u(x), Du(x), tD^2u(x) \right) dt \right) u_{x^i x^j} + F(x, u(x), Du(x), 0)$$

$$- \sum_{i,j=1}^{d} \left(\int_0^1 \frac{\partial F}{\partial r_{ij}} \left(x, u(x), Dv(x), tD^2u(x) \right) dt \right) v_{x^i x^j} - F(x, u(x), Dv(x), 0)$$

$$= F\left(x, u(x), Du(x), D^2u(x) \right) - \left(\sum_{i,j=1}^{d} a^{ij}(x) v_{x^i x^j} + F\left(x, u(x), Dv(x), 0 \right) \right),$$

$$\qquad (2.3.10)$$

with

$$a^{ij}(x) = \int_0^1 \frac{\partial F}{\partial r_{ij}} \left(x, u(x), Dv(x), tD^2u(x) \right) dt \qquad (2.3.11)$$

(this again comes from the integral of a total derivative with respect to t). Here by assumption

$$\lambda \, |\xi|^2 \le \sum_{i,j=1}^{d} a^{ij}(x)\xi^i\xi^j \quad \text{for all } x \in \Omega, \xi \in \mathbb{R}^d. \tag{2.3.12}$$

We now look for an appropriate auxiliary function v with

$$Mv := \sum a^{ij}(x)v_{x^ix^j} + F(x, u(x), Dv(x), 0) \le 0. \tag{2.3.13}$$

We now suppose that for $\delta := \operatorname{diam}(\Omega)$, Ω is contained in the strip $\{0 < x^1 < \delta\}$. We now try

$$v(x) = \max_{\partial\Omega} u^+ + \frac{\mu_2}{\lambda}\left(e^{(\mu_1+1)\delta} - e^{(\mu_1+1)x^1}\right) \tag{2.3.14}$$

$(u^+(x) = \max(0, u(x)))$.

Then

$$Mv = -\frac{\mu_2}{\lambda}\left(\mu_1 + 1\right)^2 a^{11}(x)e^{(\mu_1+1)x^1} + F(x, u(x), Dv(x), 0)$$
$$\le -\mu_2\left(\mu_1 + 1\right)^2 e^{(\mu_1+1)x^1} + \mu_2\mu_1\left(\mu_1 + 1\right)e^{(\mu_1+1)x^1} + \mu_2$$
$$\le 0$$

by (2.3.6), (2.3.12). This establishes (2.3.13). Equation (2.3.10) then implies, even under the assumption $F[u] \ge 0$ in place of $F[u] = 0$,

$$Lw \ge 0.$$

By definition of v, we also have

$$w = u - v \le 0 \quad \text{on } \partial\Omega.$$

Theorem 2.1.1 thus implies

$$u \le v \quad \text{in } \Omega,$$

and (2.3.7) follows with $c = e^{(\mu_1+1)\operatorname{diam}(\Omega)} - 1$. More precisely, under the assumption $F[u] \ge 0$, we have proved the inequality

$$\sup_{\Omega} u \le \max_{\partial\Omega} u^+ + c\frac{\mu_2}{\lambda}, \tag{2.3.15}$$

but the inequality in the other direction of course follows analogously, i.e.,

$$\inf_{\Omega} u \ge \min_{\partial\Omega} u^- - c\frac{\mu_2}{\lambda} \tag{2.3.16}$$

$(u^-(x) := \min(0, u(x)))$. $\square$

Theorem 2.3.2 is of interest even in the linear case. Let us look once more at the simple equation

$$f''(x) + \kappa f(x) = 0 \quad \text{for } x \in (0, \pi),$$
$$f(0) = f(\pi) = 0,$$

with constant κ. We may apply Theorem 2.3.2 with $\lambda = 1$, $\mu_1 = 0$,

$$\mu_2 = \begin{cases} \kappa \sup_{(0,\pi)} |f| & \text{for } \kappa > 0, \\ 0 & \text{for } \kappa \leq 0. \end{cases}$$

It follows that

$$\sup_{(0,\pi)} |f| \leq c\kappa \sup_{(0,\pi)} |f| \,;$$

i.e., if

$$\kappa < \frac{1}{c},$$

we must have $f \equiv 0$. More generally, in place of κ, one may take any function $c(x)$ with $c(x) \leq \kappa$ on $(0, \pi)$ and consider $f''(x) + c(x)f(x) = 0$, without affecting the preceding conclusion. In particular, this allows us to weaken the sign condition $c(x) \leq 0$. The sharpest possible result here is that $f \equiv 0$ if κ is smaller than the smallest eigenvalue λ_1 of $\frac{d^2}{dx^2}$ on $(0, \pi)$, i.e., 1. This analogously generalizes to other linear elliptic equations, e.g.,

$$\Delta f(x) + \kappa f(x) = 0 \quad \text{in } \Omega,$$
$$f(y) = 0 \quad \text{on } \partial\Omega.$$

Theorem 2.3.2 does imply such a result, but not with the optimal bound λ_1.

A reference for the present chapter is Gilbarg–Trudinger [9].

Summary and Perspectives

The maximum principle yields examples of so-called a priori estimates, i.e., estimates that hold for any solution of a given differential equation or class of equations, depending on the given data (boundary values, right-hand side, etc.), without the need to know the solution in advance or without even having to guarantee in advance that a solution exists. Conversely, such a priori estimates often constitute an important tool in many existence proofs. Maximum principles are characteristic for solutions of elliptic (and parabolic) PDEs, and they are not restricted to linear equations. Often, they are even the most important tool for studying certain nonlinear elliptic PDEs.

Exercises

2.1 Let $\Omega_1, \Omega_2 \subset \mathbb{R}^d$ be disjoint open sets such that $\bar{\Omega}_1 \cap \bar{\Omega}_2$ contains a smooth hypersurface T, e.g.,

$$\Omega_1 := \{(x^1, \ldots, x^d) : |x| < 1, x^1 > 0\},$$
$$\Omega_2 := \{(x^1, \ldots, x^d) : |x| < 1, x^1 < 0\},$$
$$T \ = \{(x^1, \ldots, x^d) : |x| < 1, x^1 = 0\}.$$

Let $u \in C^0(\bar{\Omega}_1 \cup \bar{\Omega}_2) \cap C^2(\Omega_1) \cap C^2(\Omega_2)$ be harmonic on Ω_1 and on Ω_2, i.e.,

$$\Delta u(x) = 0, \quad x \in \Omega_1 \cup \Omega_2.$$

Does this imply that u is harmonic on $\Omega_1 \cup \Omega_2 \cup T$?

2.2 Let Ω be open in $\mathbb{R}^2 = \{(x, y)\}$. For a nonconstant solution $u \in C^2(\Omega)$ of the differential equation

$$u_{xy} = 0 \quad \text{in } \Omega,$$

is it possible to assume an interior maximum in Ω?

2.3 Let Ω be open and bounded in $\mathbb{R}^d$. On

$$\Omega \times [0, \infty) \subset \mathbb{R}^{d+1} = \{(x^1, \ldots, x^d, t)\},$$

we consider the heat equation

$$u_t = \Delta u, \quad \text{where } \Delta = \sum_{i=1}^{d} \frac{\partial^2}{(\partial x^i)^2}.$$

Show that for bounded solutions $u \in C^2(\Omega \times (0, \infty)) \cap C^0(\bar{\Omega} \times [0, \infty))$,

$$\sup_{\Omega \times [0, \infty)} u \le \sup_{(\bar{\Omega} \times \{0\}) \cup (\partial\Omega \times [0, \infty))} u.$$

2.4 Let $u : \Omega \to \mathbb{R}$ be harmonic, $\Omega' \subset\subset \Omega \subset \mathbb{R}^d$. We then have, for all i, j between 1 and d,

$$\sup_{\Omega'} |u_{x^i x^j}| \le \left(\frac{2d}{\text{dist}(\Omega', \partial\Omega)} \right)^2 \sup_{\Omega} |u|.$$

Prove this inequality. Write down and demonstrate an analogous inequality for derivatives of arbitrary order!

2.5 Let $\Omega \subset \mathbb{R}^d$ be open and bounded. Let $u \in C^2(\Omega) \cap C^0(\bar{\Omega})$ satisfy

$$\Delta u = u^3, \quad x \in \Omega,$$
$$u \equiv 0, \quad x \in \partial\Omega.$$

Show that $u \equiv 0$ in Ω.

2.6 Prove a version of the maximum principle of Alexandrov and Bakelman for operators

$$Lu = \sum_{i,j=1}^{n} a^{ij}(x)u_{x^i x^j}(x),$$

assuming in place of ellipticity only that $\det(a^{ij}(x))$ is positive in Ω.

2.7 Control the maximum and minimum of the solution u of an elliptic Monge–Ampère equation

$$\det(u_{x^i x^j}(x)) = f(x)$$

in a bounded domain Ω.

2.8 Let $u \in C^2(\Omega)$ be a solution of the Monge–Ampère equation

$$\det(u_{x^i x^j}(x)) = f(x)$$

in the domain Ω with positive f. Suppose there exists $x_0 \in \Omega$ where the Hessian of u is positive definite. Show that the equation then is elliptic at u in all of Ω.

2.9 Let $\mathbb{R}^2 := \{(x^1, x^2)\}, \Omega := \mathring{B}(0, R_2) \setminus B(0, R_1)$ with $R_2 > R_1 > 0$. The function $\phi(x^1, x^2) := a + b\log(|x|)$ is harmonic in Ω for all a, b. Let $u \in C^2(\Omega) \cap C^0(\bar{\Omega})$ be subharmonic, i.e.,

$$\Delta u \geq 0, \quad x \in \Omega.$$

Show that

$$M(r) \leq \frac{M(R_1)\log(\frac{R_2}{r}) + M(R_2)\log(\frac{r}{R_1})}{\log(\frac{R_2}{R_1})}$$

with

$$M(r) := \max_{\partial B(0,r)} u(x)$$

and $R_1 \leq r \leq R_2$.

2.10 Let

$$u_1 := \frac{1}{2} + \frac{1}{2}(x^2 + y^2),$$

$$u_2 := \frac{3}{2} - \frac{1}{2}(x^2 + y^2).$$

Show that u_1 and u_2 solve the Monge–Ampère equation

$$u_{xx}u_{yy} - u_{xy}^2 = 1$$

and

$$u_1 = u_2 = 1 \quad \text{on } \partial B(0, 1).$$

Is this compatible with the uniqueness result for the Dirichlet problem for nonlinear elliptic PDEs?

2.11 Let $\Omega_T := \Omega \times (0, T)$, and suppose $u \in C^2(\Omega_T) \cap C^0(\bar{\Omega}_T)$ satisfies

$$u_t = \Delta u + u^2 \quad \text{in } \Omega_T,$$
$$u(x, t) > c > 0 \qquad \text{for } (x, t) \in (\Omega \times \{0\}) \cup (\partial\Omega \times [0, T)).$$

Show that
(a) $u > c$ for all $(x, t) \in \bar{\Omega}_T$.
(b) If in addition $u(x, t) = u(x, 0)$ for all $x \in \partial\Omega$ and all t, then $T < \infty$.

3. Existence Techniques I: Methods Based on the Maximum Principle

3.1 Difference Methods: Discretization of Differential Equations

The basic idea of the difference methods consists in replacing the given differential equation by a difference equation with step size h and trying to show that for $h \to 0$, the solutions of the difference equations converge to a solution of the differential equation. This is a constructive method that in particular is often applied for the numerical (approximative) computation of solutions of differential equations. In order to show the essential aspects of this method in a setting that is as simple as possible, we consider only the Laplace equation

$$\Delta u = 0 \tag{3.1.1}$$

in a bounded domain in Ω in $\mathbb{R}^d$. We cover $\mathbb{R}^d$ with an orthogonal grid of mesh size $h > 0$; i.e., we consider the points or vertices

$$\left(x^1, \ldots, x^d\right) = (n_1 h, \ldots, n_d h) \tag{3.1.2}$$

with $n_1, \ldots, n_d \in \mathbb{Z}$. The set of these vertices is called $\mathbb{R}_h^d$, and we put

$$\bar{\Omega}_h := \Omega \cap \mathbb{R}_h^d. \tag{3.1.3}$$

We say that $x = (n_1 h, \ldots, n_d h)$ and $y = (m_1 h, \ldots, m_d h)$ (all $n_i, m_j \in \mathbb{Z}$) are neighbors if

$$\sum_{i=1}^{d} |n_i - m_i| = 1, \tag{3.1.4}$$

or equivalently,

$$|x - y| = h. \tag{3.1.5}$$

The straight lines between neighboring vertices are called edges. A connected union of edges for which every vertex is contained in at most two edges is called an edge path (see Figure 3.1).

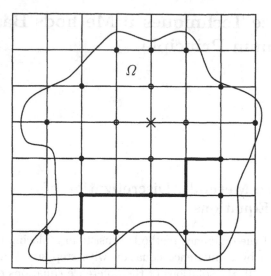

Figure 3.1. x (cross) and its neighbors (open dots) and an edge path in $\bar{\Omega}_h$ (heavy line) and vertices from Γ_h (solid dots).

The boundary vertices of $\bar{\Omega}_h$ are those vertices of $\bar{\Omega}_h$ for which not all their neighbors are contained in $\bar{\Omega}_h$. Let Γ_h be the set of boundary vertices. Vertices in $\bar{\Omega}_h$ that are not boundary vertices are called interior vertices. The set of interior vertices is called Ω_h.

We suppose that Ω_h is discretely connected, meaning that any two vertices in Ω_h can be connected by an edge path in Ω_h. We consider a function

$$u : \bar{\Omega}_h \to \mathbb{R}$$

and put, for $i = 1, \ldots, d$, $x = (x^1, \ldots, x^d) \in \Omega_h$,

$$u_i(x) := \frac{1}{h} \left(u(x^1, \ldots, x^{i-1}, x^i + h, x^{i+1}, \ldots, x^d) - u(x^1, \ldots, x^d) \right),$$

$$u_{\bar{i}}(x) := \frac{1}{h} \left(u(x^1, \ldots, x^d) - u(x^1, \ldots, x^{i-1}, x^i - h, x^{i+1}, \ldots, x^d) \right). \quad (3.1.6)$$

Thus, u_i and $u_{\bar{i}}$ are the forward and backward difference quotients in the ith coordinate direction. Analogously, we define higher-order difference quotients, e.g.,

$$u_{i\bar{i}}(x) = u_{\bar{i}i}(x) = (u_{\bar{i}})_i(x)$$

$$= \frac{1}{h^2} \left(u(x^1, \ldots, x^i + h, \ldots, x^d) - 2u(x^1, \ldots, x^d) \right.$$

$$\left. + u(x^1, \ldots, x^i - h, \ldots, x^d) \right). \quad (3.1.7)$$

If we wish to emphasize the dependence on the mesh size h, we write $u^h, u_i^h, u_{i\bar{i}}^h$ in place of $u, u_i, u_{i\bar{i}}$, etc.

The main reason for considering difference quotients, of course, is that for functions that are differentiable up to the appropriate order, for $h \to 0$, the difference quotients converge to the corresponding derivatives. For example, for $u \in C^2(\Omega)$,

$$\lim_{h \to 0} u_{i\bar{i}}^h(x_h) = \frac{\partial^2}{(\partial x^i)^2} u(x), \tag{3.1.8}$$

if $x_h \in \Omega_h$ tends to $x \in \Omega$ for $h \to 0$. Consequently, we approximate the Laplace equation

$$\Delta u = 0 \quad \text{in } \Omega$$

by the difference equation

$$\Delta_h u^h := \sum_{i=1}^d u_{i\bar{i}}^h = 0 \quad \text{in } \Omega_h, \tag{3.1.9}$$

and we call this equation the discrete Laplace equation. Our aim now is to solve the Dirichlet problem for the discrete Laplace equation

$$\Delta_h u^h = 0 \quad \text{in } \Omega_h,$$
$$u^h = g^h \quad \text{on } \Gamma_h, \tag{3.1.10}$$

and to show that under appropriate assumptions, the solutions u^h converge for $h \to 0$ to a solution of the Dirichlet problem

$$\Delta u = 0 \quad \text{in } \Omega,$$
$$u = g \quad \text{on } \partial\Omega, \tag{3.1.11}$$

where g^h is a discrete approximation of g. Considering the values of u^h at the vertices of Ω_h as unknowns, (3.1.10) leads to a linear system with the same number of equations as unknowns. Those equations that come from vertices all of whose neighbors are interior vertices themselves are homogeneous, while the others are inhomogeneous.

It is a remarkable and useful fact that many properties of the Laplace equation continue to hold for the discrete Laplace equation. We start with the *discrete maximum principle:*

Theorem 3.1.1: *Suppose*

$$\Delta_h u^h \geq 0 \quad \text{in } \Omega_h,$$

where Ω_h, as always, is supposed to be discretely connected. Then

$$\max_{\Omega_h} u^h = \max_{\Gamma_h} u^h. \tag{3.1.12}$$

If the maximum is assumed at an interior point, then u^h has to be constant.

Proof: Let x_0 be an interior vertex, and let $x_1, \ldots, x_{2d}$ be its neighbors. Then

$$\Delta_h u^h(x) = \frac{1}{h^2} \left(\sum_{\alpha=1}^{2d} u^h(x_\alpha) - 2d u^h(x_0) \right). \tag{3.1.13}$$

If $\Delta_h u^h(x) \geq 0$, then

$$u^h(x_0) \leq \frac{1}{2d} \sum_{\alpha=1}^{2d} u^h(x_\alpha), \tag{3.1.14}$$

i.e., $u^h(x_0)$ is not bigger than the arithmetic mean of the values of u^h at the neighbors of x_0. This implies

$$u^h(x_0) \leq \max_{\alpha=1,\ldots,2d} u^h(x_\alpha), \tag{3.1.15}$$

with equality only if

$$u^h(x_0) = u^h(x_\alpha) \quad \text{for all } \alpha \in \{1, \ldots, 2d\}. \tag{3.1.16}$$

Thus, if u assumes an interior maximum at a vertex x_0, it does so at all neighbors of x_0 as well, and repeating this reasoning, then also at all neighbors of neighbors, etc. Since Ω_h is discretely connected by assumption, u_h has to be constant in $\bar{\Omega}_h$. This is the strong maximum principle, which in turn implies the weak maximum principle (3.1.12). □

Corollary 3.1.1: *The discrete Dirichlet problem*

$$\Delta_h u^h = 0 \quad \text{in } \Omega_h,$$
$$u^h = g^h \quad \text{on } \Gamma^h,$$

for given g^h has at most one solution.

Proof: This follows in the usual manner by applying the maximum principle to the difference of two solutions. □

It is remarkable that in the discrete case this uniqueness result already implies an existence result:

Corollary 3.1.2: *The discrete Dirichlet problem*

$$\Delta_h u^h = 0 \quad \text{in } \Omega_h,$$
$$u^h = g^h \quad \text{on } \Gamma^h,$$

admits a unique solution for each $g^h : \Gamma_h \to \mathbb{R}$.

Proof: As already observed, the discrete problem constitutes a finite system of linear equations with the same number of equations and unknowns. Since by Corollary 3.1.1, for homogeneous boundary data $g^h = 0$, the homogeneous solution $u^h = 0$ is the unique solution, the fundamental theorem of linear algebra implies the existence of a solution for an arbitrary right-hand side, i.e., for arbitrary g^h.   □

The solution of the discrete Poisson equation

$$\Delta_h u^h = f^h \quad \text{in } \Omega^h \tag{3.1.17}$$

with given f^h is similarly simple; here, without loss of generality, we consider only the homogeneous boundary condition

$$u^h = 0 \quad \text{on } \Gamma^h, \tag{3.1.18}$$

because an inhomogeneous condition can be treated by adding a solution of the corresponding discrete Laplace equation.

In order to represent the solution, we shall now construct a Green function $G^h(x, y)$. For that purpose, we consider a particular f^h in (3.1.17), namely,

$$f^h(x) = \begin{cases} 0 & \text{for } x \neq y, \\ \frac{1}{h^2} & \text{for } x = y, \end{cases}$$

for given $y \in \Omega_h$. Then $G^h(x, y)$ is defined as the solution of (3.1.17), (3.1.18) for that f^h. The solution for an arbitrary f^h is then obtained as

$$u^h(x) = h^2 \sum_{y \in \Omega_h} G^h(x, y) f^h(y). \tag{3.1.19}$$

In order to show that solutions of the discrete Laplace equation $\Delta_h u^h = 0$ in Ω_h for $h \to 0$ converge to a solution of the Laplace equation $\Delta u = 0$ in Ω we need estimates for the u^h that do not depend on h. It turns out that as in the continuous case, such estimates can be obtained with the help of the maximum principle. Namely, for the symmetric difference quotient

$$\begin{aligned} u_{\bar i}(x) &:= \frac{1}{2h} \big(u(x^1, \ldots, x^{i-1}, x^i + h, x^{i+1}, \ldots, x^d) \\ &\quad - u(x^1, \ldots, x^{i-1}, x^i - h, x^{i+1}, \ldots, x^d) \big) \\ &= \frac{1}{2} \big(u_i(x) + u_{\bar i}(x) \big) \end{aligned} \tag{3.1.20}$$

we may prove in complete analogy with Corollary 1.2.7 the following result:

Lemma 3.1.1: *Suppose that in Ω_h,*

$$\Delta_h u^h(x) = f^h(x). \tag{3.1.21}$$

Let $x_0 \in \Omega_h$, and suppose that x_0 and all its neighbors have distance greater than or equal to R from Γ_h. Then

$$\left| u_i^h(x_0) \right| \leq \frac{d}{R} \max_{\Omega_h} \left| u^h \right| + \frac{R}{2} \max_{\Omega_h} \left| f^h \right|. \tag{3.1.22}$$

Proof: Without loss of generality $i = 1$, $x_0 = 0$. We put

$$\mu := \max_{\Omega_h} \left| u^h \right|, \quad M := \max_{\Omega_h} \left| f^h \right|.$$

We consider once more the auxiliary function

$$v^h(x) := \frac{\mu}{R^2} |x|^2 + x^1(R - x^1) \left(\frac{d\mu}{R^2} + \frac{M}{2} \right).$$

Because of

$$\Delta_h |x|^2 = \sum_{i=1}^{d} \frac{1}{h^2} \left((x^i + h)^2 + (x^i - h)^2 - 2(x^i)^2 \right) = 2d,$$

we have again

$$\Delta_h v^h(x) = -M$$

as well as

$$v^h(0, x^2, \ldots, x^d) \geq 0 \quad \text{for all } x^2, \ldots, x^d,$$
$$v^h(x) \geq \mu \quad \text{for } |x| \geq R, \quad 0 \leq x^1 \leq R.$$

Furthermore, for $\bar{u}^h(x) := \frac{1}{2}(u^h(x^1, \ldots, x^d) - u^h(-x^1, x^2, \ldots, x^d))$,

$$\left| \Delta_h \bar{u}^h(x) \right| \leq M \quad \text{for those } x \in \Omega_h, \text{ for which this expression is}$$
$$\text{defined},$$
$$\bar{u}^h(0, x^2, \ldots, x^d) = 0 \quad \text{for all } x^2, \ldots, x^d,$$
$$\left| \bar{u}^h(x) \right| \leq \mu \quad \text{for } |x| \geq R, \quad x^1 \geq 0.$$

On the discretization B_h^+ of the half-ball $B^+ := \{ |x| \leq R, x^1 > 0 \}$, we thus have

$$\Delta_h \left(v^h \pm \bar{u}^h \right) \leq 0$$

as well as

$$v^h \pm \bar{u}^h \geq 0 \quad \text{on the discrete boundary of } B_h^+$$

(in order to be precise, here one should take as the discrete boundary all vertices in the exterior of $\mathring{B}^+$ that have at least one neighbor in $\mathring{B}^+$). The maximum principle (Theorem 3.1.1) yields

$$\left|\bar{u}^h\right| \leq v^h \quad \text{in } B_h^+,$$

and hence

$$\left|u_i^h(0)\right| = \frac{1}{h}\left|\bar{u}^h(h, 0, \ldots, 0)\right| \leq \frac{1}{h}v^h(h, 0, \ldots, 0)$$
$$= \frac{d\mu}{R} + \frac{R}{2}M + \frac{\mu}{R^2}(1-d)h.$$

$\square$

For solutions of the discrete Laplace equation

$$\Delta_h u^h = 0 \quad \text{in } \Omega_h, \tag{3.1.23}$$

we then inductively get estimates for higher-order difference quotients, because if u^h is a solution, so are all difference quotients $u_i^h, u_{\bar{i}}^h, u_i^h u_{ii}^h, u_{\bar{i}\bar{i}}^h$, etc. For example, from (3.1.22) we obtain for a solution of (3.1.23) that if x_0 is far enough from the boundary Γ_h, then

$$\left|u_{\bar{i}\bar{i}}^h(x_0)\right| \leq \frac{d}{R} \max_{\Omega_h} \left|u_{\bar{i}}^h\right| \leq \frac{d^2}{R^2} \max_{\bar{\Omega}_h} \left|u^h\right| = \frac{d^2}{R^2} \max_{\Gamma_h} \left|u^h\right|. \tag{3.1.24}$$

Thus, by induction, we can bound difference quotients of any order, and we obtain the following theorem:

Theorem 3.1.2: *If all solutions u^h of*

$$\Delta_h u^h = 0 \quad \text{in } \Omega_h$$

are bounded independently of h (i.e., $\max_{\Gamma_h} \left|u^h\right| \leq \mu$), then in any subdomain $\tilde{\Omega} \subset\subset \Omega$, some subsequence of u^h converges to a harmonic function as $h \to 0$.

Convergence here first means convergence with respect to the supremum norm, i.e.,

$$\lim_{n \to 0} \max_{x \in \Omega_n} \left|u_n(x) - u(x)\right| = 0,$$

with harmonic u. By the preceding considerations, however, the difference quotients of u_n converge to the corresponding derivatives of u as well. $\square$

We wish to briefly discuss some aspects of difference equations that are important in numerical analysis. There, for theoretical reasons, one assumes that one already knows the existence of a smooth solution of the differential equation under consideration, and one wants to approximate that solution by solutions of difference equations. For that purpose, let L be an elliptic differential operator and consider discrete operators L_h that are applied to the restriction of a function u to the lattice Ω_h.

Definition 3.1.1: *The difference scheme L_h is called* consistent *with L if*

$$\lim_{h \to 0} (Lu - L_h u) = 0$$

for all $u \in C^2(\bar{\Omega})$.

The scheme L_h is called convergent *to L if the solutions u, u^h of*

$$Lu = f \quad \text{in } \Omega, u = \varphi \text{ on } \partial\Omega,$$
$$L_h u^h = f^h \quad \text{in } \Omega_h, \text{where } f^h \text{ is the restriction of } f \text{ to } \Omega_h,$$
$$u^h = \varphi^h \quad \text{on } \Gamma_h, \text{ where } \varphi^h \text{ is the restriction to } \Omega_h \text{ of a}$$
$$\text{continuous extension of } \varphi,$$

satisfy

$$\lim_{h \to 0} \max_{x \in \Omega_h} |u^h(x) - u(x)| = 0.$$

In order to see the relation between convergence and consistency we consider the "global error"

$$\sigma(x) := u^h(x) - u(x)$$

and the "local error"

$$s(x) := L_h u(x) - Lu(x)$$

and compute, for $x \in \Omega_h$,

$$L_h \sigma(x) = L_h u^h(x) - L_h u(x) = f^h(x) - Lu(x) - s(x)$$
$$= -s(x), \text{ since } f^h(x) = f(x) = Lu(x).$$

Since

$$\lim_{h \to 0} \sup_{x \in \Gamma_h} |\sigma(x)| = 0,$$

the problem essentially is

$$L_h \sigma(x) = -s(x) \quad \text{in } \Omega_h,$$
$$\sigma(x) = 0 \qquad \text{on } \Gamma_h.$$

In order to deduce the convergence of the scheme from its consistency, one thus needs to show that if $s(x)$ tends to 0, so does the solution $\sigma(x)$, and in fact uniformly. Thus, the inverses L_h^{-1} have to remain bounded in a sense that we shall not make precise here. This property is called *stability*.

In the spirit of these notions, let us show the following simple convergence result:

Theorem 3.1.3: *Let $u \in C^2(\bar{\Omega})$ be a solution of*

$$\Delta u = f \quad in \ \Omega,$$
$$u = \varphi \quad on \ \partial\Omega.$$

Let u^h be the solution

$$\Delta_h u^h = f^h \quad in \ \Omega_h,$$
$$u^h = \varphi^h \quad on \ \Gamma_h,$$

where f^h, φ^h are defined as above. Then

$$\max_{x \in \Omega_h} \left| u^h(x) - u(x) \right| \to 0 \quad for \ h \to 0.$$

Proof: Taylor's formula implies that the second-order difference quotients (which depend on the mesh size h) satisfy

$$u_{i\bar{i}}(x) = \frac{\partial^2 u}{(\partial x^i)^2} \left(x^1, \ldots, x^{i-1}, x^i + \delta^i, x^{i+1}, \ldots, x^d \right),$$

with $-h \leq \delta^i \leq h$. Since $u \in C^2(\bar{\Omega})$, we have

$$\sup_{|\delta^i| \leq h} \left(\frac{\partial^2 u}{(\partial x^i)^2}(x^1, \ldots, x^i + \delta^i, \ldots, x^d) - \frac{\partial^2 u}{(\partial x^i)^2}(x^1, \ldots, x^i, \ldots, x^d) \right) \to 0$$

for $h \to 0$, and thus the above local error satisfies

$$\sup |s(x)| \to 0 \quad for \ h \to 0.$$

Now let Ω be contained in a ball $B(x_0, R)$; without loss of generality $x_0 = 0$.

The maximum principle then implies, through comparison with the function $R^2 - |x|^2$, that a solution v of

$$\Delta_h v = \eta \quad in \ \Omega_h,$$
$$v = 0 \quad on \ \Gamma_h,$$

satisfies the estimate

$$|v(x)| \leq \frac{\sup |\eta|}{2d} \left(R^2 - |x|^2 \right).$$

Thus, the global error satisfies

$$\sup |\sigma(x)| \leq \frac{R^2}{2d} \sup |s(x)|,$$

hence the desired convergence. $\square$

3.2 The Perron Method

Let us first recall the notion of a subharmonic function from Section 1.2, since this will play a crucial role:

Definition 3.2.1: *Let $\Omega \subset \mathbb{R}^d$, $f : \Omega \to [-\infty, \infty)$ upper semicontinuous in Ω, $f \not\equiv -\infty$. The function f is called subharmonic in Ω if for all $\Omega' \subset\subset \Omega$, the following property holds:*

> *If u is harmonic in Ω', and $f \le u$ on $\partial\Omega'$, then also $f \le u$ in Ω'.*

The next lemma likewise follows from the results of Section 1.2:

Lemma 3.2.1:

 (i) *Strong maximum principle: Let v be subharmonic in Ω. If there exists $x_0 \in \Omega$ with $v(x_0) = \sup_\Omega v(x)$, then v is constant. In particular, if $v \in C^0(\bar{\Omega})$, then $v(x) \le \max_{\partial\Omega} v(y)$ for all $x \in \mathbb{R}$.*
 (ii) *If $v_1, \ldots, v_n$ are subharmonic, so is $v := \max(v_1, \ldots, v_n)$.*
 (iii) *If $v \in C^0(\bar{\Omega})$ is subharmonic and $B(y, R) \subset\subset \Omega$, then the harmonic replacement $\bar{v}$ of v, defined by*

$$\bar{v}(x) := \begin{cases} v(x) & \text{for } x \in \Omega \setminus B(y, R), \\ \frac{R^2 - |x-y|^2}{dw_d R} \int_{\partial B(y,R)} \frac{v(z)}{|z-x|^d} do(z) & \text{for } x \in B(y, R), \end{cases}$$

is subharmonic in Ω (and harmonic in $B(y, R)$).

Proof:

 (i) This is the strong maximum principle for subharmonic functions. Although we have not written it down explicitly, it is a direct consequence of Theorem 1.2.2 and Lemma 1.2.1.
 (ii) Let $\Omega' \subset\subset \Omega$, u harmonic on $\partial\Omega'$, $v \le u$ on $\partial\Omega'$. Then also

$$v_i \le u \quad \text{on } \partial\Omega' \quad \text{for } i = 1, \ldots, n,$$

and hence, since v_i is subharmonic,

$$v_i \le u \quad \text{on } \Omega'.$$

This implies

$$v_i \le u \quad \text{on } \Omega',$$

showing that v is subharmonic.

(iii) First $v \leq \bar{v}$, since v is subharmonic. Let $\Omega' \subset\subset \Omega$, u harmonic on Ω', $\bar{v} \leq u$ on $\partial\Omega'$. Since $v \leq \bar{v}$, also $v \leq u$ on $\partial\Omega'$, and thus, since v is subharmonic, $v \leq u$ on Ω' and thus $\bar{v} \leq u$ on $\Omega' \setminus \mathring{B}(y, R)$. Therefore, also $\bar{v} \leq u$ on $\Omega' \cap \partial B(y, R)$. Since $\bar{v}$ is harmonic, hence subharmonic on $\Omega' \cap B(y, R)$, we get $\bar{v} \leq u$ on $\Omega' \cap B(y, R)$. Altogether, we obtain $\bar{v} \leq u$ on Ω'. This shows that $\bar{v}$ is subharmonic.

□

For the sequel, let φ be a bounded function on Ω (not necessarily continuous).

Definition 3.2.2: *A subharmonic function $u \in C^0(\bar{\Omega})$ is called a subfunction with respect to φ if*

$$u \leq \varphi \quad \text{for all } x \in \partial\Omega.$$

Let S_φ be the set of all subfunctions with respect to φ. (Analogously, a superharmonic function $u \in C^0(\bar{\Omega})$ is called superfunction with respect to φ if $u \geq \varphi$ on $\partial\Omega$.)

The key point of the Perron method is contained in the following theorem:

Theorem 3.2.1: *Let*

$$u(x) := \sup_{v \in S_\varphi} v(x). \tag{3.2.1}$$

Then u is harmonic.

Remark: If $w \in C^2(\Omega) \cap C^0(\bar{\Omega})$ is harmonic on Ω, and if $w = \varphi$ on $\partial\Omega$, the maximum principle implies that for all subfunctions $v \in S_\varphi$, we have $v \leq w$ in Ω and hence

$$w(x) = \sup_{v \in S_\varphi} v(x).$$

Thus, w satisfies an extremal property. The idea of the Perron method (and the content of Theorem 3.2.1) is that, conversely, each supremum in S_φ yields a harmonic function.

Proof of Theorem 3.2.1: First of all, u is well-defined, since by the maximum principle $v \leq \sup_{\partial\Omega} \varphi < \infty$ for all $v \in S_\varphi$. Now let $y \in \Omega$ be arbitrary. By (3.2.1) there exists a sequence $\{v_n\} \subset S_\varphi$ with $\lim_{n \to \infty} v_n(y) = u(y)$. Replacing v_n by $\max(v_1, \ldots, v_n, \inf_{\partial\Omega} \varphi)$, we may assume without loss of generality that $(v_n)_{n \in \mathbb{N}}$ is a monotonically increasing, bounded sequence. We now choose R with $B(y, R) \subset\subset \Omega$ and consider the harmonic replacements $\bar{v}_n$ for $B(y, R)$. The maximum principle implies that $(\bar{v}_n)_{n \in \mathbb{N}}$ likewise is a monotonically increasing sequence of subharmonic functions that are even harmonic in $B(y, R)$. By the Harnack convergence theorem (Corollary 1.2.10), the sequence $(\bar{v}_n)$ converges uniformly on $B(y, R)$ towards some v that is harmonic on $B(y, R)$. Furthermore,

$$\lim_{n \to \infty} \bar{v}_n(y) = v(y) = u(y), \tag{3.2.2}$$

since $u \geq \bar{v}_n \geq v_n$ and $\lim_{n \to \infty} v_n(y) = u(y)$. By (3.2.1), we then have $v \leq u$ in $B(y, R)$. We now show that $v \equiv u$ in $B(y, R)$. Namely, if

$$v(z) < u(z) \quad \text{for some } z \in B(y, R), \tag{3.2.3}$$

by (3.2.1), we may find $\tilde{u} \in S_\varphi$ with

$$v(z) < \tilde{u}(z). \tag{3.2.4}$$

Now let

$$w_n := \max(v_n, \tilde{u}). \tag{3.2.5}$$

In the same manner as above, by the Harnack convergence theorem (Corollary 1.2.10), $\bar{w}_n$ converges uniformly on $B(y, R)$ towards some w that is harmonic on $B(y, R)$. Since $w_n \geq v_n$ and $w_n \in S_\varphi$, the maximum principle implies

$$v \leq w \leq u \quad \text{in } B(y, R). \tag{3.2.6}$$

By (3.2.2) we then have

$$w(y) = v(y), \tag{3.2.7}$$

and with the help of the strong maximum principle for harmonic functions (Corollary 1.2.3), we conclude that

$$w \equiv v \text{ in } B(y, R). \tag{3.2.8}$$

This is a contradiction, because by (3.2.4),

$$w(z) = \lim_{n \to \infty} \bar{w}_n(z) = \lim_{n \to \infty} \overline{\max(v_n(z), \tilde{u}(z))} \geq \tilde{u}(z) > v(z) = w(z).$$

Therefore, u is harmonic in Ω. $\square$

Theorem 3.2.1 tells us that we obtain a harmonic function by taking the supremum of all subfunctions of a bounded function y. It is not clear at all, however, that the boundary values of u coincide with y. Thus, we now wish to study the question of when the function $u(x) := \sup_{v \in S_\varphi} v(x)$ satisfies

$$\lim_{x \to \xi \in \partial\Omega} u(x) = \varphi(\xi).$$

For that purpose, we shall need the concept of a barrier.

Definition 3.2.3: *(a) Let $\xi \in \partial\Omega$. A function $\beta \in C^0(\overline{\Omega})$ is called a barrier at ξ with respect to Ω if*
 (i) $\beta > 0$ in $\overline{\Omega} \setminus \{\xi\}$; $\beta(\xi) = 0$,
 (ii) β is superharmonic in Ω.
(b) $\xi \in \partial\Omega$ is called regular if there exists a barrier β at ξ with respect to Ω.

Remark: The regularity is a local property of the boundary $\partial\Omega$: Let β be a local barrier at $\xi \in \partial\Omega$; i.e., there exists an open neighborhood $U(\xi)$ such that β is a barrier at ξ with respect to $U \cap \Omega$. If then $B(\xi, \rho) \subset\subset U$ and $m := \inf_{U \setminus B(\xi,\rho)} \beta$, then

$$\tilde{\beta} := \begin{cases} m & \text{for } x \in \overline{\Omega} \setminus B(\xi, \rho), \\ \min(m, \beta(x)) & \text{for } x \in \overline{\Omega} \cap B(\xi, \rho), \end{cases}$$

is a barrier at ξ with respect to Ω.

Lemma 3.2.2: *Suppose $u(x) := \sup_{v \in S_\varphi} v(x)$ in Ω. If ξ is a regular point of $\partial\Omega$, and φ is continuous at ξ, we have*

$$\lim_{x \to \xi} u(x) = \varphi(\xi). \tag{3.2.9}$$

Proof: Let $M := \sup_{\partial\Omega} |\varphi|$. Since ξ is regular, there exists a barrier β, and the continuity of y at ξ implies that for every $\varepsilon > 0$ there exists $\delta > 0$ and a constant $c = c(\varepsilon)$ such that

$$|\varphi(x) - \varphi(\xi)| < \varepsilon \quad \text{for } |x - \xi| < \delta, \tag{3.2.10}$$
$$c\beta(x) \geq 2M \quad \text{for } |x - \xi| \geq \delta \tag{3.2.11}$$

(the latter holds, since $\inf_{|x-\xi| \geq \delta} \beta(x) =: m > 0$ by definition of β). The functions

$$\varphi(\xi) + \varepsilon + c\beta(x),$$
$$\varphi(\xi) - \varepsilon - c\beta(x),$$

then are super- and subfamilies, respectively, with respect to φ, by (3.2.10), (3.2.11). By definition of u thus

$$\varphi(\xi) - \varepsilon - c\beta(x) \leq u(x),$$

and since superfunctions dominate subfunctions, we also have

$$u(x) \leq \varphi(\xi) + \varepsilon + c\beta(x).$$

Hence, altogether,

$$|u(x) - \varphi(\xi)| \leq \varepsilon + c\beta(x). \tag{3.2.12}$$

Since $\lim_{x \to \xi} \beta(x) = 0$, it follows that $\lim_{x \to \xi} u(x) = \varphi(\xi)$. $\qquad \square$

Theorem 3.2.2: *Let $\Omega \subset \mathbb{R}^d$ be bounded. The Dirichlet problem*

$$\Delta u = 0 \quad in \ \Omega,$$
$$u = \varphi \quad on \ \partial\Omega,$$

is solvable for all continuous boundary values φ if and only if all points $\xi \in \partial\Omega$ are regular.

Proof: If φ is continuous and $\partial\Omega$ is regular, then $u := \sup_{v \in S_\varphi} v$ solves the Dirichlet problem by Theorem 3.2.2. Conversely, if the Dirichlet problem is solvable for all continuous boundary values, we consider $\xi \in \partial\Omega$ and $\varphi(x) := |x - \xi|$. The solution u of the Dirichlet problem for that $\varphi \in C^0(\partial\Omega)$ then is a barrier at ξ with respect to Ω, since $u(\xi) = \varphi(\xi) = 0$ and since $\min_{\partial\Omega} \varphi(x) = 0$, by the strong maximum principle $u(x) > 0$, so that ξ is regular.   □

3.3 The Alternating Method of H.A. Schwarz

The idea of the alternating method consists in deducing the solvability of the Dirichlet problem on a union $\Omega_1 \cup \Omega_2$ from the solvability of the Dirichlet problems on Ω_1 and Ω_2. Of course, only the case $\Omega_1 \cap \Omega_2 \neq \emptyset$ is of interest here.

     In order to exhibit the idea, we first assume that we are able to solve the Dirichlet problem on Ω_1 and Ω_2 for arbitrary piecewise continuous boundary data without worrying whether or how the boundary values are assumed at their points of discontinuity. We shall need the following notation (see Figure 3.2):

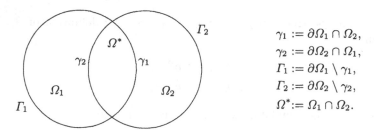

$$\gamma_1 := \partial\Omega_1 \cap \Omega_2,$$
$$\gamma_2 := \partial\Omega_2 \cap \Omega_1,$$
$$\Gamma_1 := \partial\Omega_1 \setminus \gamma_1,$$
$$\Gamma_2 := \partial\Omega_2 \setminus \gamma_2,$$
$$\Omega^* := \Omega_1 \cap \Omega_2.$$

Figure 3.2.

Then $\partial\Omega = \Gamma_1 \cup \Gamma_2$, and since we wish to consider sets Ω_1, Ω_2 that are overlapping, we assume $\partial\Omega^* = \gamma_1 \cup \gamma_2 \cup (\Gamma_1 \cap \Gamma_2)$. Thus, let boundary values φ by given on $\partial\Omega = \Gamma_1 \cup \Gamma_2$. We put

$$\varphi_i := \varphi|_{\Gamma_i} \quad (i = 1, 2),$$
$$m := \inf_{\partial\Omega} \varphi,$$
$$M := \sup_{\partial\Omega} \varphi.$$

We exclude the trivial case $\varphi = \text{const}$. Let $u_1 : \Omega_1 \to \mathbb{R}$ be harmonic with boundary values

$$u_1|_{\Gamma_1} = \varphi_1, \quad u_1|_{\gamma_1} = M. \tag{3.3.1}$$

Next, let $u_2 : \Omega_2 \to \mathbb{R}$ be harmonic with boundary values

$$u_2|_{\Gamma_2} = \varphi_2, \quad u_2|_{\gamma_2} = u_1|_{\gamma_2}. \tag{3.3.2}$$

Unless $\varphi_1 \equiv M$, by the strong maximum principle,

$$u_1 < M \quad \text{in } \Omega_1;^1 \tag{3.3.3}$$

hence in particular,

$$u_2|_{\gamma_2} < M, \tag{3.3.4}$$

and by the strong maximum principle, also

$$u_2 < M \quad \text{in } \Omega_2, \tag{3.3.5}$$

and thus in particular,

$$u_2|_{\gamma_1} < u_1|_{\gamma_1}. \tag{3.3.6}$$

If $\varphi_1 \equiv M$, then by our assumption that $\varphi \equiv \text{const}$ is excluded, $\varphi_2 \not\equiv M$, and (3.3.6) likewise holds by the maximum principle. Since by (3.3.2), u_1 and u_2 coincide on the partition of the boundary of Ω^*, by the maximum principle again

$$u_2 < u_1 \quad \text{in } \Omega^*.$$

Inductively, for $n \in \mathbb{N}$, let

$$u_{2n+1} : \Omega_1 \to \mathbb{R},$$
$$u_{2n+2} : \Omega_2 \to \mathbb{R},$$

be harmonic with boundary values

$$u_{2n+1}|_{\Gamma_1} = \varphi_1, \quad u_{2n+1}|_{\gamma_1} = u_{2n}|_{\gamma_1}, \tag{3.3.7}$$
$$u_{2n+2}|_{\Gamma_2} = \varphi_2, \quad u_{2n+2}|_{\gamma_2} = u_{2n+1}|_{\gamma_2}. \tag{3.3.8}$$

From repeated application of the strong maximum principle, we obtain

---

[1] The boundary values here are not continuous as in the maximum principle, but they can easily be approximated by continuous ones satisfying the same bounds. This easily implies that the maximum principle continues to hold in the present situation.

$$u_{2n+3} < u_{2n+2} < u_{2n+1} \quad \text{on } \Omega^*, \tag{3.3.9}$$

$$u_{2n+3} < u_{2n+1} \qquad\qquad \text{on } \Omega_1, \tag{3.3.10}$$

$$u_{2n+4} < u_{2n+2} \qquad\qquad \text{on } \Omega_2. \tag{3.3.11}$$

Thus, our sequences of functions are monotonically decreasing. Since they are also bounded from below by m, they converge to some limit

$$u : \Omega \to \mathbb{R}.$$

The Harnack convergence theorem (1.2.10)) then implies that u is harmonic on Ω_1 and Ω_2, hence also on $\Omega = \Omega_1 \cup \Omega_2$. This can also be directly deduced from the maximum principle: For simplicity, we extend u_n to all of Ω by putting

$$u_{2n+1} := u_{2n} \qquad \text{on } \Omega_2 \setminus \Omega^*,$$

$$u_{2n+2} := u_{2n+1} \quad \text{on } \Omega_1 \setminus \Omega^*.$$

Then u_{2n+1} is obtained from u_{2n} by harmonic replacement on Ω_1, and analogously, u_{2n+2} is obtained from u_{2n+1} by harmonic replacement on Ω_2. We write this symbolically as

$$u_{2n+1} = P_1 u_{2n}, \tag{3.3.12}$$

$$u_{2n+2} = P_2 u_{2n+1}. \tag{3.3.13}$$

For example, on Ω_1 we then have

$$u = \lim_{n \to \infty} u_{2n} = \lim_{n \to \infty} P_1 u_{2n}. \tag{3.3.14}$$

By the maximum principle, the uniform convergence of the boundary values (in order to get this uniform convergence, we may have to restrict ourselves to an arbitrary subdomain $\Omega_1' \subset\subset \Omega_1$) implies the uniform convergence of the harmonic extensions. Consequently, the harmonic extension of the limit of the boundary values equals the limit of the harmonic extensions, i.e.,

$$P_1 \lim_{n \to \infty} u_{2n} = \lim_{n \to \infty} P_1 u_{2n}. \tag{3.3.15}$$

Equation (3.3.14) thus yields

$$u = P_1 u, \tag{3.3.16}$$

meaning that on Ω_1, u coincides with the harmonic extension of its boundary values, i.e., is harmonic. For the same reason, u is harmonic on Ω_2.

We now assume that the boundary values φ are continuous, and that all boundary points of Ω_1 and Ω_2 are regular. Then first of all it is easy to see that u assumes its boundary values φ on $\partial\Omega \setminus (\Gamma_1 \cap \Gamma_2)$ continuously. To verify this, we carry out the same alternating process with harmonic functions $v_{2n-1} : \Omega_1 \to \mathbb{R}$, $v_{2n} : \Omega_2 \to \mathbb{R}$ starting with boundary values

$$v_1|_{\Gamma_1} = \varphi_1, \quad v_1|_{\gamma_1} = m \tag{3.3.17}$$

in place of (3.3.1). The resulting sequence $(v_n)_{n \in \mathbb{N}}$ then is monotonically increasing, and the maximum principle implies

$$v_n < u_n \text{ in } \Omega \quad \text{for all } n. \tag{3.3.18}$$

Since we assume that $\partial \Omega_1$ and $\partial \Omega_2$ are regular and φ is continuous, u_n and v_n then are continuous at every $x \in \partial \Omega \setminus (\Gamma_1 \cap \Gamma_2)$. The monotonicity of the sequence (u_n), the fact that $u_n(x) = v_n(x) = \varphi(x)$ for $x \in \partial \Omega \setminus (\Gamma_1 \cap \Gamma_2)$ for all n, and (3.3.18) then imply that $u = \lim_{n \to \infty} u_n$ at x as well.

The question whether u is continuous at $\partial \Omega_1 \cap \partial \Omega_2$ is more difficult, as can be expected already from the observation that the chosen boundary values for u_1 typically are discontinuous there even for continuous φ. In order to be able to treat that issue here in an elementary manner, we add the hypotheses that the boundaries of Ω_1 and Ω_2 are of class C^1 in some neighborhood of their intersection, and that they intersect at a nonzero angle. Under this hypotheses, we have the following lemma:

Lemma 3.3.1: *There exists some $q < 1$, depending only on Ω_1 and Ω_2, with the following property: If $w : \overline{\Omega_1} \to \mathbb{R}$ is harmonic in Ω_1, and continuous on the closure $\overline{\Omega_1}$, and if*

$$w = 0 \quad \text{on } \Gamma_1,$$
$$|w| \leq 1 \quad \text{on } \gamma_1,$$

then

$$|w| \leq q \quad \text{on } \gamma_2, \tag{3.3.19}$$

and a corresponding result holds if the roles of Ω_1 and Ω_2 are interchanged.

The *proof* will be given in Section 3.4 below.

With the help of this lemma we may now modify the alternating method in such a manner that we also get continuity on $\partial \Omega_1 \cap \partial \Omega_2$. For that purpose, we choose an arbitrary continuous extension $\bar{\varphi}$ of φ to γ_1, and in place of (3.3.1), for u_1 we require the boundary condition

$$u_1|_{\Gamma_1} = \varphi_1, \quad u_1|_{\gamma_1} = \bar{\varphi}, \tag{3.3.20}$$

and otherwise carry through the same procedure as above. Since the boundaries $\partial \Omega_1$, $\partial \Omega_2$ are assumed regular, all u_n then are continuous up to the boundary. We put

$$M_{2n+1} := \max_{\gamma_2} |u_{2n+1} - u_{2n-1}|,$$

$$M_{2n+2} := \max_{\gamma_1} |u_{2n+2} - u_{2n}|.$$

On γ_2, we then have

$$u_{2n+2} = u_{2n+1}, \; u_{2n} = u_{2n-1},$$

hence

$$u_{2n+2} - u_{2n} = u_{2n+1} - u_{2n-1},$$

and analogously on γ_1,

$$u_{2n+3} - u_{2n+1} = u_{2n+2} - u_{2n}.$$

Thus applying the lemma with $w = \frac{(u_{2n+3} - u_{2n+1})}{M_{2n+2}}$, we obtain

$$M_{2n+3} \leq q M_{2n+2}$$

and analogously

$$M_{2n+2} \leq q M_{2n+1}.$$

Thus M_n converges to 0 at least as fast as the geometric series with coefficient $q < 1$. This implies the uniform convergence of the series

$$u_1 + \sum_{n=1}^{\infty} (u_{2n+1} - u_{2n-1}) = \lim_{n \to \infty} u_{2n+1}$$

on $\bar{\Omega}_1$, and likewise the uniform convergence of the series

$$u_2 + \sum_{n=1}^{\infty} (u_{2n+2} - u_{2n}) = \lim_{n \to \infty} u_{2n}$$

on $\bar{\Omega}_2$. The corresponding limits again coincide in Ω^*, and they are harmonic on Ω_1, respectively Ω_2, so that we again obtain a harmonic function u on Ω. Since all the u_n are continuous up to the boundary and assume the boundary values given by φ on $\partial\Omega$, u then likewise assumes these boundary values continuously.

We have proved the following theorem:

Theorem 3.3.1: *Let Ω_1 and Ω_2 be bounded domains all of whose boundary points are regular for the Dirichlet problem. Suppose that $\Omega_1 \cap \Omega_2 \neq \emptyset$ and that Ω_1 and Ω_2 are of class C^1 in some neighborhood of $\partial\Omega_1 \cap \partial\Omega_2$, and that they intersect there at a nonzero angle. Then the Dirichlet problem for the Laplace equation on $\Omega := \Omega_1 \cup \Omega_2$ is solvable for any continuous boundary values.*

$\square$

3.4 Boundary Regularity

Our first task is to present the *proof of Lemma 3.3.1*:
In the sequel, with $r := |x - y| \neq 0$, we put

$$\Phi(r) := -dw_d\Gamma(r) = \begin{cases} \ln\frac{1}{r} & \text{for } d = 2, \\ \frac{1}{d-2}\frac{1}{r^{d-2}} & \text{for } d \geq 3. \end{cases} \qquad (3.4.1)$$

We then have for all $\nu \in \mathbb{R}^n$,

$$\frac{\partial}{\partial\nu}\Phi(r) = \nabla\Phi \cdot \nu = -\frac{1}{r^d}(x - y) \cdot \nu. \qquad (3.4.2)$$

We consider the situation depicted in Figure 3.3.

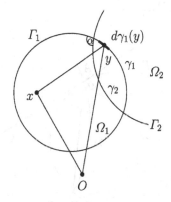

Figure 3.3.

That is, $x \in \Omega_1$; $y \in \gamma_1$, $\alpha \neq 0, \pi, \partial\Omega_1, \partial\Omega_2 \in C^1$. Let $d\gamma_1(y)$ be an infinitesimal boundary portion of γ_1 (see Figure 3.4).

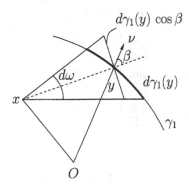

Figure 3.4.

Let $d\omega$ be the infinitesimal spatial angle at which the boundary piece $d\gamma_1(y)$ is seen from x. We then have

$$d\gamma_1(y) \cos \beta = |x - y|^{d-1} d\omega \tag{3.4.3}$$

and $\cos \beta = \frac{y-x}{|y-x|}$. This and (3.4.2) imply

$$h(x) := \int_{\gamma_1} \frac{\partial}{\partial \nu} \Phi(r) d\gamma_1(y) = \int_{\gamma_1} d\omega. \tag{3.4.4}$$

The geometric meaning of (3.4.4) is that $\int_{\gamma_1} \frac{\partial \Phi}{\partial \nu}(r) d\gamma_1(y)$ describes the spatial angle at which the boundary piece γ_1 is seen at x. Since derivatives of harmonic functions are harmonic as well, (3.4.4) yields a function h that is harmonic on Ω_1 and continuous on $\partial \Omega_1 \setminus (\Gamma_1 \cap \Gamma_2)$. In order to make the proof of Lemma 3.3.1 geometrically as transparent as possible, from now on, we only consider the case $d = 2$ and point out that the proof in the case $d \geq 3$ proceeds analogously.

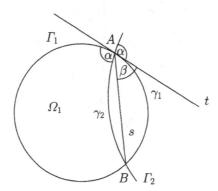

Figure 3.5.

Let A and B be the two points where Γ_1 and Γ_2 intersect (Figure 3.5). Then h is not continuous at A and B, because

$$\lim_{\substack{x \to A \\ x \in \Gamma_1}} h(x) = \beta, \tag{3.4.5}$$

$$\lim_{\substack{x \to A \\ x \in \gamma_1}} h(x) = \beta + \pi, \tag{3.4.6}$$

$$\lim_{\substack{x \to A \\ x \in \gamma_2}} h(x) = \alpha + \beta. \tag{3.4.7}$$

Let

$$\rho(x) := \pi \quad \text{for } x \in \gamma_1$$

and

$$\rho(x) := 0 \quad \text{for } x \in \Gamma_1.$$

Then $h|_{\partial\Omega_1} - \rho$ is continuous on all of $\partial\Omega_1$, because

$$\lim_{\substack{x \to A \\ x \in \Gamma_1}} (h(x) - \rho(x)) = \lim_{\substack{x \to A \\ x \in \Gamma_1}} h(x) - 0 = \beta,$$

$$\lim_{\substack{x \to A \\ x \in \gamma_1}} (h(x) - \rho(x)) = \lim_{\substack{x \to A \\ x \in \gamma_1}} h(x) - \pi = \beta + \pi - \pi = \beta.$$

By assumption, there then exists a function $u \in C^2(\Omega_1) \cap C^0(\bar{\Omega}_1)$ with

$$\Delta u = 0 \qquad \text{in } \Omega_1,$$
$$u = h|_{\partial\Omega_1} - \rho \quad \text{on } \partial\Omega_1.$$

For

$$v(x) := \frac{h(x) - u(x)}{\pi} \tag{3.4.8}$$

we have

$$\Delta v = 0 \quad \text{for } x \in \Omega_1,$$
$$v(x) = 0 \quad \text{for } x \in \Gamma_1,$$
$$v(x) = 1 \quad \text{for } x \in \gamma_1.$$

The strong maximum principle thus implies

$$v(x) < 1 \quad \text{for all } x \in \Omega_1, \tag{3.4.9}$$

and in particular,

$$v(x) < 1 \quad \text{for all } x \in \gamma_2. \tag{3.4.10}$$

Now

$$\lim_{\substack{x \to A \\ x \in \gamma_2}} v(x) = \frac{1}{\pi} \left(\lim_{\substack{x \to A \\ x \in \gamma_2}} h(x) - \beta \right) = \frac{\alpha}{\pi} < 1, \tag{3.4.11}$$

since $\alpha < \pi$ by assumption. Analogously, $\lim_{\substack{x \to B \\ x \in \gamma_2}} v(x) < 1$, and hence since $\bar{\gamma}_2$ is compact,

$$v(x) < q < 1 \quad \text{for all } x \in \bar{\gamma}_2 \tag{3.4.12}$$

for some $q > 0$. We put $m := v - w$ and obtain

$$m(x) = 0 \quad \text{for } x \in \Gamma_1,$$
$$m(x) \geq 0 \quad \text{for } x \in \gamma_1.$$

Since m is continuous in $\partial\Omega_1 \setminus (\Gamma_1 \cap \Gamma_2)$, and $\partial\Omega_1$ is regular, it follows that

$$\lim_{x \to x_0} m(x) = m(x_0) \quad \text{for all } x_0 \in \partial\Omega_1 \setminus (\Gamma_1 \cap \Gamma_2).$$

By the maximum principle, $m(x) \geq 0$ for all $x \in \Omega_1$, and since also

$$\lim_{x \to A} m(x) = \lim_{x \to A} v(x) - w(A) = \lim_{x \to A} v(x) \geq 0 \quad (w \text{ is continuous}),$$

we have for all $x \in \bar\gamma_2$,

$$w(x) \leq v(x) < q < 1. \tag{3.4.13}$$

The analogous considerations for $M := v + w$ yield the inequality

$$-w(x) \leq v(x) < q < 1; \tag{3.4.14}$$

hence, altogether,

$$|w(x)| < q < 1 \quad \text{for all } x \in \bar\gamma_2.$$

$\square$

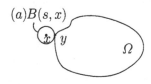

(a)$B(s, x)$

(b)

Figure 3.6.

We now wish to present a sufficient condition for the regularity of a boundary point $y \in \partial\Omega$:

Definition 3.4.1: Ω *satisfies an exterior sphere condition at* $y \in \partial\Omega$ *if there exists* $x_0 \in \mathbb{R}^n$ *with* $\bar{B}(\rho, x_0) \cap \bar\Omega = \{y\}$.

Examples: (a) All convex regions and all regions of class C^2 satisfy an exterior sphere condition at every boundary point. (See Figure 3.6(a).)

(b) At inward cusps, the exterior sphere condition does not hold. (See Figure 3.6(b).)

Lemma 3.4.1: *If Ω satisfies an exterior sphere condition at y, then $\partial\Omega$ is regular at y.*

Proof:

$$\beta(x) := \begin{cases} \dfrac{1}{\rho^{d-2}} - \dfrac{1}{|x-x_0|^{d-2}} & \text{for } d \geq 3, \\[2mm] \ln \dfrac{|x-x_0|}{\rho} & \text{for } d = 2, \end{cases}$$

yields a barrier at y. Namely, $\beta(y) = 0$, and β is harmonic in $\mathbb{R}^n \setminus \{x_0\}$, hence in particular in Ω. Since for $x \in \bar\Omega \setminus \{y\}$, $|x - x_0| > \varrho$, also $\beta(x) > 0$ for all $x \in \bar\Omega \setminus \{y\}$.

$\square$

We now wish to present Lebesgue's example of a nonregular boundary point, constructing a domain with a sufficiently pointed inward cusp.

Let $\mathbb{R}^3 = \{(x, y, z)\}$, $x \in [0, 1]$, $\rho^2 := y^2 + z^2$,

$$u(x, y, z) := \int_0^1 \frac{x_0}{\sqrt{(x_0 - x)^2 + \rho^2}} dx_0 = v(x, \rho) - 2x \ln \rho$$

with

$$v(x, \rho) = \sqrt{(1 - x)^2 + \rho^2} - \sqrt{x^2 + \rho^2}$$
$$+ x \ln \left| \left(1 - x + \sqrt{(1 - x)^2 + \rho^2} \right) \left(x + \sqrt{x^2 + \rho^2} \right) \right|.$$

We have

$$\lim_{\substack{(x, \rho) \to 0 \\ x > 0}} v(x, \rho) = 1.$$

The limiting value of $-2x \ln \rho$, however, crucially depends on the sequence (x, ρ) converging to 0. For example, if $\rho = |x|^n$, we have

$$-2x \ln \rho = -2nx \ln |x| \xrightarrow{x \to 0} 0.$$

On the other hand, if $\rho = e^{-\frac{k}{2x}}$, $k, x > 0$, we have

$$\lim_{(x, \rho) \to 0} (-2x \ln \rho) = k > 0.$$

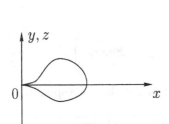

Figure 3.7.

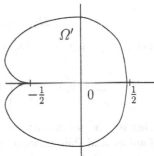

Figure 3.8.

The surface $\rho = e^{-\frac{k}{2x}}$ has an "infinitely pointed" cusp at 0. (See Figure 3.7.)

Considering u as a potential, this means that the equipotential surfaces of u for the value $1 + k$ come together at 0, in such a manner that $f'(0) = 0$ if the equipotential surface is given by $\rho = f(x)$. With Ω as an equipotential surface for $1 + k$, then u solves the exterior Dirichlet problem, and by reflection at the ball $(x - \frac{1}{2})^2 + y^2 + z^2 = \frac{1}{4}$, one obtains a region Ω' as in Figure 3.8).

Depending on the manner, in which one approaches the cusp, one obtains different limiting values, and this shows that the solution of the potential problem cannot be continuous at $(x, y, z) = (-\frac{1}{2}, 0, 0)$, and hence $\partial \Omega'$ is not regular at $(-\frac{1}{2}, 0, 0)$.

Summary

The maximum principle is the decisive tool for showing the convergence of various approximation schemes for harmonic functions. The difference methods replace the Laplace equation, a differential equation, by difference equations on a discrete grid, i.e., by finite-dimensional linear systems. The maximum principle implies uniqueness, and since we have a finite-dimensional system, then it also implies the existence of a solution, as well as the control of the solution by its boundary values.

The Perron method constructs a harmonic function with given boundary values as the supremum of all subharmonic functions with those boundary values. Whether this solution is continuous at the boundary depends on the geometry of the boundary, however.

The alternating method of H.A. Schwarz obtains a solution on the union of two overlapping domains by alternately solving the Dirichlet problem on each of the two domains with boundary values in the overlapping part coming from the solution of the previous step on the other domain.

Exercises

3.1 Employing the notation of Section 3.1, let $x_0 \in \Omega_h \subset \mathbb{R}_h^2$ have neighbors $x_1, \ldots, x_4$. Let $x_5, \ldots, x_8$ be those points in $\mathbb{R}^3$ that are neighbors of exactly two of the points $x_1, \ldots, x_4$. We put

$$\tilde{\Omega}_h := \{x_0 \in \Omega_h : x_1, \ldots, x_8 \in \bar{\Omega}_h).$$

For $u : \bar{\Omega}_h \to \mathbb{R}$, $x_0 \in \tilde{\Omega}_h$, we put

$$\tilde{\Delta}_h u(x_0) = \frac{1}{6h^2} \left(4 \sum_{\alpha=1}^{4} u(x_\alpha) + \sum_{\beta=5}^{8} u(x_\beta) - 20u(x_0) \right).$$

Discuss the solvability of the Dirichlet problem for the corresponding Laplace and Poisson equations.

3.2 Let $x_0 \in \Omega_h$ have neighbors $x_1, \ldots, x_{2d}$. We consider a difference operator Lu for $u : \bar{\Omega}_h \to \mathbb{R}$,

$$Lu(x_0) = \sum_{\alpha=0}^{2d} b_\alpha u(x_\alpha),$$

satisfying the following assumptions:

$$b_\alpha \geq 0 \quad \text{for } \alpha = 1, \ldots, 2d, \ \sum_{\alpha=1}^{2d} b_\alpha > 0, \ \sum_{\alpha=0}^{2d} b_\alpha \leq 0.$$

Prove the weak maximum principle: $Lu \geq 0$ in Ω_h implies

$$\max_{\Omega_h} u \leq \max_{\Gamma_h} u.$$

3.3 Under the assumptions of Section 3.2, assume in addition

$$b_\alpha > 0 \quad \text{for } \alpha = 1, \ldots, 2d,$$

and let Ω_h be discretely connected. Show that if a solution of $Lu \geq 0$ assume its maximum at a point of Ω_h, it has to be constant.

3.4 Carry out the details of the alternating method for the union of three domains.

3.5 Let u be harmonic on the domain Ω, $x_0 \in \Omega$, $B(x_0, R) \subset \Omega, 0 \leq r \leq \rho \leq R, \rho^2 = rR$. Then

$$\int_{|\vartheta|=1} u(x_0 + r\vartheta)u(x_0 + R\vartheta)d\vartheta = \int_{|\vartheta|=1} u^2(x_0 + \rho\vartheta)d\vartheta.$$

Conclude that if u is constant in some neighborhood of x_0, it is constant on all of Ω.

4. Existence Techniques II: Parabolic Methods. The Heat Equation

4.1 The Heat Equation: Definition and Maximum Principles

Let $\Omega \in \mathbb{R}^d$ be open, $(0, T) \subset \mathbb{R} \cup \{\infty\}$,

$$\Omega_T := \Omega \times (0, T),$$
$$\partial^* \Omega_T := \left(\bar{\Omega} \times \{0\} \right) \cup \left(\partial \Omega \times \overline{(0, T)} \right). \quad \text{(See Figure 4.1.)}$$

We call $\partial^* \Omega_T$ the reduced boundary of Ω_T.

For each fixed $t \in (0, T)$ let $u(x, t) \in C^2(\Omega)$, and for each fixed $x \in \Omega$ let $u(x, t) \in C^1((0, T))$. Moreover, let $f \in C^0(\partial^* \Omega_T)$, $u \in C^0(\bar{\Omega}_T)$. We say that u solves the heat equation with boundary values f if

$$\begin{aligned} u_t(x, t) &= \Delta_x u(x, t) && \text{for } (x, t) \in \Omega_T, \\ u(x, t) &= f(x, t) && \text{for } (x, t) \in \partial^* \Omega_T. \end{aligned} \quad (4.1.1)$$

Written out with a less compressed notation, the differential equation is

$$\frac{\partial}{\partial t} u(x, t) = \sum_{i=1}^{d} \frac{\partial^2}{\partial x_i^2} u(x, t).$$

Equation (4.1.1) is a linear, parabolic partial differential equation of second order. The reason that here, in contrast to the Dirichlet problem for harmonic functions, we are prescribing boundary values only at the reduced boundary is that for a solution of a parabolic equation, the values of u on $\Omega \times \{T\}$ are already determined by its values on $\partial^* \Omega_T$, as we shall see in the sequel.

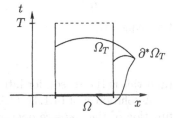

Figure 4.1.

The heat equation describes the evolution of temperature in heat-conducting media and is likewise important in many other diffusion processes. For example, if we have a body in $\mathbb{R}^3$ with given temperature distribution at time t_0 and if we keep the temperature on its surface constant, this determines its temperature distribution uniquely at all times $t > t_0$. This is a heuristic reason for prescribing the boundary values in (4.1.1) only at the reduced boundary.

Replacing t by $-t$ in (4.1.1) does not transform the heat equation into itself. Thus, there is a distinction between "past" and "future". This is likewise heuristically plausible.

In order to gain some understanding of the heat equation, let us try to find solutions with separated variables, i.e., of the form

$$u(x,t) = v(x)w(t). \tag{4.1.2}$$

Inserting this ansatz into (4.1.1), we obtain

$$\frac{w_t(t)}{w(t)} = \frac{\Delta v(x)}{v(x)}. \tag{4.1.3}$$

Since the left-hand side of (4.1.3) is a function of t only, while the right-hand side is a function of x, each of them has to be constant. Thus

$$\Delta v(x) = -\lambda v(x), \tag{4.1.4}$$

$$w_t(t) = -\lambda w(t), \tag{4.1.5}$$

for some constant λ. We consider the case where we assume homogeneous boundary conditions on $\partial \Omega \times [0, \infty)$, i.e.,

$$u(x,t) = 0 \quad \text{for } x \in \partial \Omega,$$

or equivalently,

$$v(x) = 0 \quad \text{for } x \in \partial \Omega. \tag{4.1.6}$$

From (4.1.4) we then get through multiplication by v and integration by parts

$$\int_\Omega |Dv(x)|^2 dx = -\int_\Omega v(x)\Delta v(x)dx = \lambda \int_\Omega v(x)^2 dx.$$

Consequently,

$$\lambda \geq 0$$

(and this is the reason for introducing the minus sign in (4.1.4) and (4.1.5)).

A solution v of (4.1.4), (4.1.6) that is not identically 0 is called an eigenfunction of the Laplace operator, and λ an eigenvalue. We shall see in Section 9.5 that the eigenvalues constitute a discrete sequence $(\lambda_n)_{n \in \mathbb{N}}$, $\lambda_n \to \infty$ for $n \to \infty$. Thus, a nontrivial solution of (4.1.4), (4.1.6) exists precisely if $\lambda = \lambda_n$, for some $n \in \mathbb{N}$. The solution of (4.1.5) then is simply given by

$$w(t) = w(0)e^{-\lambda t}.$$

So, if we denote an eigenfunction for the eigenvalue λ_n by v_n, we obtain the solution

$$u(x,t) = v_n(x)w(0)e^{-\lambda_n t}$$

of the heat equation (4.1.1), with the homogeneous boundary condition

$$u(x,t) = 0 \quad \text{for } x \in \partial \Omega$$

and the initial condition

$$u(x,0) = v_n(x)w(0).$$

This seems to be a rather special solution. Nevertheless, in a certain sense this is the prototype of a solution. Namely, because (4.1.1) is a linear equation, any linear combination of solutions is a solution itself, and so we may take sums of such solutions for different eigenvalues λ_n. In fact, as we shall demonstrate in Section 9.5, any L^2-function on Ω, and thus in particular any continuous function f on $\bar{\Omega}$, assuming Ω to be bounded, that vanishes on $\partial \Omega$, can be expanded as

$$f(x) = \sum_{n \in \mathbb{N}} \alpha_n v_n(x), \tag{4.1.7}$$

where the $v_n(x)$ are the eigenfunctions of Δ, normalized via

$$\int_\Omega v_n(x)^2 dx = 1$$

and mutually orthogonal:

$$\int_\Omega v_n(x)v_m(x)dx = 0 \quad \text{for } n \neq m.$$

Then α_n can be computed as

$$\alpha_n = \int_\Omega v_n(x)f(x)dx.$$

We then have an expansion for the solution of

$$
\begin{aligned}
u_t(x,t) &= \Delta u(x,t) && \text{for } x \in \Omega, t \geq 0, \\
u(x,t) &= 0 && \text{for } x \in \partial \Omega, t \geq 0, \\
u(x,0) &= f(x) && \left(= \sum_n \alpha_n v_n(x) \right), \quad \text{for } x \in \Omega,
\end{aligned}
\tag{4.1.8}
$$

namely,

$$u(x,t) = \sum_{n \in \mathbb{N}} \alpha_n e^{-\lambda_n t} v_n(x). \tag{4.1.9}$$

Since all the λ_n are nonnegative, we see from this representation that all the "modes" $\alpha_n v_n(x)$ of the initial values f are decaying in time for a solution of the heat equation. In this sense, the heat equation regularizes or smoothes out its initial values. In particular, since thus all factors $e^{-\lambda_n t}$ are less than or equal to 1 for $t \geq 0$, the series (4.1.9) converges in $L^2(\Omega)$, because (4.1.7) does.

If instead of the heat equation we considered the backward heat equation

$$u_t = -\Delta u,$$

then the analogous expansion would be $u(x,t) = \sum_n \alpha_n e^{\lambda_n t} v_n(x)$, and so the modes would grow, and differences would be exponentially enlarged, and in fact, in general, the series will no longer converge for positive t. This expresses the distinction between "past" and "future" built into the heat equation and alluded to above.

If we write

$$q(x,y,t) := \sum_{n \in \mathbb{N}} e^{-\lambda_n t} v_n(x) v_n(y), \tag{4.1.10}$$

and if we can use the results of Section 9.5 to show the convergence of this series, we may represent the solution $u(x,t)$ of (4.1.8) as

$$\begin{aligned}
u(x,t) &= \sum_{n \in \mathbb{N}} e^{-\lambda_n t} v_n(x) \int_\Omega v_n(y) f(y) dy \quad \text{by (4.1.9)} \\
&= \int_\Omega q(x,y,t) f(y) dy.
\end{aligned} \tag{4.1.11}$$

Instead of demonstrating the convergence of the series (4.1.10) and that $u(x,t)$ given by (4.1.9) is smooth for $t > 0$ and permits differentiation under the sum, in this chapter we shall pursue a different strategy to construct the "heat kernel" $q(x,y,t)$ in Section 4.3.

For $x,y \in \mathbb{R}^n$, $t, t_0 \in \mathbb{R}$, $t \neq t_0$, we define the heat kernel at (y,t_0) as

$$\Lambda(x,y,t,t_0) := \frac{1}{(4\pi |t - t_0|)^{\frac{d}{2}}} e^{\frac{|x-y|^2}{4(t_0 - t)}}.$$

We then have

$$\Lambda_t(x,y,t,t_0) = -\frac{d}{2(t - t_0)} \Lambda(x,y,t,t_0) + \frac{|x-y|^2}{4(t_0 - t)^2} \Lambda(x,y,t,t_0),$$

$$\Lambda_{x_i}(x,y,t,t_0) = \frac{x^i - y^i}{2(t_0 - t)} \Lambda(x,y,t,t_0),$$

$$\Lambda_{x_i x_i}(x,y,t,t_0) = \frac{(x^i - y^i)^2}{4(t_0 - t)^2} \Lambda(x,y,t,t_0) + \frac{1}{2(t_0 - t)} \Lambda(x,y,t,t_0),$$

i.e.,

$$\Delta_x \Lambda(x,y,t,t_0) = \frac{|x-y|^2}{4(t_0-t)^2} \Lambda(x,y,t,t_0) + \frac{d}{2(t_0-t)} \Lambda(x,y,t,t_0)$$
$$= \Lambda_t(x,y,t,t_0).$$

The heat kernel thus is a solution of (4.1.1). The heat kernel Λ is similarly important for the heat equation as the fundamental solution Γ is for the Laplace equation.

We first wish to derive a representation formula for solutions of the (homogeneous and inhomogeneous) heat equation that will permit us to compute the values of u at time T from the values of u and its normal derivative on $\partial^* \Omega_T$. For that purpose, we shall first assume that u solves the equation

$$u_t(x,t) = \Delta u(x,t) + \varphi(x,t) \quad \text{in } \Omega_T$$

for some bounded integrable function $\varphi(x,t)$ and that $\Omega \subset \mathbb{R}^d$ is bounded and such that the divergence theorem holds. Let v satisfy $v_t = -\Delta v$ on Ω_T. Then

$$\int_{\Omega_T} v\varphi\, dx\, dt = \int_{\Omega_T} v(u_t - \Delta u)\, dx\, dt$$

$$= \int_\Omega \left(\int_0^T v(x,t) u_t(x,t)\, dt \right) dx - \int_0^T \left(\int_\Omega v\Delta u\, dx \right) dt$$

$$= \int_\Omega \left[v(x,T)u(x,T) - v(x,0)u(x,0) - \int_0^T v_t(x,t)u(x,t)dt \right] dx$$

$$- \int_0^T \left(\int_\Omega u\Delta v\, dx \right) dt - \int_0^T \int_{\partial\Omega} \left(v\frac{\partial u}{\partial \nu} - u\frac{\partial v}{\partial \nu} \right) do\, dt$$

$$= \int_{\Omega \times \{T\}} vu\, dx - \int_{\Omega \times \{0\}} vu\, dx - \int_0^T \int_{\partial\Omega} \left(v\frac{\partial u}{\partial \nu} - u\frac{\partial v}{\partial \nu} \right) do\, dt.$$
$$(4.1.12)$$

For $v(x,t) := \Lambda(x,y,T+\varepsilon,t)$ with $T > 0$ and $y \in \Omega^d$ fixed we then have, because of $v_t = -\Delta v$,

$$\int_{\Omega \times \{T\}} \Lambda u\, dx = \int_{\Omega_T} \Lambda\varphi\, dx\, dt + \int_{\Omega \times \{0\}} \Lambda u\, dx$$
$$+ \int_0^T \left(\int_{\partial\Omega} \left(\Lambda\frac{\partial u}{\partial \nu} - u\frac{\partial \Lambda}{\partial \nu} \right) do \right) dt.$$
$$(4.1.13)$$

For $\varepsilon \to 0$, the term on the left-hand side becomes

$$\lim_{\varepsilon \to 0} \int_\Omega \Lambda(x,y,T+\varepsilon,T)u(x,T)dx = u(y,T).$$

Furthermore, $\Lambda(x, y, T + \varepsilon, t)$ is uniformly continuous in ε, x, t for $\varepsilon \geq 0$, $x \in \partial\Omega$, and $0 \leq t \leq T$ or for $x \in \Omega$, $t = 0$. Thus (4.1.13) implies, letting $\varepsilon \to 0$,

$$u(y, T) = \int_{\Omega_T} \Lambda(x, y, T, t)\varphi(x, t)\, dx\, dt + \int_{\Omega} \Lambda(x, y, T, 0)u(x, 0)\, dx$$
$$+ \int_0^T \left(\int_{\partial\Omega} \left(\Lambda(x, y, T, t)\frac{\partial u(x, t)}{\partial \nu} - u(x, t)\frac{\partial \Lambda(x, y, T, t)}{\partial \nu} \right) do \right) dt. \quad (4.1.14)$$

This formula, however, does not yet solve the initial boundary value problem, since in (4.1.14), in addition to $u(x, t)$ for $x \in \partial\Omega$, $t > 0$, and $u(x, 0)$, also the normal derivative $\frac{\partial u}{\partial \nu}(x, t)$ for $x \in \partial\Omega$, $t > 0$, enters. Thus we should try to replace $\Lambda(x, y, T, t)$ by a kernel that vanishes on $\partial\Omega \times (0, \infty)$. This is the task that we shall address in Section 4.3. Here, we shall modify the construction in a somewhat different manner. Namely, we do not replace the kernel, but change the domain of integration so that the kernel becomes constant on its boundary. Thus, for $\mu > 0$, we let

$$M(y, T; \mu) := \left\{ (x, s) \in \mathbb{R}^d \times \mathbb{R}, s \leq T : \frac{1}{(4\pi(T - s))^{\frac{d}{2}}} e^{-\frac{|x-y|^2}{4(T-s)}} \geq \mu \right\}.$$

For any $y \in \Omega, T > 0$, we may find $\mu_0 > 0$ such that for all $\mu > \mu_0$,

$$M(y, T; \mu) \subset \Omega \times [0, T].$$

We always have

$$(y, T) \in M(y, T; \mu),$$

and in fact, $M(y, T; \mu) \cap \{s = T\}$ consists of the single point (y, T). For t falling below T, $M(y, T; \mu) \cap \{s = t\}$ is a ball in $\mathbb{R}^d$ with center (y, t) whose radius first grows but then starts to shrink again if t is decreased further, until it becomes 0 at a certain value of t.

We then perform the above computation on $M(y, T; \mu)$ $(\mu > \mu_0)$ in place of Ω_T, with

$$v(x, t) := \Lambda(x, y, T + \varepsilon, t) - \mu,$$

and as before, we may perform the limit $\varepsilon \searrow 0$. Then

$$v(x, t) = 0 \quad \text{for } (x, t) \in \partial M(y, T; \mu),$$

so that the corresponding boundary term disappears.

Here, we are interested only in the homogeneous heat equation, and so, we put $\varphi = 0$. We then obtain the representation formula

$$u(y,T) = -\int_{\partial M(y,T;\mu)} u(x,t)\frac{\partial \Lambda}{\partial \nu_x}(x,y,T,t)do(x,t)$$

$$= \mu \int_{\partial M(y,T;\mu)} u(x,t)\frac{|x-y|}{2(T-t)}do(x,t), \qquad (4.1.15)$$

since

$$\frac{\partial \Lambda}{\partial \nu_x} = -\frac{|x-y|}{2(T-t)}\Lambda = -\frac{|x-y|}{2(T-t)}\mu \quad \text{on } \partial M(y,T;\mu).$$

In general, the maximum principles for parabolic equations are qualitatively different from those for elliptic equations. Namely, one often gets stronger conclusions in the parabolic case.

Theorem 4.1.1: *Let u be as in the assumptions of (4.1.1). Let $\Omega \subset \mathbb{R}^d$ be open and bounded and*

$$\Delta u - u_t \geq 0 \quad \text{in } \Omega_T. \qquad (4.1.16)$$

We then have

$$\sup_{\bar{\Omega}_T} u = \sup_{\partial^* \Omega_T} u. \qquad (4.1.17)$$

(If $T < \infty$, we can take max in place of sup.)

Proof: Without loss of generality $T < \infty$.

(i) Suppose first

$$\Delta u - u_t > 0 \quad \text{in } \Omega_T. \qquad (4.1.18)$$

For $0 < \varepsilon < T$, by continuity of u and compactness of $\bar{\Omega}_{T-\varepsilon}$, there exists $(x_0,t_0) \in \bar{\Omega}_{T-\varepsilon}$ with

$$u(x_0,t_0) = \max_{\bar{\Omega}_{T-\varepsilon}} u. \qquad (4.1.19)$$

If we had $(x_0,t_0) \in \Omega_{T-\varepsilon}$, then $\Delta u(x_0,t_0) \leq 0$, $\nabla u(x_0,t_0) = 0$, $u_t(x_0,t_0) = 0$ would lead to a contradiction; hence we must have $(x_0,t_0) \in \partial\Omega_{T-\varepsilon}$. For $t = T-\varepsilon$ and $x \in \Omega$, we would get $\Delta u(x_0,t_0) \leq 0$, $u_t(x_0,t_0) \geq 0$, likewise contradicting (4.1.18). Thus we conclude that

$$\max_{\bar{\Omega}_{T-\varepsilon}} u = \max_{\partial^* \Omega_{T-\varepsilon}} u, \qquad (4.1.20)$$

and for $\varepsilon \to 0$, (4.1.20) yields the claim, since u is continuous.

(ii) If we have more generally $\Delta u - u_t \geq 0$, we let $v := u - \varepsilon t$, $\varepsilon > 0$. We have

$$v_t = u_t - \varepsilon \leq \Delta u - \varepsilon = \Delta v - \varepsilon < \Delta v,$$

and thus by (i),

$$\max_{\bar{\Omega}_T} u = \max_{\bar{\Omega}_T}(v + \varepsilon t) \leq \max_{\bar{\Omega}_T} v + \varepsilon T = \max_{\partial^* \Omega_T} v + \varepsilon T \leq \max_{\partial^* \Omega_T} u + \varepsilon T,$$

and $\varepsilon \to 0$ yields the claim.

$\square$

Theorem 4.1.1 directly leads to a uniqueness result:

Corollary 4.1.1: *Let u, v be solutions of (4.1.1) with $u = v$ on $\partial^* \Omega_T$, where $\Omega \subset \mathbb{R}^d$ is bounded. Then $u = v$ on $\bar{\Omega}_T$.*

Proof: We apply Theorem 4.1.1 to $u - v$ and $v - u$.

$\square$

This uniqueness holds only for bounded Ω, however. If, e.g., $\Omega = \mathbb{R}^d$, uniqueness holds only under additional assumptions on the solution u.

Theorem 4.1.2: *Let $\Omega = \mathbb{R}^d$ and suppose*

$$
\begin{aligned}
\Delta u - u_t &\geq 0 && \text{in } \Omega_T, \\
u(x,t) &\leq M e^{\lambda |x|^2} && \text{in } \Omega_T \text{ for } M, \lambda > 0, \\
u(x,0) &= f(x) && x \in \Omega = \mathbb{R}^d.
\end{aligned}
\tag{4.1.21}
$$

Then

$$\sup_{\bar{\Omega}_T} u \leq \sup_{\mathbb{R}^d} f. \tag{4.1.22}$$

Remark: This maximum principle implies the uniqueness of solutions of the differential equation

$$
\begin{aligned}
\Delta u &= u_t && \text{on } \Omega_T = \mathbb{R}^d \times (0,T), \\
u(x,0) &= f(x) && \text{for } x \in \mathbb{R}^d, \\
u(x,t) &\leq M e^{\lambda |x|^2} && \text{for } (x,t) \in \Omega_T.
\end{aligned}
$$

The condition (4.1.21) is a condition for the growth of u at infinity. If this condtion does not hold, there are counterexamples for uniqueness. For example, let us choose

$$u(x,t) := \sum_{n=0}^{\infty} \frac{g^n(t)}{(2n)!} x^{2n}$$

with

$$g(t) := \begin{cases} e^{\frac{-1}{t^k}} & t > 0, \text{ for some } k > 1, \\ 0 & t = 0, \end{cases}$$

$$v(x,t) := 0 \quad \text{for all } (x,t) \in \mathbb{R} \times (0,\infty).$$

Then u and v are solutions of (4.1.1) with $f(x) = 0$. For further details we refer to the book of F. John [10].

Proof of Theorem 4.1.2: Since we can divide the interval $(0,T)$ into subintervals of length $\tau < \frac{1}{4\lambda}$, it suffices to prove the claim for $T < \frac{1}{4\lambda}$, because we shall then get

$$\sup_{\mathbb{R}^d \times [0,k\tau]} u \leq \sup_{\mathbb{R}^d \times [0,(k-1)\tau]} u \leq \cdots \leq \sup_{\mathbb{R}^d} f(x).$$

Thus let $T < \frac{1}{4\lambda}$. We may then find $\varepsilon > 0$ with

$$T + \varepsilon < \frac{1}{4\lambda}. \tag{4.1.23}$$

For fixed $y \in \mathbb{R}^d$ and $\delta > 0$, we consider

$$(x,t) := u(x,t) - \delta\Lambda(x,y,t,T+\varepsilon), \quad 0 \leq t \leq T. \tag{4.1.24}$$

It follows that

$$v_t^\delta - \Delta v^\delta = u_t - \Delta u \leq 0, \tag{4.1.25}$$

since Λ is a solution of the heat equation. For $\Omega^\rho := B(y,\rho)$, we thus obtain from Theorem 4.1.1

$$v^\delta(y,t) \leq \max_{\partial^*\Omega^\rho} v^\delta. \tag{4.1.26}$$

Moreover,

$$v^\delta(x,0) \leq u(x,0) \leq \sup_{\mathbb{R}^d} f, \tag{4.1.27}$$

and for $|x - y| = \rho$,

$$v^\delta(x,t) \leq Me^{\lambda|x|^2} - \delta\frac{1}{(4\pi(T+\varepsilon-t))^{\frac{d}{2}}} \exp\left(\frac{\rho^2}{4(T+\varepsilon-t)}\right)$$

$$\leq Me^{\lambda(|y|+\rho)^2} - \delta\frac{1}{(4\pi(T+\varepsilon))^{\frac{d}{2}}} \exp\left(\frac{\rho^2}{4(T+\varepsilon)}\right).$$

Because of (4.1.23), for sufficiently large ρ, the second term has a larger exponent than the first, and so the whole expression can be made arbitrarily negative; in particular, we can achieve that it is not larger than $\sup_{\mathbb{R}^d} f$. Consequently,

$$v^\delta \leq \sup_{\mathbb{R}^d} f \quad \text{on } \partial^* \Omega^\rho. \tag{4.1.28}$$

Thus, (4.1.26) and (4.1.28) yield

$$v^\delta(y,t) = u(y,t) - \delta \Lambda(y,y,t,T+\varepsilon) = u(y,t) - \delta \frac{1}{(4\pi(T+\varepsilon-t))^{\frac{d}{2}}}$$
$$\leq \sup_{\mathbb{R}^d} f.$$

The conclusion follows by letting $\delta \to 0$. $\qquad\qquad\square$

Actually, we can use the representation formula (4.1.12) to obtain a strong maximum principle for the heat equation, in the same manner as the mean value formula could be used to obtain Corollary 1.2.3:

Theorem 4.1.3: *Let $\Omega \subset \mathbb{R}^d$ be open and bounded and*

$$\Delta u - u_t = 0 \quad \text{in } \Omega_T,$$

with the regularity properties specified at the beginning of this section. Then if there exists some $(x_0, t_0) \in \Omega \times (0, T]$ with

$$u(x_0, t_0) = \max_{\Omega_T} u \quad (\text{or with } u(x_0, t_0) = \min_{\Omega_T} u),$$

then u is constant in $\bar{\Omega}_{t_0}$.

Proof: The proof is the same as that of Lemma 1.2.1, using the representation formula (4.1.12). (Note that by applying (4.1.12) to the function $u \equiv 1$, we obtain

$$\mu \int_{\partial M(y,T;\mu)} \frac{|x-y|}{2(T-t)} do(x,t) = 1,$$

and so a general u that solves the heat equation is indeed represented as some average. Also, $M(y, T; \mu_2) \subset M(y, T; \mu_1)$ for $\mu_1 \leq \mu_2$, and as $\mu \to \infty$, the sets $M(y, T; \mu)$ shrink to the point (y, T).) $\qquad\qquad\square$

Of course, the maximum principle also holds for subsolutions, i.e., if

$$\Delta u - u_t \geq 0 \quad \text{in } \Omega_T.$$

In that case, we get the inequality "$\leq$" in place of "$=$" in (4.1.12), which is what is required for the proof of the maximum principle. Likewise, the statement with the minimum holds for solutions of

$$\Delta u - u_t \leq 0.$$

Slightly more generally, we even have

Corollary 4.1.2: *Let $\Omega \subset \mathbb{R}^d$ be open and bounded and*

$$\Delta u(x,t) + c(x,t)u(x,t) - u_t(x,t) \geq 0 \quad in \ \Omega_T,$$

with some bounded function

$$c(x,t) \leq 0 \quad in \ \Omega_T. \tag{4.1.29}$$

Then if there exists some $(x_0, t_0) \in \Omega \times (0, T]$ with

$$u(x_0, t_0) = \max_{\overline{\Omega}_T} u \geq 0, \tag{4.1.30}$$

then u is constant in $\bar{\Omega}_{t_0}$.

Proof: Our scheme of proof still applies because, since c is nonpositive, at a nonnegative maximum point (x_0, t_0) of u, $c(x_0, t_0)u(x_0, t_0) \leq 0$ which strengthens the inequality used in the proof.   □

Again, we obtain a minimum principle when we reverse all signs.
For use in §5.1 below, we now derive a parabolic version of E.Hopf's boundary point lemma 2.1.2. Compared with §2.1, we shall reverse here the scheme of proof, that is, deduce the boundary point lemma from the strong maximum principle instead of the other way around. This is possible because here we consider less general differential operators than the ones in §2.1 so that we could deduce our maximum principle from the representation formula. Of course, one can also deduce general Hopf type maximum principles in the parabolic case, in a manner analogous to §2.1, but we do not pursue that here as it will not yield conceptually or technically new insights.

Lemma 4.1.1: *Suppose the function c is bounded and satisfies $c(x,t) \leq 0$ in Ω_T. Let u solve the differential inequality*

$$\Delta u(x,t) + c(x,t)u(x,t) - u_t(x,t) \geq 0 \quad in \ \Omega_T,$$

and let $(x_0, t_0) \in \partial^* \Omega_T$. Moreover, assume

(i) u is continuous at (x_0, t_0);
(ii) $u(x_0, t_0) \geq 0$ if $c(x) \not\equiv 0$;
(iii) $u(x_0, t_0) > u(x,t)$ for all $(x,t) \in \Omega_T$;
(iv) there exists a ball $\mathring{B}((y,t_1), R) \subset \Omega_T$ with $(x_0, t_0) \in \partial B((y,t_1), R)$.

We then have, with $r := |(x,t) - (y,t_1)|$,

$$\frac{\partial u}{\partial r}(x_0, t_0) > 0, \tag{4.1.31}$$

provided that this derivative (in the direction of the exterior normal of Ω_T) exists.

Proof: With the auxiliary function

$$v(x) := e^{-\gamma(|x-y|^2+(t-t_1)^2)} - e^{-\gamma R^2},$$

the proof proceeds as the one of Lemma 2.1.2, employing this time the maximum principle Theorem 4.1.3. □

I do not know of any good recent book that gives a detailed and systematic presentation of parabolic differential equations. Some older, but still useful, references are [6], [16].

4.2 The Fundamental Solution of the Heat Equation. The Heat Equation and the Laplace Equation

We first consider the so-called fundamental solution

$$K(x, y, t) = \Lambda(x, y, t, 0) = \frac{1}{(4\pi t)^{\frac{d}{2}}} e^{-\frac{|x-y|^2}{4t}}, \qquad (4.2.1)$$

and we first observe that for all $x \in \mathbb{R}^d$, $t > 0$,

$$\int_{\mathbb{R}^d} K(x, y, t) dy = \frac{1}{(4\pi t)^{\frac{d}{2}}} d\omega_d \int_0^\infty e^{-\frac{r^2}{4t}} r^{d-1} dr = \frac{1}{\pi^{\frac{d}{2}}} d\omega_d \int_0^\infty e^{-s^2} s^{d-1} ds$$

$$= \frac{1}{\pi^{\frac{d}{2}}} \int_{\mathbb{R}^d} e^{-|y|^2} dy = 1. \qquad (4.2.2)$$

For bounded and continuous $f : \mathbb{R}^d \to \mathbb{R}$, we consider the convolution

$$u(x, t) = \int_{\mathbb{R}^d} K(x, y, t) f(y) dy = \frac{1}{(4\pi t)^{\frac{d}{2}}} \int_{\mathbb{R}^d} e^{-\frac{|x-y|^2}{4t}} f(y) dy. \qquad (4.2.3)$$

Lemma 4.2.1: *Let $f : \mathbb{R}^d \to \mathbb{R}$ be bounded and continuous. Then*

$$u(x, t) = \int_{\mathbb{R}^d} K(x, y, t) f(y) dy$$

is of class C^∞ on $\mathbb{R}^d \times (0, \infty)$, and it solves the heat equation

$$u_t = \Delta u. \qquad (4.2.4)$$

Proof: That u is of class C^∞ follows, by differentiating under the integral (which is permitted by standard theorems), from the C^∞ property of $K(x, y, t)$. Consequently, we also obtain

$$\frac{\partial}{\partial t} u(x, t) = \int_{\mathbb{R}^d} \frac{\partial}{\partial t} K(x, y, t) f(y) dy = \int_{\mathbb{R}^d} \Delta_x K(x, y, t) f(y) dy = \Delta_x u(x, t).$$

$\square$

Lemma 4.2.2: *Under the assumptions of Lemma 4.2.1, we have for every $x \in \mathbb{R}^d$,*

$$\lim_{t \to 0} u(x, t) = f(x).$$

Proof:

$$
\begin{aligned}
|f(x) - u(x,t)| &= \left| f(x) - \int_{\mathbb{R}^d} K(x,y,t) f(y) dy \right| \\
&= \left| \int_{\mathbb{R}^d} K(x,y,t)(f(x) - f(y)) dy \right| \text{ with (4.2.2)} \\
&= \left| \frac{1}{(4\pi t)^{\frac{d}{2}}} \int_0^\infty e^{-\frac{r^2}{4t}} r^{d-1} \int_{S^{d-1}} (f(x) - f(x+r\xi)) \, do(\xi) \, dr \right| \\
&= \left| \frac{1}{\pi^{\frac{d}{2}}} \int_0^\infty e^{-s^2} s^{d-1} \int_{S^{d-1}} \left(f(x) - f(x + 2\sqrt{t}s\xi) \right) do(\xi) \, ds \right| \\
&= \left| \cdots \int_0^M \cdots + \cdots \int_M^\infty \cdots \right| \\
&\leq \sup_{y \in B(x, 2\sqrt{t}M)} |f(x) - f(y)| + 2 \sup_{\mathbb{R}^d} |f| \frac{d\omega_d}{\pi^{\frac{d}{2}}} \int_M^\infty e^{-s^2} s^{d-1} ds.
\end{aligned}
$$

Given $\varepsilon > 0$, we first choose M so large that the second summand is less than $\varepsilon/2$, and we then choose $t_0 > 0$ so small that for all t with $0 < t < t_0$, the first summand is less than $\varepsilon/2$ as well. This implies the continuity. $\square$

By (4.2.3), we have thus found a solution of the initial value problem

$$
\begin{aligned}
u_t(x,t) - \Delta u(x,t) &= 0 \quad \text{for } x \in \mathbb{R}^d, \quad t > 0, \\
u(x,0) &= f(x),
\end{aligned}
$$

for the heat equation. By Theorem 4.1.2 this is the only solution that grows at most exponentially.

According to the physical interpretation, $u(x,t)$ is supposed to describe the evolution in time of the temperature for initial values $f(x)$. We should note, however, that in contrast to physically more realistic theories, we here obtain an infinite propagation speed as for any positive time $t > 0$, the temperature $u(x,t)$ at the point x is influenced by the initial values at all arbitrarily far away points y, although the strength decays exponentially with the distance $|x - y|$.

In the case where f has compact support K, i.e., $f(x) = 0$ for $x \notin K$, the function from (4.2.3) satisfies

$$
|u(x,t)| \leq \frac{1}{(4\pi t)^{\frac{d}{2}}} e^{-\frac{\text{dist}(x,K)^2}{4t}} \int_K |f(y)| \, dy, \tag{4.2.5}
$$

which goes to 0 as $t \to \infty$.

Remark: (4.2.5) yields an explicit exponential rate of convergence!

More generally, one is interested in the initial boundary value problem for the inhomogeneous heat equation:

Let $\Omega \subset \mathbb{R}^d$ be a domain, and let $\varphi \in C^0(\Omega \times [0,\infty))$, $f \in C^0(\Omega)$, $g \in C^0(\partial\Omega \times (0,\infty))$ be given. We wish to find a solution of

$$\frac{\partial u(x,t)}{\partial t} - \Delta u(x,t) = \varphi(x,t) \quad \text{in } \Omega \times (0,\infty),$$
$$u(x,0) = f(x) \quad \text{in } \Omega, \tag{4.2.6}$$
$$u(x,t) = g(x,t) \quad \text{for } x \in \partial\Omega, \quad t \in (0,\infty).$$

In order for this problem to make sense, one should require a compatibility condition between the initial and the boundary values: $f \in C^0(\bar{\Omega})$, $g \in C^0(\partial\Omega \times [0,\infty))$, and

$$f(x) = g(x,0) \quad \text{for } x \in \partial\Omega. \tag{4.2.7}$$

We want to investigate the connection between this problem and the Dirichlet problem for the Laplace equation, and for that purpose, we consider the case where $\varphi \equiv 0$ and $g(x,t) = g(x)$ is independent of t. For the following consideration whose purpose is to serve as motivation, we assume that $u(x,t)$ is differentiable sufficiently many times up to the boundary. (Of course, this is an issue that will need a more careful study later on.) We then compute

$$\left(\frac{\partial}{\partial t} - \Delta\right)\frac{1}{2}u_t^2 = u_t u_{tt} - u_t \Delta u_t - \sum_{i=1}^{d} u_{x^i t}^2 = u_t \frac{\partial}{\partial t}(u_t - \Delta u) - \sum_{i=1}^{d} u_{x^i t}^2$$

$$= -\sum_{i=1}^{d} u_{x^i t}^2 \leq 0. \tag{4.2.8}$$

According to Theorem 4.1.1,

$$v(t) := \sup_{x \in \Omega} \left|\frac{\partial u(x,t)}{\partial t}\right|^2$$

then is a nonincreasing function of t.

We now consider

$$E(u(\cdot,t)) = \frac{1}{2}\int_\Omega \sum_{i=1}^{d} u_{x^i}^2 \, dx$$

and compute

$$\frac{\partial}{\partial t}E(u(\cdot,t)) = \int_\Omega \sum_{i=1}^{d} u_{t x^i} u_{x^i} \, dx$$

$$= -\int_\Omega u_t \Delta u \, dx, \text{ since } u_t(x,t) = \frac{\partial}{\partial t}g(x) = 0 \quad \text{for } x \in \partial\Omega$$

$$= -\int_\Omega u_t^2 \, dx \leq 0. \tag{4.2.9}$$

With (4.2.8), we then conclude that

$$\frac{\partial^2}{\partial t^2} E(u(\cdot, t)) = -\int_\Omega \frac{\partial}{\partial t} u_t^2 dx = -\int_\Omega \Delta u_t^2 dx + 2 \int_\Omega \sum_{i=1}^d u_{x^i t}^2 dx$$

$$= -\int_{\partial \Omega} \frac{\partial}{\partial \nu} u_t^2 do(x) + 2 \int_\Omega \sum_{i=1}^d u_{x^i t}^2 dx.$$

Since $u_t^2 \geq 0$ in Ω, $u_t^2 = 0$ on $\partial\Omega$, we have on $\partial\Omega$,

$$\frac{\partial}{\partial \nu} u_t^2 \leq 0.$$

It follows that

$$\frac{\partial^2}{\partial t^2} E(u(\cdot, t)) \geq 0. \tag{4.2.10}$$

Thus $E(u(\cdot, t))$ is a monotonically nonincreasing and convex function of t. In particular, we obtain

$$\frac{\partial}{\partial t} E(u(\cdot, t)) \leq \alpha := \lim_{t \to \infty} \frac{\partial}{\partial t} E(u(\cdot, t)) \leq 0. \tag{4.2.11}$$

Since $E(u(\cdot, t)) \geq 0$ for all t, we must have $\alpha = 0$, because otherwise for sufficiently large T,

$$E(u(\cdot, T)) = E(u(\cdot, 0)) + \int_0^T \frac{\partial}{\partial t} E(u(\cdot, t)) dt \leq E(u(\cdot, 0)) + \alpha T < 0.$$

Thus it follows that

$$\lim_{t \to \infty} \int_\Omega u_t^2 dx = 0. \tag{4.2.12}$$

In order to get pointwise convergence as well, we have to utilize the maximum principle once more. We extend $u_t^2(x, 0)$ from Ω to all of $\mathbb{R}^d$ as a nonnegative, continuous function l with compact support and put

$$v(x, t) := \int_{\mathbb{R}^d} \frac{1}{(4\pi t)^{\frac{d}{2}}} e^{-\frac{|x-y|^2}{4t}} l(y) dy. \tag{4.2.13}$$

We then have

$$v_t - \Delta v = 0,$$

and since $l \geq 0$, also

$$v > 0,$$

and thus in particular

$$v \geq u_t^2 \quad \text{on } \partial\Omega.$$

Thus $w := u_t^2 - v$ satisfies

$$\frac{\partial}{\partial t} w - \Delta w \leq 0 \quad \text{in } \Omega, \tag{4.2.14}$$

$$w \leq 0 \quad \text{on } \partial\Omega,$$

$$w(x,0) = 0 \quad \text{for } x \in \Omega, t = 0.$$

Theorem 4.1.1 then implies

$$w(x,t) \leq 0,$$

i.e.,

$$u_t^2(x,t) \leq v(x,t) \quad \text{for all } x \in \Omega, t > 0. \tag{4.2.15}$$

Since l has compact support, from Lemma 4.2.2

$$\lim_{t \to \infty} v(x,t) = 0 \quad \text{for all } x \in \Omega,$$

and thus also

$$\lim_{t \to \infty} u_t^2(x,t) = 0 \quad \text{for all } x \in \Omega. \tag{4.2.16}$$

We thus conclude that provided that our regularity assumptions are valid the time derivative of a solution of our initial boundary value theorem with boundary values that are constant in time goes to 0 as $t \to \infty$. Thus, if we can show that $u(x,t)$ converges for $t \to \infty$ with respect to x in C^2, the limit function u_∞ needs to satisfy

$$\Delta u_\infty = 0,$$

i.e., be harmonic. If we can even show convergence up to the boundary, then u_∞ satisfies the Dirichlet condition

$$u_\infty(x) = g(x) \quad \text{for } x \in \partial\Omega.$$

From the remark about (4.2.5), we even see that $u_t(x,t)$ converges to 0 exponentially in t.

If we know already that the Dirichlet problem

$$\Delta u_\infty = 0 \quad \text{in } \Omega,$$

$$u_\infty = g \quad \text{on } \partial\Omega, \tag{4.2.17}$$

admits a solution, it is easy to show that any solution $u(x,t)$ of the heat equation with appropriate boundary values converges to u_∞. Namely, we even have the following result:

Theorem 4.2.1: *Let Ω be a bounded domain in $\mathbb{R}^d$, and let $g(x,t)$ be continuous on $\partial\Omega \times (0, \infty)$, and suppose*

$$\lim_{t\to\infty} g(x,t) = g(x) \quad \text{uniformly in } x \in \partial\Omega. \tag{4.2.18}$$

Let $F(x,t)$ be continuous on $\Omega \times (0, \infty)$, and suppose

$$\lim_{t\to\infty} F(x,t) = F(x) \quad \text{uniformly in } x \in \Omega. \tag{4.2.19}$$

Let $u(x,t)$ be a solution of

$$\Delta u(x,t) - \frac{\partial}{\partial t}u(x,t) = F(x,t) \quad \text{for } x \in \Omega, \quad 0 < t < \infty,$$
$$u(x,t) = g(x,t) \quad \text{for } x \in \partial\Omega, \quad 0 < t < \infty. \tag{4.2.20}$$

Let $v(x)$ be a solution of

$$\Delta v(x) = F(x) \quad \text{for } x \in \Omega,$$
$$v(x) = g(x) \quad \text{for } x \in \partial\Omega. \tag{4.2.21}$$

We then have

$$\lim_{t\to\infty} u(x,t) = v(x) \quad \text{uniformly in } x \in \Omega. \tag{4.2.22}$$

Proof: We consider the difference

$$w(x,t) = u(x,t) - v(x). \tag{4.2.23}$$

Then

$$\Delta w(x,t) - \frac{\partial}{\partial t}w(x,t) = F(x,t) - F(x) \quad \text{in } \Omega \times (0, \infty),$$
$$w(x,t) = g(x,t) - g(x) \quad \text{in } \partial\Omega \times (0, \infty), \tag{4.2.24}$$

and the claim follows from the following lemma:

Lemma 4.2.3: *Let Ω be a bounded domain in $\mathbb{R}^d$, let $\phi(x,t)$ be continuous on $\Omega \times (0, \infty)$, and suppose*

$$\lim_{t\to\infty} \phi(x,t) = 0 \quad \text{uniformly in } x \in \Omega. \tag{4.2.25}$$

Let $\gamma(x,t)$ be continuous on $\partial\Omega \times (0, \infty)$, and suppose

$$\lim_{t\to\infty} \gamma(x,t) = 0 \quad \text{uniformly in } x \in \partial\Omega. \tag{4.2.26}$$

Let $w(x,t)$ be a solution of

$$\Delta w(x,t) - \frac{\partial}{\partial t} w(x,t) = \phi(x,t) \quad in \ \Omega \times (0,\infty),$$
$$w(x,t) = \gamma(x,t) \quad in \ \partial\Omega \times (0,\infty). \tag{4.2.27}$$

Then

$$\lim_{t\to\infty} w(x,t) = 0 \quad uniformly \ in \ x \in \Omega. \tag{4.2.28}$$

Proof: We choose $R > 0$ such that

$$2x^1 < R \quad \text{for all } x = (x^1, \dots, x^d) \in \Omega, \tag{4.2.29}$$

and consider

$$k(x) := e^R - e^{x^1}. \tag{4.2.30}$$

Then

$$\Delta k = -e^{x^1}.$$

With $\kappa := \inf_{x \in \Omega} e^{x^1}$, we thus have

$$\Delta k \leq -\kappa. \tag{4.2.31}$$

We consider, with constants η, c_0, τ to be determined, and with

$$\kappa_0 := \inf_{x \in \Omega} k(x), \quad \kappa_1 := \sup_{x \in \Omega} k(x),$$

the expression

$$m(x,t) := \eta \frac{k(x)}{\kappa} + \eta \frac{k(x)}{\kappa_0} + c_0 \frac{k(x)}{\kappa_0} e^{-\frac{\kappa}{\kappa_1}(t-\tau)} \tag{4.2.32}$$

in $\Omega \times [\tau, \infty)$.
Then

$$\Delta m(x,t) - \frac{\partial}{\partial t} m(x,t)$$
$$< -\eta - \eta \frac{\kappa}{\kappa_0} - c_0 \frac{\kappa}{\kappa_0} e^{-\frac{\kappa}{\kappa_1}(t-\tau)} + c_0 \frac{\kappa_1}{\kappa_0} \frac{\kappa}{\kappa_1} e^{-\frac{\kappa}{\kappa_1}(t-\tau)} < -\eta. \tag{4.2.33}$$

Furthermore,

$$m(x,\tau) > c_0 \quad \text{for } x \in \Omega, \tag{4.2.34}$$
$$m(x,t) > \eta \quad \text{for } (x,t) \in \partial\Omega \times [\tau, \infty). \tag{4.2.35}$$

By our assumptions (4.2.25), (4.2.26), for every η, there exists some $\tau = \tau(\eta)$ with

$$|\phi(x,t)| < \eta \quad \text{for } x \in \Omega, \quad t \geq \tau, \qquad (4.2.36)$$
$$|\gamma(x,t)| < \eta \quad \text{for } x \in \partial\Omega, \quad t \geq \tau. \qquad (4.2.37)$$

In (4.2.32) we now put

$$\tau = \tau(\eta), \quad c_0 = \sup_{x \in \Omega} |w(x,\tau)|.$$

Then

$$m(x,\tau) \pm w(x,\tau) \geq 0 \quad \text{for } x \in \Omega \text{ by } (4.2.34),$$
$$m(x,t) \pm w(x,t) \geq 0 \quad \text{for } x \in \partial\Omega, t \geq \tau,$$

$$\text{by } (4.2.35), (4.2.37), (4.2.27);$$

$$\left(\Delta - \frac{\partial}{\partial t}\right)(m(x,t) \pm w(x,t)) \leq 0 \quad \text{for } x \in \Omega, t \geq \tau,$$

$$\text{by } (4.2.33), (4.2.36), (4.2.27).$$

It follows from Theorem 4.1.1 (observe that it is irrelevant that our functions are defined only on $\Omega \times [\tau, \infty)$ instead of $\Omega \times [0, \infty)$, and initial values are given on $\Omega \times \{\tau\}$) that

$$|w(x,t)| \leq m(x,t) \quad \text{for } x \in \Omega, \quad t > \tau,$$

$$\leq \eta\left(\frac{\kappa_1}{\kappa} + \frac{\kappa_1}{\kappa_0}\right) + c_0 \frac{\kappa_1}{\kappa_0} e^{-\frac{\kappa}{\kappa_1}(t-\tau)},$$

and this becomes smaller than any given $\varepsilon > 0$ if $\eta > 0$ from (4.2.36), (4.2.37) is sufficiently small and $t > \tau(\eta)$ is sufficiently large. $\qquad \square$

4.3 The Initial Boundary Value Problem for the Heat Equation

In this section, we wish to study the initial boundary value problem for the inhomogeneous heat equation

$$u_t(x,t) - \Delta u(x,t) = \varphi(x,t) \quad \text{for } x \in \Omega, t > 0,$$
$$u(x,t) = g(x,t) \quad \text{for } x \in \partial\Omega, t > 0, \qquad (4.3.1)$$
$$u(x,0) = f(x) \quad \text{for } x \in \Omega,$$

with given (continuous and smooth) functions φ, g, f. We shall need some preparations.

Lemma 4.3.1: *Let Ω be a bounded domain of class C^2 in $\mathbb{R}^d$. Then for every $\alpha < \frac{d}{2} + 1$, $T > 0$ there exists a constant $c = c(\alpha, T, d, \Omega)$ such that for all $x_0, x \in \partial\Omega$, $0 < t \leq T$, letting ν denote the exterior normal of $\partial\Omega$, we have*

$$\left|\frac{\partial K}{\partial \nu_x}(x, x_0, t)\right| \leq ct^{-\alpha}|x - x_0|^{-d+2\alpha}.$$

Proof:

$$\frac{\partial}{\partial \nu_x} K(x, x_0, t) = \frac{1}{(4\pi t)^{\frac{d}{2}}} \frac{\partial}{\partial \nu_x} e^{-\frac{|x-x_0|^2}{4t}} = -\frac{1}{(4\pi t)^{\frac{d}{2}}} \frac{(x-x_0)\cdot \nu_x}{2t} e^{-\frac{|x-x_0|^2}{4t}}.$$

As we are assuming that the boundary of Ω is a manifold of class C^2, and since $x, x_0 \in \partial\Omega$, and ν_x is normal to $\partial\Omega$, we have

$$|(x - x_0)\cdot \nu_x| \le c_1 |x - x_0|^2$$

with a constant c_1 depending on the geometry of $\partial\Omega$. Thus

$$\left| \frac{\partial}{\partial \nu_x} K(x, x_0, t) \right| \le c_2 t^{-\frac{d}{2}-1} |x - x_0|^2 e^{-\frac{|x-x_0|^2}{4t}} \tag{4.3.2}$$

with some constant c_2. With a parameter $\beta > 0$, we now consider the function

$$\psi(s) := s^\beta e^{-s} \quad \text{for } s > 0. \tag{4.3.3}$$

Inserting $s = \frac{|x-x_0|^2}{4t}$, $\beta = \frac{d}{2} + 1 - \alpha$, we obtain from (4.3.3)

$$e^{-\frac{|x-x_0|^2}{4t}} \le c_3 |x - x_0|^{-d-2+2\alpha} t^{\frac{d}{2}+1-\alpha}, \tag{4.3.4}$$

with c_3 depending on β, i.e., on d and α. Inserting (4.3.4) into (4.3.2) yields the assertion. $\qquad\square$

Lemma 4.3.2: *Let $\Omega \subset \mathbb{R}^d$ be a bounded domain of class C^2 with exterior normal ν, and let $\gamma \in C^0(\partial\Omega \times [0, T])$ $(T > 0)$. We put*

$$v(x, t) := -\int_0^t \int_{\partial\Omega} \frac{\partial K}{\partial \nu_y}(x, y, \tau)\gamma(y, t - \tau)do(y)d\tau. \tag{4.3.5}$$

We then have

$$v \in C^\infty(\Omega \times [0, T]),$$
$$v(x, 0) = 0 \quad \text{for all } x \in \Omega, \tag{4.3.6}$$

and for all $x_0 \in \partial\Omega$, $0 < t \le T$,

$$\lim_{x \to x_0} v(x, t) = \frac{\gamma(x_0, t)}{2} - \int_0^t \int_{\partial\Omega} \frac{\partial K}{\partial \nu_y}(x_0, y, \tau)\gamma(y, t - \tau)do(y)d\tau. \tag{4.3.7}$$

Proof: First of all, Lemma 4.3.1, with $\alpha = \frac{3}{4}$, implies that the integral in (4.3.5) indeed exists. The C^∞-regularity of v with respect to x then follows from the corresponding regularity of the kernel K by the change of variables $\sigma = t - \tau$. Equation (4.3.6) is obvious as well. It remains to verify the jump relation (4.3.7). For that purpose, it obviously suffices to investigate

$$-\int_0^{\tau_0}\int_{\partial\Omega\cap B(x_0,\delta)}\frac{\partial K}{\partial\nu_y}(x,y,\tau)\gamma(y,t-\tau)do(y)d\tau \qquad (4.3.8)$$

for arbitrarily small $\tau_0 > 0$, $\delta > 0$. In particular, we may assume that δ_0 and τ are chosen such that for any given $\varepsilon > 0$, we have for $y \in \partial\Omega$, $|y - x_0| < \delta$, and $0 \leq \tau < \tau_0$,

$$|\gamma(x_0,t) - \gamma(y,t-\tau)| < \varepsilon.$$

Thus, we shall have an error of magnitude controlled by ε if in place of (4.3.8), we evaluate the integral

$$-\int_0^{\tau_0}\int_{\partial\Omega\cap B(x_0,\delta)}\frac{\partial K}{\partial\nu_y}(x,y,\tau)\gamma(x_0,t)do(y)d\tau. \qquad (4.3.9)$$

Extracting the factor $\gamma(x_0,t)$ it remains to show that

$$-\lim_{x\to x_0}\int_0^{\tau_0}\int_{\partial\Omega\cap B(x_0,\delta)}\frac{\partial K}{\partial\nu_y}(x,y,\tau)do(y)d\tau = \frac{1}{2}+O(\delta). \qquad (4.3.10)$$

Also, we observe that since γ is continuous, it suffices to show that (4.3.10) holds uniformly in x_0 if x approaches $\partial\Omega$ in the direction normal to $\partial\Omega$. In other words, letting $\nu(x_0)$ denote the exterior normal vector of $\partial\Omega$ at x_0, we may assume

$$x = x_0 - \mu\nu(x_0).$$

In that case, $\mu^2 = |x - x_0|^2$, and since $\partial\Omega$ is of class C^2, for $y \in \partial\Omega$,

$$|x - y|^2 = |y - x_0|^2 + \mu^2 + O\left(|y - x_0|^2\,|x - x_0|\right).$$

The term $O\left(|y - x_0|^2\,|x - x_0|\right)$ here is a higher-order term that does not influence the validity of our subsequent limit processes, and so we shall omit it in the sequel for the sake of simplicity. Likewise, for $y \in \partial\Omega$,

$$(x - y)\cdot\nu_y = (x - x_0)\cdot\nu_y + (x_0 - y)\cdot\nu_y = -\mu + O\left(|x_0 - y|^2\right),$$

and the term $O(|x_0 - y|^2)$ may be neglected again.

Thus we approximate

$$\frac{\partial K}{\partial\nu_y}(x,y,\tau) = \frac{1}{(4\pi\tau)^{\frac{d}{2}}}\frac{(x - y)\cdot\nu_y}{2\tau}e^{-\frac{|x-y|^2}{4\tau}}$$

by

$$\frac{1}{(4\pi\tau)^{\frac{d}{2}}}\frac{(-\mu)}{2\tau}e^{-\frac{|x_0-y|^2}{4\tau}}e^{-\frac{\mu^2}{4\tau}}.$$

This means that we need to estimate the expression

$$\int_0^{\tau_0} \int_{\partial\Omega \cap B(x_0,\delta)} \frac{1}{2(4\pi)^{\frac{d}{2}}} \frac{\mu}{\tau^{\frac{d}{2}+1}} e^{-\frac{|x_0-y|^2}{4\tau}} e^{-\frac{\mu^2}{4\tau}} do(y) d\tau.$$

We introduce polar coordinates with center x_0 and put $\sigma = |x_0 - y|$. We then obtain, again up to a higher-order error term,

$$\mu \text{Vol}(S^{d-2}) \frac{1}{2(4\pi)^{\frac{d}{2}}} \int_0^{\tau_0} \frac{1}{\tau^{\frac{d}{2}+1}} e^{-\frac{\mu^2}{4\tau}} \int_0^\delta e^{-\frac{r^2}{4\tau}} r^{d-2} dr \, d\tau,$$

where S^{d-2} is the unit sphere in $\mathbb{R}^{d-1}$

$$= \frac{\mu \text{Vol}(S^{d-2})}{4\pi^{\frac{d}{2}}} \int_0^{\tau_0} \frac{1}{\tau^{\frac{3}{2}}} e^{-\frac{\mu^2}{4\tau}} \int_0^{\frac{\delta}{2\tau^{\frac{1}{2}}}} e^{-s^2} s^{d-2} ds \, d\tau$$

$$= \frac{\text{Vol}(S^{d-2})}{2\pi^{\frac{d}{2}}} \int_{\frac{\mu^2}{4\tau_0}}^\infty \frac{1}{\sigma^{\frac{1}{2}}} e^{-\sigma} \int_0^{\frac{\delta\sigma^{\frac{1}{2}}}{\mu}} e^{-s^2} s^{d-2} ds \, d\sigma.$$

In this integral we may let μ tend to 0 and obtain as limit

$$\frac{\text{Vol}(S^{d-2})}{2\pi^{\frac{d}{2}}} \int_0^\infty \frac{1}{\sigma^{\frac{1}{2}}} e^{-\sigma} \int_0^\infty e^{-s^2} s^{d-2} ds \, d\sigma = \frac{1}{2}. \qquad (4.3.11)$$

By our preceding considerations, this implies (4.3.10).

Equation (4.3.11) is shown with the help of the gamma function

$$\Gamma(x) = \int_0^\infty e^{-t} t^{x-1} dt \quad \text{for } x > 0.$$

We have

$$\Gamma(x+1) = x\Gamma(x) \quad \text{for all } x > 0,$$

and because of $\Gamma(1) = 1$, then

$$\Gamma(n+1) = n! \quad \text{for } n \in \mathbb{N}.$$

Moreover,

$$\int_0^\infty s^n e^{-s^2} ds = \frac{1}{2} \Gamma\left(\frac{n+1}{2}\right) \quad \text{for all } n \in \mathbb{N}.$$

In particular,

$$\Gamma\left(\frac{1}{2}\right) = 2 \int_0^\infty e^{-s^2} ds = \sqrt{\pi}$$

and

$$\pi^{\frac{d}{2}} = \int_{\mathbb{R}^d} e^{-|x|^2} dx = \text{Vol}(S^{d-1}) \int_0^\infty e^{-r^2} r^{d-1} dr = \frac{1}{2} \text{Vol}(S^{d-1}) \Gamma\left(\frac{d}{2}\right);$$

hence

$$\text{Vol}(S^{d-1}) = \frac{2\pi^{\frac{d}{2}}}{\Gamma\left(\frac{d}{2}\right)}.$$

With these formulae, the integral (4.3.11) becomes

$$\frac{2\pi^{\frac{d-1}{2}}}{\Gamma\left(\frac{d-1}{2}\right)} \frac{1}{2\pi^{\frac{d}{2}}} \Gamma\left(\frac{1}{2}\right) \cdot \frac{1}{2} \Gamma\left(\frac{d-1}{2}\right) = \frac{1}{2}.$$

□

In an analogous manner, one proves the following lemma:

Lemma 4.3.3: *Under the assumptions of Lemma 4.3.2, for*

$$w(x,t) := \int_0^t \int_{\partial\Omega} K(x,y,\tau)\gamma(y,t-\tau)\, do(y)\, d\tau \qquad (4.3.12)$$

($x \in \Omega$, $0 \le t \le T$), we have

$$w \in C^\infty(\Omega \times [0,T]),$$
$$w(x,0) = 0 \quad \text{for } x \in \Omega. \qquad (4.3.13)$$

The function w extends continuously to $\bar{\Omega} \times [0,T]$, and for $x_0 \in \partial\Omega$ we have

$$\lim_{x \to x_0} \nabla_x w(x,t) \cdot \nu(x_0) = \frac{\gamma(x_0,t)}{2} + \int_0^t \int_{\partial\Omega} \frac{\partial K}{\partial \nu_{x_0}}(x_0,y,\tau)\gamma(y,t-\tau)\, do(y)\, d\tau.$$

$$(4.3.14)$$

□

We now want to try first to find a solution of

$$\Delta u - \frac{\partial}{\partial t} u = 0 \qquad \text{in } \Omega \times (0,\infty),$$
$$u(x,0) = 0 \qquad \text{for } x \in \Omega,$$
$$u(x,t) = g(x,t) \quad \text{for } x \in \partial\Omega,\ t > 0, \qquad (4.3.15)$$

by Lemma 4.3.2.
We try

$$u(x,t) = -\int_0^t \int_{\partial\Omega} \frac{\partial K}{\partial \nu_y}(x,y,t-\tau)\gamma(y,\tau)\, do(y)\, d\tau, \qquad (4.3.16)$$

with a function $\gamma(x, t)$ yet to be determined. As a consequence of (4.3.7), (4.3.15), γ has to satisfy, for $x_0 \in \partial\Omega$,

$$g(x_0, t) = \frac{1}{2}\gamma(x_0, t) - \int_0^t \int_{\partial\Omega} \frac{\partial K}{\partial\nu_y}(x_0, y, t - \tau)\gamma(y, \tau) \, do(y) \, d\tau,$$

i.e.,

$$\gamma(x_0, t) = 2g(x_0, t) + 2 \int_0^t \int_{\partial\Omega} \frac{\partial K}{\partial\nu_y}(x_0, y, t - \tau)\gamma(y, \tau) \, do(y) \, d\tau. \quad (4.3.17)$$

This is a fixed-point equation for γ, and one may attempt to solve it by iteration; i.e., for $x_0 \in \partial\Omega$,

$$\gamma_0(x_0, t) = 2g(x_0, t),$$

$$\gamma_n(x_0, t) = 2g(x_0, t) + 2 \int_0^t \int_{\partial\Omega} \frac{\partial K}{\partial\nu_y}(x_0, y, t - \tau)\gamma_{n-1}(y, \tau) \, do(y) d\tau$$

for $n \in \mathbb{N}$. Recursively, we obtain

$$\gamma_n(x_0, t) = 2g(x_0, t) + 2 \int_0^t \int_{\partial\Omega} \sum_{\nu=1}^n S_\nu(x_0, y, t - \tau)g(y, \tau) \, do(y) d\tau \quad (4.3.18)$$

with

$$S_1(x_0, y, t) = 2\frac{\partial K}{\partial\nu_y}(x_0, y, t),$$

$$S_{\nu+1}(x_0, y, t) = 2 \int_0^t \int_{\partial\Omega} S_\nu(x_0, z, t - \tau)\frac{\partial K}{\partial\nu_y}(z, y, \tau) \, do(z) \, d\tau.$$

In order to show that this iteration indeed yields a solution, we have to verify that the series

$$S(x_0, y, t) = \sum_{\nu=1}^\infty S_\nu(x_0, y, t)$$

converges.

Choosing once more $\alpha = \frac{3}{4}$ in Lemma 4.3.1, we obtain

$$|S_1(x_0, y, t)| \leq ct^{-3/4} |x_0 - y|^{-(d-1)+\frac{1}{2}}.$$

Iteratively, we get

$$|S_n(x_0, y, t)| \leq c_n t^{-1+\frac{n}{4}} |x_0 - y|^{-(d-1)+\frac{n}{2}}.$$

We now choose $n = \max(4, 2(d - 1))$ so that both exponents are positive. If now

$$|S_m(x_0, y, t)| \leq \beta_m t^\alpha \quad \text{for some constant } \beta_m \text{ and some } \alpha \geq 0,$$

then

$$|S_{m+1}(x_0, y, t)| \leq c\beta_0\beta_m \int_0^t (t-\tau)^\alpha \tau^{-3/4} \, d\tau,$$

where the constant c comes from Lemma 4.3.1 and

$$\beta_0 := \sup_{y \in \partial\Omega} \int_{\partial\Omega} |z-y|^{-(d-1)+\frac{1}{2}} \, do(z).$$

Furthermore,

$$\int_0^t (t-\tau)^\alpha \tau^{-3/4} d\tau = \frac{\Gamma(1+\alpha)\Gamma\left(\frac{1}{4}\right)}{\Gamma\left(\frac{5}{4}+\alpha\right)} t^{\alpha+1/4},$$

where on the right-hand side we have the gamma function introduced above.
Thus

$$|S_{n+\nu}(x_0, y, t)| \leq \beta_n (c\beta_0)^\nu t^{\alpha+\nu/4} \prod_{\mu=1}^\nu \frac{\Gamma\left(\alpha + \frac{3}{4} + \mu/4\right)\Gamma\left(\frac{1}{4}\right)}{\Gamma(\alpha+1+\mu/4)}.$$

Since the gamma function grows factorially as a function of its arguments, this implies that

$$\sum_{\nu=1}^\infty S_\nu(x_0, y, t)$$

converges absolutely and uniformly on $\partial\Omega \times \partial\Omega \times [0, T]$ for every $T > 0$. We thus have the following result:

Theorem 4.3.1: *The initial boundary value problem for the heat equation on a bounded domain $\Omega \subset \mathbb{R}^d$ of class C^2, namely,*

$$\Delta u(x, t) - \frac{\partial}{\partial t} u(x, t) = 0 \qquad \text{in } \Omega \times (0, \infty),$$

$$u(x, 0) = 0 \qquad \text{in } \Omega,$$

$$u(x, t) = g(x, t) \qquad \text{for } x \in \partial\Omega, \quad t > 0,$$

with given continuous g, admits a unique solution. That solution can be represented as

$$u(x, t) = -\int_0^t \int_{\partial\Omega} \Sigma(x, y, t-\tau) g(y, \tau) \, do(y) \, d\tau, \tag{4.3.19}$$

where

$$\Sigma(x, y, t) = 2\frac{\partial K}{\partial \nu_y}(x, y, t) + 2\int_0^t \int_{\partial\Omega} \frac{\partial K}{\partial \nu_z}(x, z, t-\tau) \sum_{\nu=1}^\infty S_\nu(z, y, \tau) \, do(z) \, d\tau.$$

$$\tag{4.3.20}$$

Proof: Since the series $\sum_{\nu=1}^{\infty} S_\nu$ converges,

$$\gamma(x_0, t) = 2g(x_0, t) + 2 \int_0^t \int_{\partial\Omega} \sum_{\nu=1}^{\infty} S_\nu(x_0, y, t-\tau)g(y, \tau) \, do(y) \, d\tau$$

is a solution of (4.3.17). Inserting this into (4.3.16), we obtain (4.3.20). Here, one should note that

$$t^{-3/4} |y - x|^{-(d-1)+\frac{1}{2}} \sum_{\nu=1}^{\infty} S_\nu(x_0, y, \tau),$$

and hence also $\Sigma(x, y, t)$ converges absolutely and uniformly on $\partial\Omega \times \partial\Omega \times [0, T]$ for every $T > 0$. Thus, we may differentiate term by term under the integral and show that u solves the heat equation. The boundary values are assumed by construction, and it is clear that u vanishes at $t = 0$. Uniqueness follows from Theorem 4.1.1.   □

Definition 4.3.1: *Let $\Omega \subset \mathbb{R}^d$ be a domain. A function $q(x, y, t)$ that is defined for $x, y \in \bar\Omega$, $t > 0$, is called the heat kernel of Ω if*

(i)

$$\left(\Delta_x - \frac{\partial}{\partial t}\right) q(x, y, t) = 0 \quad \text{for } x, y \in \Omega, \, t > 0, \tag{4.3.21}$$

(ii)

$$q(x, y, t) = 0 \quad \text{for } x \in \partial\Omega, \tag{4.3.22}$$

(iii) and for all continuous $f : \Omega \to \mathbb{R}$

$$\lim_{t \to 0} \int_\Omega q(x, y, t)f(x)dx = f(y) \quad \text{for all } y \in \Omega. \tag{4.3.23}$$

Corollary 4.3.1: *Any bounded domain $\Omega \subset \mathbb{R}^d$ of class C^2 has a heat kernel, and this heat kernel is of class C^1 on $\bar\Omega$ with respect to the spatial variables y. The heat kernel is positive in Ω, for all $t > 0$.*

Proof: For each $y \in \Omega$, by Theorem 4.3.1 we solve the boundary value problem for the heat equation with initial values 0 and

$$g(x, t) = -K(x, y, t).$$

The solution is called $\mu(x, y, t)$, and we put

$$q(x, y, t) := K(x, y, t) + \mu(x, y, t). \tag{4.3.24}$$

Obviously, $q(x, y, t)$ satisfies (i) und (ii), and since

$$\lim_{t \to 0} \mu(x, y, t) = 0,$$

and $K(x, y, t)$ satisfies (iii), then so does $q(x, y, t)$.

Lemma 4.3.3 implies that q can be extended to $\bar{\Omega}$ as a continuously differentiable function of the spatial variables.

That $q(x, y, t) > 0$ for all $x, y \in \Omega, t > 0$ follows from the strong maximum principle (Theorem 4.1.3). Namely,

$$q(x, y, t) = 0 \quad \text{for } x \in \partial\Omega,$$
$$q(x, y, t) = 0 \quad \text{for } x, y, \in \Omega, x \neq y,$$

while (iii) implies

$$q(x, y, t) > 0 \quad \text{if } |x - y| \text{ and } t > 0 \text{ are sufficiently small.}$$

Thus, $q \geq 0$ and $q \neq 0$, and so by Theorem 4.1.3,

$$q > 0 \quad \text{in } \Omega \times \Omega \times (0, \infty).$$

$\square$

Lemma 4.3.4 (Duhamel principle): *For all functions u, v on $\Omega \times [0, T]$ with the appropriate regularity conditions, we have*

$$\int_0^T \int_\Omega \left\{ v(x, t) \left(\Delta u(x, T - t) + u_t(x, T - t) \right) \right.$$

$$\left. - u(x, T - t) \left(\Delta v(x, t) - v_t(x, t) \right) \right\} dx \, dt$$

$$= \int_0^T \int_{\partial\Omega} \left\{ \frac{\partial u}{\partial \nu}(y, T - t) v(y, t) - \frac{\partial v}{\partial \nu}(y, t) u(y, T - t) \right\} do(y) \, dt$$

$$+ \int_\Omega \left\{ u(x, 0) v(x, T) - u(x, T) v(x, 0) \right\} dx. \tag{4.3.25}$$

Proof: Same as the proof of (4.1.12). $\square$

Corollary 4.3.2: *If the heat kernel $q(z, w, T)$ of Ω is of class C^1 on $\bar{\Omega}$ with respect to the spatial variables, then it is symmetric with respect to z and w, i.e.,*

$$q(z, w, T) = q(w, z, T) \quad \text{for all } z, w \in \Omega, \ T > 0. \tag{4.3.26}$$

Proof: In (4.3.25), we put $u(x, t) = q(x, z, t)$, $v(x, t) = q(x, w, t)$. The double integrals vanish by properties (i) and (ii) of Definition 4.3.1. Property (iii) of Definition 4.3.1 then yields $v(z, T) = u(w, T)$, which is the asserted symmetry. $\square$

Theorem 4.3.2: *Let $\Omega \subset \mathbb{R}^d$ be a bounded domain of class C^2 with heat kernel $q(x, y, t)$ according to Corollary 4.3.1, and let*

$$\varphi \in C^0(\bar{\Omega} \times [0, \infty)), \quad g \in C^0(\partial\Omega \times (0, \infty)), \quad f \in C^0(\Omega).$$

Then the initial boundary value problem

$$u_t(x,t) - \Delta u(x,t) = \varphi(x,t) \quad \text{for } x \in \Omega, t > 0,$$
$$u(x,t) = g(x,t) \quad \text{for } x \in \partial\Omega, t > 0,$$
$$u(x,0) = f(x) \quad \text{for } x \in \Omega, \tag{4.3.27}$$

admits a unique solution that is continuous on $\bar{\Omega} \times [0,\infty) \setminus \partial\Omega \times \{0\}$ and is represented by the formula

$$u(x,t) = \int_0^t \int_\Omega q(x,y,t-\tau)\varphi(y,\tau)dy\,d\tau$$
$$+ \int_\Omega q(x,y,t)f(y)dy - \int_0^t \int_{\partial\Omega} \frac{\partial q}{\partial\nu_y}(x,y,t-\tau)g(y,\tau)do(y)d\tau. \tag{4.3.28}$$

Proof: Uniqueness follows from the maximum principle. We split the existence problem into two subproblems:
We solve

$$v_t(x,t) - \Delta v(x,t) = 0 \quad \text{for } x \in \Omega, t > 0,$$
$$v(x,t) = g(x,t) \quad \text{for } x \in \partial\Omega, t > 0, \tag{4.3.29}$$
$$v(x,0) = f(x) \quad \text{for } x \in \Omega,$$

i.e., the homogeneous equation with the prescribed initial and boundary conditions, and

$$w_t(x,t) - \Delta w(x,t) = \varphi(x,t) \quad \text{for } x \in \Omega, t > 0,$$
$$w(x,t) = 0 \quad \text{for } x \subset \partial\Omega, t > 0, \tag{4.3.30}$$
$$w(x,0) = 0 \quad \text{for } x \in \Omega,$$

i.e., the inhomogeneous equation with vanishing initial and boundary values. The solution of (4.3.27) is then given by

$$u = v + w.$$

We first address (4.3.29), and we claim that the solution v can be represented as

$$v(x,t) = \int_\Omega q(x,y,t)f(y)dy - \int_0^t \int_{\partial\Omega} \frac{\partial q}{\partial\nu_y}(x,y,t-\tau)g(y,\tau)do(y)d\tau.$$

The facts that v solves the heat equation and the initial condition $v(x,0) = f(x)$ follow from the corresponding properties of q. Moreover, $q(x,y,t) = K(x,y,t) + \mu(x,y,t)$ with $\mu(x,y,t)$ coming from the proof of Corollary 4.3.1. By Theorem 4.3.1, this μ can be represented as

$$\mu(x,y,t) = \int_0^t \int_{\partial\Omega} \Sigma(x,z,t-\tau)K(z,y,\tau)do(z)\,d\tau, \tag{4.3.31}$$

and by Lemma 4.3.3, we have for $y \in \partial\Omega$,

$$\frac{\partial \mu}{\partial \nu_y}(x,y,t) = \frac{\Sigma(x,y,t)}{2} + \int_0^t \int_{\partial\Omega} \Sigma(x,z,t-\tau)\frac{\partial K}{\partial \nu_y}(z,y,\tau)do(z)\,d\tau.$$

(4.3.32)

This means that the second integral on the right-hand side of (4.3.28) is precisely of the type (4.3.19), and thus, by the considerations of Theorem 4.3.1, v indeed satisfies the boundary condition $v(x,t) = g(x,t)$ for $x \in \partial\Omega$, because the first integral vanishes on the boundary.

We now turn to (4.3.30). For every $\tau > 0$, we let $z(x,t,\tau)$ be the solution of

$$\begin{aligned} z_t(x,t;\tau) - \Delta z(x,t,\tau) &= 0 && \text{for } x \in \Omega, t > \tau, \\ z(x,t;\tau) &= 0 && \text{for } x \in \partial\Omega, t > \tau, \\ z(x,\tau;\tau) &= \varphi(x,\tau) && \text{for } x \in \Omega. \end{aligned}$$

(4.3.33)

This is a special case of (4.3.29), which we already know how to solve, except that the initial conditions are not prescribed at $t = 0$, but at $t = \tau$. This case, however, is trivially reduced to the case of initial conditions at $t = 0$ by replacing t by $t - \tau$, i.e., considering $\zeta(x,t;\tau) = z(x,t+\tau;\tau)$. Thus, (4.3.33) can be solved.

We then put

$$w(x,t) = \int_0^t z(x,t;\tau)d\tau.$$

(4.3.34)

Then

$$w_t(x,t) = \int_0^t z_t(x,t;\tau)d\tau + z(x,t;t) = \int_0^t \Delta z(x,t;\tau)d\tau + \varphi(x,t)$$
$$= \Delta w(x,t) + \varphi(x,t)$$

and

$$\begin{aligned} w(x,t) &= 0 && \text{for } x \in \partial\Omega, t > 0, \\ w(x,0) &= 0 && \text{for } x \in \Omega. \end{aligned}$$

Thus, w is a solution of (4.3.30) as required, and the proof is complete, since the representation formula (4.3.28) follows from the one for v and the one for w that, by (4.3.34), comes from integrating the one for z. The latter in turn solves (4.3.33), and so by what has been proved already, is given by

$$z(x,t;\tau) = \int_\Omega q(x,y,t-\tau)\varphi(x,\tau)dy.$$

Thus, inserting this into (4.3.34), we obtain

$$w(x,t) = \int_0^t \int_\Omega q(x,y,t-\tau)\varphi(x,\tau)dy\,d\tau. \qquad (4.3.35)$$

This completes the proof.   □

We briefly interrupt our discussion of the solution of the heat equation and record the following simple result on the heat kernel q for subsequent use:

$$\int_\Omega q(x,y,t)dy \leq 1 \qquad (4.3.36)$$

for all $t \geq 0$. To start, we have

$$\lim_{t \to 0} \int_\Omega q(x,y,t)dy = 1. \qquad (4.3.37)$$

This follows from (4.3.23) with $f \equiv 1$ and the proof of Corollary 4.3.1 which enables to replace the integration w.r.t. x in (4.3.23) by the one w.r.t. y in (4.3.37). Next, we observe that

$$\frac{\partial q}{\partial \nu_y}(x,y,t) \leq 0 \qquad (4.3.38)$$

because q is nonnegative in Ω and vanishes on the boundary $\partial \Omega$ (see (4.3.22) and Corollary 4.3.1). We then note that the solution of Theorem 4.3.2 for $\varphi \equiv 1$, $g(x,t) = t$, $f(x) = 0$ is given by $u(x,t) = t$. In the representation formula (4.3.28), using (4.3.38), this yields

$$\int_0^t \int_\Omega q(x,y,t-\tau)dy\,d\tau \leq t, \qquad (4.3.39)$$

from which (4.3.36) is derived upon a little reflection.

We now resume the discussion of the solution established in Theorem 4.3.2. We did not claim continuity of our solution at the corner $\partial \Omega \times \{0\}$, and in general, we cannot expect continuity there unless we assume a matching condition between the initial and the boundary values. We do have, however,

Theorem 4.3.3: *The solution of Theorem 4.3.2 is continuous on all of $\bar{\Omega} \times [0, \infty)$ when we have the compatibility condition*

$$g(x,0) = f(x) \quad \text{for } x \in \partial \Omega. \qquad (4.3.40)$$

Proof: While the continuity at the corner $\partial \Omega \times \{0\}$ could also be established from a refinement of our previous considerations, we provide here some independent and simpler reasoning. By the general superposition argument that we have already employed a few times (in particular in the proof of Theorem 4.3.2), it suffices to establish continuity for a solution of

$$v_t(x,t) - \Delta v(x,t) = 0 \qquad \text{for } x \in \Omega, \, t > 0,$$
$$v(x,t) = g(x,t) \quad \text{for } x \in \partial\Omega, \, t > 0,$$
$$v(x,0) = 0 \qquad \text{for } x \in \Omega, \qquad\qquad (4.3.41)$$

with a continuous g satisfying

$$g(x,0) = 0 \quad \text{for } x \in \partial\Omega, \qquad\qquad (4.3.42)$$

and for a solution of

$$w_t(x,t) - \Delta w(x,t) = 0 \qquad \text{for } x \in \Omega, \, t > 0,$$
$$w(x,t) = 0 \qquad \text{for } x \in \partial\Omega, \, t > 0,$$
$$w(x,0) = f(x) \quad \text{for } x \in \Omega, \qquad\qquad (4.3.43)$$

with a continuous f satisfying

$$f(x) = 0 \quad \text{for } x \in \partial\Omega. \qquad\qquad (4.3.44)$$

(We leave it to the reader to check the case of a solution of the inhomogeneous equation $u_t(x,t) - \Delta u(x,t) = \varphi(x,t)$ with vanishing initial and boundary values.)

To deal with the first case, we consider, for $\tau > 0$,

$$\tilde{v}_t(x,t) - \Delta\tilde{v}(x,t) = 0 \qquad \text{for } x \in \Omega, \, t > 0,$$
$$\tilde{v}(x,t) = 0 \qquad \text{for } x \in \partial\Omega, \, 0 < t \leq \tau,$$
$$\tilde{v}(x,t) = g(x,t-\tau) \quad \text{for } x \in \partial\Omega, \, t > \tau,$$
$$\tilde{v}(x,0) = 0 \qquad \text{for } x \in \Omega. \qquad\qquad (4.3.45)$$

Since, by (4.3.42), the boundary values are continuous at $t = \tau$, by the boundary continuity result of Theorem 4.3.2, $\tilde{v}(x,\tau)$ is continuous for $x \in \partial\Omega$. Also, by uniqueness, $\tilde{v}(x,t) = 0$ for $0 \leq t \leq \tau$, because both the boundary and initial values vanish there. Therefore, again by uniqueness, $v(x,t) = \tilde{v}(x,t+\tau)$ and we conclude the continuity of $v(x,0)$ for $x \in \partial\Omega$.

We can now turn to the second case. We consider some bounded C^2 domain $\tilde{\Omega}$ with $\bar{\Omega} \subset \tilde{\Omega}$. We put $f^+(x) := \max(f(x),0)$ for $x \in \Omega$ and $f(x) = 0$ for $x \in \tilde{\Omega}\backslash\Omega$. Then, because of (4.3.44), f^+ is continuous on $\tilde{\Omega}$. We then solve

$$\tilde{w}_t(x,t) - \Delta\tilde{w}(x,t) = 0 \qquad \text{for } x \in \tilde{\Omega}, \, t > 0,$$
$$\tilde{w}(x,t) = 0 \qquad \text{for } x \in \partial\tilde{\Omega}, \, t > 0,$$
$$\tilde{w}(x,0) = f^+(x) \quad \text{for } x \in \tilde{\Omega}. \qquad\qquad (4.3.46)$$

By the continuity result of Theorem 4.3.2, $\tilde{w}(x,0)$ is continuous for $x \in \tilde{\Omega}$, and therefore in particular for $x \in \partial\Omega$. Since $f^+(x) = 0$ for $x \in \partial\Omega$, $\tilde{w}(x,t) \to 0$ for $x \in \partial\Omega$ and $t \to 0$. Since the initial values of $\tilde{w}$ are non-negative,

$\tilde{w}(x,t) \geq 0$ for all $x \in \tilde{\Omega}$ and $t \geq 0$ by the maximum principle (Theorem 4.1.1). In particular, $\tilde{w}(x,t) \geq w(x,t)$ for $x \in \partial\Omega$ since $w(x,t) = 0$ there. Since also $\tilde{w}(x,0) = f^+(x) \geq f(x) = w(x,0)$, the maximum principle implies $\tilde{w}(x,t) \geq w(x,t)$ for all $x \in \tilde{\Omega}, t \geq 0$. Altogether, $w(x,0) \leq 0$ for $x \in \partial\Omega$. Doing the same reasoning with $f^-(x) := \min(f(x),0)$, we conclude that also $w(x,0) \geq 0$ for $x \in \partial\Omega$, that is, altogether, $w(x,0) = 0$ for $x \in \partial\Omega$. This completes the proof. $\qquad\square$

Remark: Theorem 4.3.2 does not claim that u is twice differentiable with respect to x, and in fact, this need not be true for a φ that is merely continuous. However, one may still justify the equation

$$u_t(x,t) - \Delta u(x,t) = \varphi(x,t).$$

We shall return to the analogous issue in the elliptic case in Sections 10.1 and 11.1. In Section 11.1, we shall verify that u is twice continuously differentiable with respect to x if we assume that φ is Hölder continuous.

Here, we shall now concentrate on the case $\varphi = 0$ and address the regularity issue both in the interior of Ω and at its boundary. We recall the representation formula (4.1.14) for a solution of the heat equation on Ω,

$$u(x,t) = \int_{\Omega} K(x,y,t)u(y,0)\,dy$$
$$+ \int_0^t \int_{\partial\Omega} \left(K(x,y,t-\tau)\frac{\partial u(y,\tau)}{\partial \nu} - \frac{\partial K}{\partial \nu_y}(x,y,t-\tau)u(y,\tau) \right) do(y)\,d\tau.$$
$$(4.3.47)$$

We put $K(x,y,s) = 0$ for $s \leq 0$ and may then integrate the second integral from 0 to ∞ instead of from 0 to t. Then $K(x,y,s)$ is of class C^∞ for $x,y \in \mathbb{R}^d$, $s \in \mathbb{R}$, except at $x = y, s = 0$. We thus have the following theorem:

Theorem 4.3.4: *Any solution $u(x,t)$ of the heat equation in a domain Ω is of class C^∞ with respect to $x \in \Omega$, $t > 0$.*

Proof: Since we do not know whether the normal derivative $\frac{\partial u}{\partial \nu}$ exists on $\partial\Omega$ and is continuous there, we cannot apply (4.3.47) directly. Instead, for given $x \in \Omega$, we consider some ball $B(x,r)$ contained in Ω. We then apply (4.3.47) on $\mathring{B}(x,r)$ in place of Ω. Since $\partial B(x,r)$ in Ω is contained in Ω, and u as a solution of the heat equation is of class C^1 there, the normal derivative $\frac{\partial u}{\partial \nu}$ on $\partial B(x,r)$ causes no problem, and the assertion is obtained. $\qquad\square$

In particular, the heat kernel $q(x,y,t)$ of a bounded C^2-domain Ω is of class C^∞ with respect to $x,y \in \Omega$, $t > 0$. This also follows directly from (4.3.24), (4.3.31), (4.3.20), and the regularity properties of $\Sigma(x,y,t)$ established in Theorem 4.3.1. From these solutions it also follows that $\frac{\partial q}{\partial \nu_y}(x,y,t)$ for $y \in \partial\Omega$ is of class C^∞ with respect to $x \in \Omega$, $t > 0$. Thus, one can also use the representation formula (4.3.28) for deriving regularity properties. Putting $q(x, y, s) = 0$ for $s < 0$, we may again extend the second integral in (4.3.28) from 0 to ∞, and we then obtain by integrating by parts, assuming that the boundary values are differentiable with respect to t,

$$
\frac{\partial}{\partial t} u(x, t) = \int_\Omega \frac{\partial}{\partial t} q(x, y, t) f(y) dy
$$
$$
- \int_0^\infty \int_{\partial\Omega} \frac{\partial q}{\partial \nu_y}(x, y, t - \tau) \frac{\partial}{\partial \tau} g(y, \tau) \, do(y) \, d\tau
$$
$$
+ \lim_{\tau \to 0} \int_{\partial\Omega} \frac{\partial g}{\partial \nu_y}(x, y, t - \tau) g(y, \tau) \, do(y). \tag{4.3.48}
$$

Since $q(x, y, t) = 0$ for $x \in \partial\Omega$, $y \in \Omega$, $t > 0$, also $\frac{\partial q}{\partial \nu_y}(x, y, t - \tau) = 0$ for $x, y \in \partial\Omega, \tau < t$ and

$$
\frac{\partial}{\partial t} q(x, y, t) = 0 \quad \text{for } x \in \partial\Omega, \ y \in \Omega, \ t > 0 \tag{4.3.49}
$$

(passing to the limit here is again justified by (4.3.31)). Since the second integral in (4.3.48) has boundary values $\frac{\partial}{\partial t} g(x, t)$, we thus have the following result:

Lemma 4.3.5: *Let u be a solution of the heat equation on the bounded C^2-domain Ω with continuous boundary values $g(x, t)$ that are differentiable with respect to t. Then u is also differentiable with respect to t, for $x \in \partial\Omega$, $t > 0$, and we have*

$$
\frac{\partial}{\partial t} u(x, t) = \frac{\partial}{\partial t} g(x, t) \quad \text{for } x \in \partial\Omega, \ t > 0. \tag{4.3.50}
$$

$\square$

We are now in position to establish the connection between the heat and Laplace equation rigorously that we had arrived at from heuristic considerations in Section 4.2.

Theorem 4.3.5: *Let $\Omega \subset \mathbb{R}^d$ be a bounded domain of class C^2, and let $f \in C^0(\Omega)$, $g \in C^0(\partial\Omega)$. Let u be the solution of Theorem 4.3.2 of the initial boundary value problem*

$$
\begin{aligned}
\Delta u(x, t) - u_t(x, t) &= 0 && \text{for } x \in \Omega, \ t > 0, \\
u(x, 0) &= f(x) && \text{for } x \in \Omega, \\
u(x, t) &= g(x) && \text{for } x \in \partial\Omega, \quad t > 0.
\end{aligned} \tag{4.3.51}
$$

Then u converges for $t \to \infty$ uniformly on $\bar{\Omega}$ towards a solution of the Dirichlet problem for the Laplace equation

$$
\begin{aligned}
\Delta u(x) &= 0 && \text{for } x \in \Omega, \\
u(x) &= g(x) && \text{for } x \in \partial\Omega.
\end{aligned} \tag{4.3.52}
$$

Proof: We write $u(x, t) = u^1(x, t) + u^2(x, t)$, where u^1 and u^2 both solve the heat equation, and u^1 has the correct initial values, i.e.,

$$u^1(x, 0) = f(x) \quad \text{for } x \in \Omega,$$

while u^2 has the correct boundary values, i.e.,

$$u^2(x, t) = g(x) \quad \text{for } x \in \partial\Omega, \ t > 0,$$

as well as

$$u^1(x, t) = 0 \quad \text{for } x \in \partial\Omega, \ t > 0,$$
$$u^2(x, 0) = 0 \quad \text{for } x \in \Omega.$$

By Lemma 4.2.3, we have

$$\lim_{t \to \infty} u^1(x, t) = 0.$$

Thus, the initial values f are irrelevant, and we may assume without loss of generality that $f \equiv 0$, i.e., $u = u^2$.

One easily sees that $q(x, y, t) > 0$ for $x, y \in \Omega$, because $q(x, y, t) = 0$ for all $x \in \partial\Omega$, and by (iii) of Definition 4.3.1, $q(x, y, t) > 0$ for $x, y \in \Omega$ and sufficiently small $t > 0$. Since q solves the heat equation, by the strong maximum principle q then is indeed positive in the interior of Ω for all $t > 0$ (see Corollary 4.3.1).

Therefore, we always have

$$\frac{\partial q}{\partial \nu_y}(x, y, t) \leq 0. \tag{4.3.53}$$

Since $q(x, y, t)$ solves the heat equation with vanishing boundary values, Lemma 4.2.3 also implies

$$\lim_{t \to \infty} q(x, y, t) = 0 \text{ uniformly in } \bar{\Omega} \times \bar{\Omega} \tag{4.3.54}$$

(utilizing the symmetry $q(x, y, t) = q(y, x, t)$ from Corollary 4.3.1). We then have for $t_2 > t_1$,

$$|u(x, t_2) - u(x, t_1)| = \left| \int_{t_1}^{t_2} \int_{\partial\Omega} \frac{\partial q}{\partial \nu_z}(x, z, t) g(z) do(z) dt \right|$$

$$\leq \max_{\partial\Omega} |g| \int_{t_1}^{t_2} \int_{\partial\Omega} \left(-\frac{\partial q}{\partial \nu_z}(x, z, t) \right) do(z) dt$$

$$= -\max_{\partial\Omega} |g| \int_{t_1}^{t_2} \int_{\Omega} \Delta_y q(x, y, t) dy \, dt$$

$$= -\max_{\partial\Omega} |g| \int_{t_1}^{t_2} \int_{\Omega} q_t(x, y, t) dy \, dt$$

$$= -\max_{\partial\Omega} |g| \int_{\Omega} \{q(x, y, t_2) - q(x, y, t_1)\} \, dy$$

$$\to 0 \quad \text{for } t_1, t_2 \to \infty \text{ by (4.3.54)}.$$

Thus $u(x,t)$ converges for $t \to \infty$ uniformly towards some limit function $u(x)$ that then also satisfies the boundary condition

$$u(x) = g(x) \quad \text{for } x \in \partial\Omega.$$

Theorem 4.3.2 also implies

$$u(x) = -\int_0^\infty \int_{\partial\Omega} \frac{\partial q}{\partial \nu_z}(x, z, t) g(z) do(z) dt.$$

We now consider the derivatives $\frac{\partial}{\partial t} u(x,t) =: v(x,t)$. Then $v(x,t)$ is a solution of the heat equation itself, namely with boundary values $v(x,t) = 0$ for $x \in \partial\Omega$ by Lemma 4.3.5. By Lemma 4.2.3, v then converges uniformly to 0 on $\bar\Omega$ for $t \to \infty$. Therefore, $\Delta u(x,t)$ converges uniformly to 0 in $\bar\Omega$ for $t \to \infty$, too. Thus, we must have

$$\Delta u(x) = 0.$$

□

As a consequence of Theorem 4.3.5, we obtain a new proof for the solvability of the Dirichlet problem for the Laplace equation on bounded domains of class C^2, i.e., a special case of Theorem 3.2.2 (together with Lemma 3.4.1):

Corollary 4.3.3: *Let $\Omega \subset \mathbb{R}^d$ be a bounded domain of class C^2, and let $g : \partial\Omega \to \mathbb{R}$ be continuous. Then the Dirichlet problem*

$$\Delta u(x) = 0 \qquad \text{for } x \in \Omega, \tag{4.3.55}$$

$$u(x) = g(x) \quad \text{for } x \in \partial\Omega, \tag{4.3.56}$$

admits a solution that is unique by the maximum principle. □

References for this Section are Chavel [3] and the sources given there.

4.4 Discrete Methods

Both for the heuristics and for numerical purposes, it can be useful to discretize the heat equation. For that, we shall proceed as in Section 3.1 and also keep the notation of that section. In addition to the spatial variables, we also need to discretize the time variable t; the corresponding step size will be denoted by k. It will turn out to be best to choose k different from the spatial grid size h.

The discretization of the heat equation

$$u_t(x,t) = \Delta u(x,t) \tag{4.4.1}$$

is now straightforward:

$$\frac{1}{k}\left(u^{h,k}(x,t+k) - u^{h,k}(x,t)\right) \tag{4.4.2}$$

$$= \Delta_h u^{h,k}(x,t)$$

$$= \frac{1}{h^2}\sum_{i=1}^{d}\left\{u^{h,k}\left(x^1,\ldots,x^{i-1},x^i+h,x^{i+1},\ldots,x^d,t\right)\right.$$

$$\left. - 2u^{h,k}\left(x^1,\ldots,x^d,t\right) + u^{h,k}\left(x^1,\ldots,x^i-h,\ldots,x^d,t\right)\right\}.$$

Thus, for discretizing the time derivative, we have selected a forward difference quotient. In order to simplify the notation, we shall mostly write u in
place of $u^{h,k}$. Choosing

$$h^2 = 2dk, \tag{4.4.3}$$

the term $u(x,t)$ drops out, and (4.4.2) becomes

$$u(x,t+k) =$$

$$\frac{1}{2d}\sum_{i=1}^{d}\left(u\left(x^1,\ldots,x^i+h,\ldots,x^d,t\right) + u\left(x^1,\ldots,x^i-h,\ldots,x^d,t\right)\right). \tag{4.4.4}$$

This means that $u(x,t+k)$ is the arithmetic mean of the values of u at the $2d$
spatial neighbors of (x,t). From this observation, one sees that if the process
stabilizes as time grows, one obtains a solution of the discretized Laplace
equation asymptotically as in the continuous case.

It is possible to prove convergence results as in Section 3.1. Here, however,
we shall not carry this out. We wish to remark, however, that the process
can become unstable if $h^2 < 2dk$. The reader may try to find some examples.
This means that if one wishes h to be small so as to guarantee accuracy of
the approximation with respect to the spatial variables, then k has to be
extremely small to guarantee stability of the scheme. This makes the scheme
impractical for numerical use.

The mean value property of (4.4.4) also suggests the following semidiscrete
approximation of the heat equation: Let $\Omega \subset \mathbb{R}^d$ be a bounded domain. For
$\varepsilon > 0$, we put $\Omega_\varepsilon := \{x \in \Omega : \operatorname{dist}(x,\partial\Omega) > \varepsilon\}$. Let a continuous function
$g : \partial\Omega \to \mathbb{R}$ be given, with a continuous extension to $\bar{\Omega} \setminus \Omega_\varepsilon$, again denoted
by g. Finally, let initial values $f : \Omega \to \mathbb{R}$ be given. We put iteratively

$$\tilde{u}(x,0) = f(x) \qquad\qquad \text{for } x \in \Omega,$$

$$\tilde{u}(x,0) = 0 \qquad\qquad \text{for } x \in \mathbb{R}^d \setminus \Omega,$$

$$u(x,nk) = \frac{1}{\omega_d \varepsilon^d}\int_{B(x,\varepsilon)}\tilde{u}(y,(n-1)k)\,dy \quad \text{for } x \in \Omega, n \in \mathbb{N},$$

and

$$\tilde{u}(x, nk) = \begin{cases} u(x, nk) & \text{for } x \in \Omega_\varepsilon, \\ g(x) & \text{for } x \in \mathbb{R}^d \setminus \Omega_\varepsilon, \end{cases} \quad n \in \mathbb{N}.$$

Thus, in the nth step, the value of the function at $x \in \Omega_\varepsilon$ is obtained as the mean of the values of the preceding step of the ball $B(x, \varepsilon)$. A solution that is time independent then satisfies a mean value property and thus is harmonic in Ω_ε according to the remark after Corollary 1.2.5.

Summary

In the present chapter we have investigated the heat equation on a domain $\Omega \in \mathbb{R}^d$,

$$\frac{\partial}{\partial t} u(x, t) - \Delta u(x, t) = 0 \quad \text{for } x \in \Omega, t > 0.$$

We prescribed initial values

$$u(x, 0) = f(x) \quad \text{for } x \in \Omega,$$

and in the case that Ω has a boundary $\partial\Omega$, also boundary values

$$u(y, t) = g(y, t) \quad \text{for } y \in \partial\Omega, t \geq 0.$$

In particular, we studied the Euclidean fundamental solution

$$K(x, y, t) = \frac{1}{(4\pi t)^{\frac{d}{2}}} e^{-\frac{|x-y|^2}{4t}},$$

and we obtained the solution of the initial value problem on $\mathbb{R}^d$ by convolution

$$u(x, t) = \int_{\mathbb{R}^d} K(x, y, t) f(y) dy.$$

If Ω is a bounded domain of class C^2, we established the existence of the heat kernel $q(x, y, t)$, and we solved the initial boundary value problem by the formula

$$u(x, t) = \int_\Omega q(x, y, t) f(y) dy - \int_0^t \int_{\partial\Omega} \frac{\partial q}{\partial \nu_z}(x, z, t - \tau) g(z, \tau) do(z) d\tau.$$

In particular, $u(x, t)$ is of class C^∞ for $x \in \Omega$, $t > 0$, because of the corresponding regularity properties of the kernel $q(x, y, t)$. The solutions satisfy a maximum principle saying that a maximum or minimum can be assumed only on $\Omega \times \{0\}$ or on $\partial\Omega \times [0, \infty)$ unless the solution is constant. Consequently, solutions are unique. If the boundary values $g(y)$ do not depend on t, then $u(x, t)$ converges for $t \to \infty$ towards a solution of the Dirichlet problem for the Laplace equation

$$\Delta u(x) = 0 \qquad \text{in } \Omega,$$
$$u(x) = g(x) \quad \text{for } x \in \partial\Omega.$$

This yields a new existence proof for that problem, although requiring stronger assumptions for the domain Ω when compared with the existence proof of Chapter 3. The present proof, on the other hand, is more constructive in the sense of giving an explicit prescription for how to reach a harmonic state from some given state f.

Exercises

4.1 Let $\Omega \subset \mathbb{R}^d$ be bounded, $\Omega_T := \Omega \times (0, T)$. Let

$$L := \sum_{i,j=1}^{d} a^{ij}(x,t)\frac{\partial^2}{\partial x^i \partial x^j} + \sum_{i=1}^{d} b^i(x,t)\frac{\partial}{\partial x^i}$$

be elliptic for all $(x,t) \in \Omega_T$, and suppose

$$u_t \leq Lu,$$

where $u \in C^0(\bar\Omega_T)$ is twice continuously differentiable with respect to $x \in \Omega$ and once with respect to $t \in (0,T)$.
Show that

$$\sup_{\Omega_T} u = \sup_{\partial^* \Omega_T} u.$$

4.2 Using the heat kernel $\Lambda(x,y,t,0) = K(x,y,t)$, derive a representation formula for solutions of the heat equation on Ω_T with a bounded $\Omega \subset \mathbb{R}^d$ and $T < \infty$.

4.3 Show that for K as in Exercise 4.2,

$$K(x,0,s+t) = \int_{\mathbb{R}^d} K(x,y,t)K(y,0,s)dy$$

(a) if $s, t > 0$;
(b) if $0 < t < -s$.

4.4 Let Σ be the grid consisting of the points (x,t) with $x = nh$, $t = mk$, $n, m \in \mathbb{Z}$, $m \geq 0$, and let v be the solution of the discrete heat equation

$$\frac{v(x,t+k) - v(x,t)}{k} - \frac{v(x+h,t) - 2v(x,t) + v(x-h,t)}{h^2} = 0$$

with $v(x,0) = f(x) \in C^0(\mathbb{R})$.
Show that for $\frac{k}{h^2} = \frac{1}{2}$,

$$v(nh, mk) = 2^{-m} \sum_{j=0}^{m} \binom{m}{j} f((n-m+2j)h).$$

Conclude from this that

$$\sup_{\Sigma} |v| \leq \sup_{R} |f|.$$

4.5 Use the method of Section 4.3 to obtain a solution of the Poisson equation on $\Omega \subset \mathbb{R}^d$, a bounded domain of class C^2, continuous boundary values $g : \partial\Omega \to \mathbb{R}$, and continuous right-hand side $\varphi : \Omega \to \mathbb{R}$, i.e., of

$$\Delta u(x) = \varphi(x) \quad \text{for } x \in \Omega,$$
$$u(x) = g(x) \quad \text{for } x \in \partial\Omega.$$

(For the regularity issue, we need to refer to Section 11.1.)

5. Reaction-Diffusion Equations and Systems

5.1 Reaction-Diffusion Equations

In this section, we wish to study the initial boundary value problem for nonlinear parabolic equations of the form

$$u_t(x,t) - \Delta u(x,t) = F(x,t,u) \quad \text{for } x \in \Omega, t > 0,$$
$$u(x,t) = g(x,t) \qquad \text{for } x \in \partial\Omega, t > 0, \qquad (5.1.1)$$
$$u(x,0) = f(x) \qquad \text{for } x \in \Omega,$$

with given (continuous and smooth) functions g, f and a Lipschitz continuous function F (in fact, Lipschitz continuity is only needed w.r.t. to u; for x and t, continuity suffices). The nonlinearity of this equation comes from the u-dependence of F. While we may consider (5.1.1) as a heat equation with a nonlinear term on the right hand side, that is, as a generalization of

$$u_t(x,t) - \Delta u(x,t) = 0 \quad \text{for } x \in \Omega, t > 0 \qquad (5.1.2)$$

(with the same boundary and initial values), in the case where F does not depend on the spatial variable x, i.e. $F = F(t,u)$, we may alternatively view (5.1.1) as a generalization of the ODE

$$u_t(t) = F(t,u) \quad \text{for } t > 0,$$
$$u(0) = u_0. \qquad (5.1.3)$$

For such equations, we have, for the case of a Lipschitz continuous F, a local existence theorem, the Picard-Lindelöf theorem. This says that for given initial value u_0, we may find some $t_0 > 0$ with the property that a unique solution exists for $0 \leq t < t_0$. When F is bounded, solutions exist for all t, as follows from an iterated application of the Picard-Lindelöf theorem. When F is unbounded, however, solutions may become infinite in finite time; a standard example is

$$u_t(t) = u^2(t) \qquad (5.1.4)$$

with positive initial value u_0. The solution is

$$u(t) = (\frac{1}{u_0} - t)^{-1} \qquad (5.1.5)$$

which for positive u_0 becomes infinite in finite time, at $t = \frac{1}{u_0}$.

We shall see in this section that this qualitative type of behavior, in particular the local (in time) existence result, carries over to the reaction-diffusion equation (5.1.1). In fact, the local existence can be shown like the Picard-Lindelöf theorem by an application of the Banach fixed point theorem; here, of course, we need to utilize also the results for the heat equation (5.1.2) established in Section 4.3. We shall thus start by establishing the local existence result:

Theorem 5.1.1: Let $\Omega \subset \mathbb{R}^d$ be a bounded domain of class C^2, and let

$$g \in C^0(\partial\Omega \times [0, t_0]), \quad f \in C^0(\bar{\Omega}),$$
$$\text{with } g(x, 0) = f(x) \quad \text{for } x \in \partial\Omega,$$

and let

$$F \in C^0(\bar{\Omega} \times [0, t_0] \times \mathbb{R})$$

be locally bounded, that is, given $\eta > 0$ and $f \in C^0(\bar{\Omega})$, there exists $M = M(\eta)$ with

$$|F(x, t, v(x))| \leq M \quad \text{for } x \in \bar{\Omega}, t \in [0, t_0], |v(x) - f(x)| \leq \eta, \quad (5.1.6)$$

and locally Lipschitz continuous w.r.t. u, that is, there exists a constant $L = L(\eta)$ with

$$|F(x, t, u_1(x)) - F(x, t, u_2(x))| \leq L|u_1(x) - u_2(x)| \quad (5.1.7)$$
$$\text{for } x \in \bar{\Omega}, t \in [0, t_0], \|u_1 - f\|_{C^0(\bar{\Omega})}, \|u_2 - f\|_{C^0(\bar{\Omega})} < \eta.$$

(Of course, (5.1.6) follows from (5.1.7), but it is convenient to list it separately.)

Then there exists some $t_1 \leq t_0$ for which the initial boundary value problem

$$u_t(x, t) - \Delta u(x, t) = F(x, t, u) \quad \text{for } x \in \Omega, 0 < t \leq t_1,$$
$$u(x, t) = g(x, t) \quad \text{for } x \in \partial\Omega, 0 < t \leq t_1,$$
$$u(x, 0) = f(x) \quad \text{for } x \in \Omega, \quad (5.1.8)$$

admits a unique solution that is continuous on $\bar{\Omega} \times [0, t_1]$.

Proof: Let $q(x, y, t)$ be the heat kernel of Ω of Corollary 4.3.1. According to (4.3.28), a solution then needs to satisfy

$$u(x, t) = \int_0^t \int_\Omega q(x, y, t - \tau) F(y, \tau, u(y, \tau)) dy\, d\tau$$

$$+ \int_\Omega q(x, y, t) f(y) dy - \int_0^t \int_{\partial\Omega} \frac{\partial q}{\partial \nu_y}(x, y, t - \tau) g(y, \tau) do(y) d\tau. \quad (5.1.9)$$

A solution of (5.1.9) then is a fixed point of

$$\Phi : v \mapsto \int_0^t \int_\Omega q(x,y,t-\tau)F(y,\tau,v(y,\tau))dy\,d\tau$$

$$+ \int_\Omega q(x,y,t)f(y)dy - \int_0^t \int_{\partial\Omega} \frac{\partial q}{\partial \nu_y}(x,y,t-\tau)g(y,\tau)do(y)d\tau \quad (5.1.10)$$

which maps $C^0(\bar\Omega \times [0,t_0])$ to itself. We consider the set

$$A := \{v \in C^0(\bar\Omega \times [0,t_1]) : \sup_{x\in\bar\Omega,0\le t\le t_1} |v(x,t) - f(x)| < \eta\}. \quad (5.1.11)$$

Here, we choose $t_1 > 0$ so small that

$$t_1 M \le \frac{\eta}{2} \quad (5.1.12)$$

and

$$t_1 L < 1. \quad (5.1.13)$$

For $v \in A$

$$|\Phi(v)(x,t) - f(x)|$$

$$\le |\int_0^t \int_\Omega q(x,y,t-\tau)F(y,\tau,v(y,\tau))dy\,d\tau|$$

$$+ |\int_\Omega q(x,y,t)f(y)dy - \int_0^t \int_{\partial\Omega} \frac{\partial q}{\partial \nu_y}(x,y,t-\tau)g(y,\tau)do(y)d\tau - f(x)|$$

$$\le tM + c_{f,g}(t) \quad (5.1.14)$$

where we have used (4.3.39) and $c_{f,g}(t)$ controls the difference of the solution $u_0(x,t)$ at time t of the heat equation with initial values f and boundary values g from its initial values, that is, $\sup_{x\in\bar\Omega} |u_0(x,t) - f(x)|$. That latter quantity can be made arbitrarily small, for example smaller than $\frac{\eta}{2}$ by choosing t sufficiently small, by continuity of the solution of the heat equation (see Theorem 4.3.3). Together with (5.1.12), we then have, by choosing t_1 sufficiently small,

$$|\Phi(v)(x,t) - f(x)| < \eta, \quad (5.1.15)$$

that is, $\Phi(v) \in A$. Thus, Φ maps the set A to itself.
We shall now show that Φ is a contraction on A: for $v, w \in A$, using (4.3.39) again, and our Lipschitz condition (5.1.7),

$$\sup_{x\in\bar\Omega,0\le t\le t_1} |\Phi(v)(x,t) - \Phi(w)(x,t)|$$

$$= \sup_{x\in\bar\Omega,0\le t\le t_1} |\int_0^t \int_\Omega q(x,y,t-\tau)(F(y,\tau,v(y,\tau)) - F(y,\tau,w(y,\tau)))dy\,d\tau|$$

$$\le t_1 L \sup_{x\in\bar\Omega,0\le t\le t_1} |v(x,t) - w(x,t)|, \quad (5.1.16)$$

with $t_1 L < 1$ by (5.1.13). Thus, Φ is a contraction on A, and the Banach fixed point theorem (see Theorem A.1 of the appendix) yields the existence of a unique fixed point in A that then is a solution of our problem (5.1.8). We still need to exclude that there exists a solution outside A, but this is simple as the next lemma shows. $\square$

Lemma 5.1.1: *Let $u_1(x,t), u_2(x,t) \in C^0(\bar{\Omega} \times [0,T])$ be solutions of (5.1.8) with $u_i(x,t) = g(x,t)$ for $x \in \partial\Omega, 0 \le t \le T$, $|u_i(x,0) - f(x)| \le \frac{\eta}{2}$ for $x \in \bar{\Omega}$, $i = 1,2$. Then there exists a constant $K = K(\eta)$ with*

$$\sup_{x \in \bar{\Omega}} |u_1(x,t) - u_2(x,t)| \le e^{Kt} \sup_{x \in \bar{\Omega}} |u_1(x,0) - u_2(x,0)| \quad \text{for } 0 \le t \le T.$$

$$(5.1.17)$$

Proof: By the representation formula (4.3.28),

$$u_1(x,t) - u_2(x,t) = \int_\Omega q(x,y,t)(u_1(y,0) - u_2(y,0))dy$$

$$+ \int_0^t \int_\Omega q(x,y,t-\tau)(F(y,\tau,u_1(y,\tau)) - F(y,\tau,u_2(y,\tau)))dy\, d\tau$$

$$(5.1.18)$$

Then, as long as $\sup_x |u_i(x,t) - f(x)| \le \eta$, we have the bound from (5.1.7)

$$|F(x,t,u_1(x,t)) - F(x,t,u_2(x,t))| \le L|u_1(x,t) - u_2(x,t)| \quad (5.1.19)$$

Using (4.3.36) and (5.1.19) in (5.1.18), we obtain

$$\sup_{x \in \bar{\Omega}} |u_1(x,t) - u_2(x,t)| \le \sup_{x \in \bar{\Omega}} |u_1(x,0) - u_2(x,0)| + \int_0^t \sup_{x \in \bar{\Omega}} |u_1(x,\tau) - u_2(x,\tau)| d\tau$$

$$(5.1.20)$$

which implies the claim by the following general calculus inequality. $\square$

Lemma 5.1.2: *Let the integrable function $\phi : [0,T] \to \mathbb{R}^+$ satisfy*

$$\phi(t) \le \phi(0) + c \int_0^t \phi(\tau)d\tau \qquad (5.1.21)$$

for all $0 \le t \le T$ and some constant c. Then for $0 \le t \le T$

$$\phi(t) \le e^{ct}\phi(0). \qquad (5.1.22)$$

Proof: From (5.1.21)

$$\frac{d}{dt}(e^{-ct} \int_0^t \phi(\tau)d\tau) \le e^{-ct}\phi(0),$$

hence

$$e^{-ct} \int_0^t \phi(\tau)d\tau \le \frac{1 - e^{-ct}}{c}\phi(0),$$

from which, with (5.1.21), the desired inequality (5.1.22) follows. $\square$

We have the following important consequence of Theorem 5.1.1, a global existence theorem:

Corollary 5.1.1: *Under the assumptions of Theorem 5.1.1, suppose that the solution $u(x, t)$ of (5.1.8) satisfies the a-priori bound*

$$\sup_{x \in \bar{\Omega}, 0 \le \tau \le t} |u(x, \tau)| \le K \qquad (5.1.23)$$

for all times t for which it exists, with some fixed constant K. Then the solution $u(x, t)$ exists for all times $0 \le t < \infty$.

Proof: Suppose the solution exists for $0 \le t \le T$. Then we apply Theorem 5.1.1 at time T instead of 0, with initial values $u(x, T)$ in place of the original initial values $u(x, 0)$ and conclude that the solution continues to exist on some interval $[0, T + t_0)$ for some $t_0 > 0$ that only depends on K. We can therefore iterate the procedure to obtain a solution for all time.   □

In order to understand the qualitative behavior of solutions of reaction-diffusion equations

$$u_t(x, t) - \Delta u(x, t) = F(t, u) \quad \text{on } \Omega_T, \qquad (5.1.24)$$

it is useful to compare them with solutions of the pure reaction equation

$$v_t(x, t) = F(t, v), \qquad (5.1.25)$$

which, when the initial values

$$v(x, 0) = v_0 \qquad (5.1.26)$$

do not depend on x, likewise is independent of the spatial variable x. It therefore satisfies the homogeneous Neumann boundary condition

$$\frac{\partial v}{\partial \nu} = 0, \qquad (5.1.27)$$

where ν, as always, is the exterior normal of the domain Ω. Therefore, comparison is easiest when we also assume that u satisfies such a Neumann condition

$$\frac{\partial u}{\partial \nu} = 0 \quad \text{on } \partial \Omega, \qquad (5.1.28)$$

instead of the Dirichlet condition of (5.1.1). We therefore investigate that situation now, even though in Chapter 4 we have not derived existence theorems for parabolic equations with Neumann boundary conditions. For such results, we refer to [6]. – We have the following general comparison result:

Lemma 5.1.3: *Let u, v be of class C^2 w.r.t. $x \in \Omega$, of class C^1 w.r.t. $t \in [0, T]$, and satisfy*

$$u_t(x,t) - \Delta u(x,t) - F(x,t,u) \geq v_t(x,t) - \Delta v(x,t) - F(x,t,v) \text{ for } x \in \Omega, 0 < t \leq T,$$

$$\frac{\partial u(x,t)}{\partial \nu} \geq \frac{\partial v(x,t)}{\partial \nu} \qquad \text{for } x \in \partial\Omega, 0 < t \leq T,$$

$$u(x,0) \geq v(x,0) \qquad \text{for } x \in \Omega,$$
$$(5.1.29)$$

with our above assumptions on F. Then

$$u(x,t) \geq v(x,t) \quad \text{for } x \in \bar{\Omega}, 0 \leq t \leq T. \tag{5.1.30}$$

Proof: $w(x,t) := u(x,t) - v(x,t)$ satisfies $w(x,0) \geq 0$ in Ω and $\frac{\partial w}{\partial \nu} \geq 0$ on $\partial\Omega \times [0,T]$, as well as

$$w_t(x,t) - \Delta w(x,t) - \frac{dF(x,t,\eta)}{du} w(x,t) \geq 0 \tag{5.1.31}$$

with $\eta := su + (1-s)v$ for some $0 < s < 1$. Lemma 4.1.1 then implies $w \geq 0$, that is, (5.1.30). □

For example, a solution of

$$u_t - \Delta u = -u^3 \quad \text{for } x \in \bar{\Omega}, t > 0 \tag{5.1.32}$$

with

$$u(x,0) = u_0(x) \quad \text{for } x \in \Omega, \quad \frac{\partial u(x,t)}{\partial \nu} = 0 \quad \text{for } x \in \partial\Omega, t > 0 \tag{5.1.33}$$

can be sandwiched between solutions of

$$v_t(t) = -v^3(t), \quad v(0) = m, \quad \text{and } w_t(t) = -w^3(t), \quad w(0) = M \tag{5.1.34}$$

with $m \leq u_0(x) \leq M$, that is, we have

$$v(t) \leq u(x,t) \leq w(t) \quad \text{for } x \in \bar{\Omega}, t > 0. \tag{5.1.35}$$

Since v and w as solutions of (5.1.34) tend to 0 for $t \to \infty$, we conclude that $u(x,t)$ (assuming that it exists for all $t \geq 0$) also tends to 0 for $t \to \infty$ uniformly in $x \in \Omega$.

We now come to one of the topics that make reaction-diffusion interesting and useful models for pattern formation, namely, travelling waves. We consider the reaction-diffusion equation in one-dimensional space

$$u_t = u_{xx} + f(u) \tag{5.1.36}$$

and look for solutions of the form

$$u(x,t) = v(x - ct) = v(s), \text{ with } s := x - ct. \tag{5.1.37}$$

This travelling wave solution moves at constant speed c, assumed to be > 0 w.l.o.g, in the increasing x-direction. In particular, if we move the coordinate system with speed c, that is, keep $x - ct$ constant, then the solution also stays constant. We do not expect such a solution for every wave speed c, but at most for particular values that then need to be determined.

A travelling wave solution $v(s)$ of (5.1.36) satisfies the ODE

$$v''(s) + cv'(s) + f(v) = 0, \text{ with } ' = \frac{d}{ds}. \tag{5.1.38}$$

When $f \equiv 0$, then a solution must be of the form $v(s) = c_0 + c_1 e^{-cs}$ and therefore becomes unbounded for $s \to -\infty$, that is for $t \to \infty$. In other words, for the heat equation, there is no non-trivial bounded travelling wave. In contrast to this, depending on the precise non-linear structure of f, such travelling waves solutions may exist for reaction-diffusion equations. This is one of the reasons why such equations are interesting.

As an example, we consider the Fisher equation in one dimension,

$$u_t = u_{xx} + u(1 - u). \tag{5.1.39}$$

This is a model for the growth of populations under limiting constraints: The term $-u^2$ on the r.h.s. limits the population size. Due to such an interpretation, one is primarily interested in non-negative solutions.

We now apply some standard concepts from dynamical systems[1] to the underlying reaction equation

$$u_t = u(1 - u). \tag{5.1.40}$$

The fixed points of this equation are $u = 0$ and $u = 1$. The first one is unstable, the second one stable. The travelling wave equation (5.1.38) then is

$$v''(s) + cv'(s) + v(1 - v) = 0. \tag{5.1.41}$$

With $w := v'$, this is converted into the first order system

$$v' = w, \quad w' = -cw - v(1 - v). \tag{5.1.42}$$

The fixed points then are $(0,0)$ and $(1,0)$. The eigenvalues of the linearization at $(0,0)$, that is, of the linear system

$$\nu' = \mu, \quad \mu' = -c\mu - \nu, \tag{5.1.43}$$

are

---

[1] Readers who are not familiar with this can consult [13].

$$\lambda_\pm = \frac{1}{2}(-c \pm \sqrt{c^2 - 4}).$$ (5.1.44)

For $c^2 \geq 4$, they are both real and negative, and so the solution of (5.1.43) yields a stable node. For $c^2 < 4$, they are conjugate complex with a negative real part, and we obtain a stable spiral. Since a stable spiral oscillates about 0, in that case, we cannot expect a non-negative solution, and so, we do not consider this case here. Also, for symmetry reasons, we may restrict ourselves to the case $c > 0$, and since we want to exclude the spiral then to $c \geq 2$. The eigenvalues of the linearization at $(1,0)$, that is, of the linear system

$$\nu' = \mu, \quad \mu' = -c\mu + \nu,$$ (5.1.45)

are

$$\lambda_\pm = \frac{1}{2}(-c \pm \sqrt{c^2 + 4});$$ (5.1.46)

they are real and of different signs, and we obtain a saddle. Thus, the stability properties are reversed when compared to (5.1.40) which, of course, results from the fact that $\frac{ds}{dt} = -c$ is negative.

For $c \geq 2$, one finds a solution with $v \geq 0$ from $(1,0)$ to $(0,0)$, that is, with $v(-\infty) = 1, v(\infty) = 0$. $v' \leq 0$ for this solution. We recall that the value of a travelling wave solution is constant when $x - ct$ is constant. Thus, in the present case, when time t advances, the values for large negative values of x which are close to 1 are propagated to the whole real line, and for $t \to \infty$, the solution becomes 1 everywhere. In this sense, the behavior of the ODE (5.1.40) where a trajectory goes from the unstable fixed point 0 to the stable fixed point 1 is translated into a travelling wave that spreads a nucleus taking the value 1 for $x = -\infty$ to the entire space.

The question for which initial conditions a solution of (5.1.39) evolves to such a travelling wave, and what the value of c then is, has been widely studied in the literature since the seminal work of Kolmogorov and his coworkers [15]. For example, they showed when $u(x,0) = 1$ for $x \leq x_1$, $0 \leq u(x,0) \leq 1$ for $x_1 \leq x \leq x_2$, $u(x,0) = 0$ for $x \geq x_2$, then the solution $u(x,t)$ evolves towards a travelling wave with speed $c = 2$. In general, the wave speed c depends on the asymptotic behavior of $u(x,0)$ for $x \to \pm\infty$.

5.2 Reaction-Diffusion Systems

In this section, we extend the considerations of the previous section to systems of coupled reaction-diffusion equations. More precisely, we wish to study the initial boundary value problems for nonlinear parabolic systems of the form

$$u_t^\alpha(x,t) - d_\alpha \Delta u^\alpha(x,t) = F^\alpha(x,t,u) \quad \text{for } x \in \Omega, t > 0, \alpha = 1, \dots, n, \quad (5.2.1)$$

for suitable initial and boundary conditions. Here, $u = (u^1, \dots, u^n)$ consists of n components, the d_α are non-negative constants, and the functions

$F^\alpha(x, t, u)$ are assumed to be continuous w.r.t. x, t and Lipschitz continuous w.r.t. u, as in the preceding section. Again, the u-dependence here is the important one.

We note that in (5.2.1), the different components u^α are only coupled through the non-linear terms $F(x, t, u)$ while the left hand side of (5.2.1) for each α only involves u^α, but no other component u^β for $\beta \neq \alpha$. Here, we allow some of the diffusion constants d_α to vanish. The corresponding equation for $u^\alpha(x, t)$ then becomes an ordinary differential equation with the spatial coordinate x assuming the role of a parameter. If we ignore the coupling with other components u^β with positive diffusion constants d_β, then such a $u^\alpha(x, t)$ evolves independently for each position x. In particular, in the absence of diffusion, it is no longer meaningful to impose a Dirichlet boundary condition. When d_α is positive, however, diffusion between the different spatial positions takes place. – We have already explained in §4.1 why the diffusion constants should not be negative.

We first observe that, when we assume that the d_α are positive, the proofs of Theorem 5.1.1 and Corollary 5.1.1 extend to the present case when we make corresponding assumptions on the initial and boundary values. The reason is that the proof of Theorem 5.1.1 only needs norm estimates coming from Lipschitz bounds, but no further detailed knowledge on the structure of the right hand side. Thus

Corollary 5.2.1: *Let the diffusion constants d_α all be positive. Under the assumptions of Theorem 5.1.1 for the right hand side components F^α, and with the same type of boundary conditions for the components u^α, suppose that the solution $u(x, t) = (u^1(x, t), \ldots, u^n(x, t))$ of (5.2.1) satisfies the a-priori bound*

$$\sup_{x \in \bar\Omega, 0 \leq \tau \leq t} |u(x, \tau)| \leq K \tag{5.2.2}$$

for all times t for which it exists, with some fixed constant K. Then the solution $u(x, t)$ exists for all times $0 \leq t < \infty$.

$\square$

For the following considerations, it will be simplest to assume homogeneous Neumann boundary conditions

$$\frac{\partial u^\alpha(x, t)}{\partial \nu} = 0 \text{ for } x \in \partial\Omega, \ t > 0, \ \alpha = 1, \ldots, n. \tag{5.2.3}$$

We also assume that F is independent of x and t, that is, $F = F(u)$.

Again, we assume that the solution $u(x, t)$ stays bounded and consequently exists for all time. We want to compare $u(x, t)$ with its spatial average $\bar u$ defined by

$$\bar u^\alpha(t) := \frac{1}{\|\Omega\|} \int_\Omega u^\alpha(x, t) dx \tag{5.2.4}$$

where $\|\Omega\|$ is the Lebesgue measure of Ω.

We also assume that the right hand side F is differentiable w.r.t. u, and

$$\sup_{x,t} \left\| \frac{dF(x,t,u(x,t))}{du} \right\| \leq L. \tag{5.2.5}$$

Finally, let

$$d := \min_{\alpha=1,\ldots,n} d_\alpha > 0 \tag{5.2.6}$$

and $\lambda_1 > 0$ be the smallest Neumann eigenvalue of Δ on Ω, according to Theorem 9.5.2 below. We then have

Theorem 5.2.1: *Assume that $u(x,t)$ is a bounded solution of (5.2.1) with homogeneous Neumann boundary conditions (5.2.3). Assume that*

$$\delta := d\lambda_1 - L > 0. \tag{5.2.7}$$

Then

$$\int_\Omega \sum_{i=1}^d u_{x^i}(x,t)^2 dx \leq c_1 e^{-\delta t} \tag{5.2.8}$$

for a constant c_1, and

$$\int_\Omega |u(x,t) - \bar{u}(t)|^2 dx \leq c_2 e^{-\delta t} \tag{5.2.9}$$

for a constant c_2.

Thus, under the conditions of the theorem, spatial oscillations decay exponentially, and the solution asymptotically behaves like its spatial average. In the next §5.3, we shall investigate situations where this does not happen.

Proof: We shall leave out the summation over the index α in our notation, that is, write $u_{x^i}^2$ or $u_{x^i} u_{x^i}$ in place of $\sum_{\alpha=1}^n u_{x^i}^\alpha u_{x^i}^\alpha$ and so on. We put, as in §4.2,

$$E(u(\cdot,t)) = \frac{1}{2} \int_\Omega \sum_{i=1}^d u_{x^i}^2 dx$$

and compute

$$\frac{\partial}{\partial t} E(u(\cdot,t)) = \int_\Omega \sum_{i=1}^d u_{tx^i} u_{x^i} dx$$

$$= \int_\Omega \sum_{i=1}^d u_{x^i} \frac{\partial(\Delta u + F(u))}{\partial x^i} dx$$

$$= -\int_\Omega (\Delta u)^2 dx + \int_\Omega \sum_{i=1}^d u_{x^i} \frac{\partial F}{\partial u} u_{x^i}, \text{ since } \frac{\partial u(x,t)}{\partial \nu} = 0 \quad \text{for } x \in \partial\Omega$$

$$\leq (-\lambda_1 + L) \int_\Omega \sum_{i=1}^d u_{x^i}^2 dx \leq 2\delta E(u(\cdot,t)), \tag{5.2.10}$$

using Corollary 9.5.1 below and (5.2.7). This differential inequality by integration readily implies (5.2.8).
By Corollary 9.5.1 again, we have

$$\lambda_1 \int_\Omega |u(x,t) - \bar{u}(t)|^2 dx \le \int_\Omega \sum_{i=1}^d u_{x^i}(x,t)^2 dx, \qquad (5.2.11)$$

and so (5.2.8) implies (5.2.9).   □

We now consider the case where all the diffusion constants d_α are equal. After rescaling, we may then assume that all $d_\alpha = 1$ so that we are looking at the system

$$u_t^\alpha(x,t) - \Delta u^\alpha(x,t) = F^\alpha(x,t,u) \quad \text{for } x \in \Omega, t > 0. \qquad (5.2.12)$$

We then have

Theorem 5.2.2: *Assume that $u(x,t)$ is a bounded solution of (5.2.12) with homogeneous Neumann boundary conditions (5.2.3). Assume that*

$$\delta = \lambda_1 - L > 0. \qquad (5.2.13)$$

Then

$$\sup_{x \in \Omega} |u(x,t) - \bar{u}(t)| \le c_3 e^{-\delta t} \qquad (5.2.14)$$

for a constant c_3.

Proof: Again, we shall leave out the summation over the index α in our notation, that is, write u_t^2 or $u_t u_t$ in place of $\sum_{\alpha=1}^n u_t^\alpha u_t^\alpha$ and so on.
As in §4.2, we compute

$$\left(\frac{\partial}{\partial t} - \Delta\right) \frac{1}{2} u_t^2 = u_t u_{tt} - u_t \Delta u_t - \sum_{i=1}^d u_{x^i t}^2 = u_t \frac{\partial}{\partial t}(u_t - \Delta u) - \sum_{i=1}^d u_{x^i t}^2$$

$$\le L u_t^2 - \sum_{i=1}^d (u_{x^i t})^2. \qquad (5.2.15)$$

Therefore, by Corollary 9.5.1,

$$\frac{\partial}{\partial t} \int_\Omega u_t^2 = \int_\Omega \left(\frac{\partial}{\partial t} u_t^2 - \Delta u_t^2\right) \le (L - \lambda_1) \int_\Omega u_t^2 \le 0 \qquad (5.2.16)$$

by (5.2.7). According to Theorem 4.1.1, therefore

$$v(t) := \sup_{x \in \Omega} \left|\frac{\partial u(x,t)}{\partial t}\right|^2$$

is a nonincreasing function of t. In particular, $\frac{\partial u(x,t)}{\partial t}$ remains uniformly bounded in t. Writing our equation for u^α as[1]

$$\Delta u^\alpha(x,t) = \frac{1}{d_\alpha}(u_t^\alpha(x,t) - F^\alpha(x,t,u)), \qquad (5.2.17)$$

we may then apply Theorem 11.1.2a below to obtain $C^{1,\sigma}$ bounds on $u(x,t)$ as a function of x that are independent of t, for some $0 < \sigma < 1$. Then, first using the Sobolev embedding theorem 9.1.1 for some $p > d$ (d here is the dimension of the domain Ω, not to be confused with the minimum of the diffusion constants), and then these pointwise, time-independent bounds on $u(x,t)$ and $\frac{\partial u(x,t)}{\partial x^i}$,

$$\sup_{x \in \Omega} |u(x,t) - \bar{u}(t)| \leq \int_\Omega |u(x,t) - \bar{u}(t)|^p dx + \int_\Omega \sum_i |\tfrac{\partial}{\partial x^i}(u(x,t) - \bar{u}(t))|^p dx$$

$$\leq \int_\Omega |u(x,t) - \bar{u}(t)|^2 dx + \int_\Omega \sum_i |\tfrac{\partial u(x,t)}{\partial x^i}|^2 dx.$$

From (5.2.8) and (5.2.9), we then obtain (5.2.14).   □

A reference for reaction-diffusion equations and systems that we have used in this chapter is [21].

5.3 The Turing Mechanism

The Turing mechanism is a reaction-diffusion system that has been proposed as a model for biological and chemical pattern formation. We discuss it here in order to show how the interaction between reaction and diffusion processes can give rise to structures that neither of the two processes is capable of creating by itself. The Turing mechanism creates instabilities w. r. t. spatial variables for temporally stable states in a system of two coupled reaction-diffusion equations with different diffusion constants. This is in contrast to the situation considered in the previous §, where we have derived conditions under which a solution asymptotically becomes spatially constant (see Theorems 5.2.1, 5.2.2). – In this section, we shall need to draw upon some results about eigenvalues of the Laplace operator that will only be established in §9.5 below (see in particular Theorem 9.5.2).
The system is of the form

$$\begin{aligned} u_t &= \Delta u + \gamma f(u,v), \\ v_t &= d\Delta v + \gamma g(u,v). \end{aligned} \qquad (5.3.1)$$

where the important parameter is the diffusion constant d that will subsequently be taken > 1. Its relation with the properties of the reaction functions

---

[1] For this step, we no longer need the assumption that the d_α are all equal, and so, we keep them in the next formula.

f, g will drive the whole process. The parameter $\gamma > 0$ is only introduced for the subsequent analysis, instead of absorbing it into the functions f and g. Here $u, v : \Omega \times \mathbb{R}^+ \to \mathbb{R}$ for some bounded domain $\Omega \subset \mathbb{R}^d$ of class C^∞, and we fix the initial values

$$u(x, 0), v(x, 0) \quad \text{for } x \in \Omega,$$

and impose Neumann boundary conditions

$$\frac{\partial u}{\partial n}(x, t) = 0 = \frac{\partial v}{\partial n}(x, t) \quad \text{for all } x \in \partial\Omega, t \geq 0.$$

One can also study Dirichlet type boundary condition, for example $u = u_0, v = v_0$ on $\partial\Omega$ where u_0, v_0 are a fixed point of the reaction system as introduced below. In fact, the easiest analysis results when we assume periodic boundary conditions.

In order to facilitate the mathematical analysis, we have rescaled the independent as well as the dependent variables compared to the biological or chemical models treated in the literature on pattern formation. We now present some such examples, again in our rescaled version. All parameters a, b, ρ, K, k in those examples are assumed to be positive.

(1) Schnakenberg reaction

$$u_t = \Delta u + \gamma(a - u + u^2 v),$$
$$v_t = d\Delta v + \gamma(b - u^2 v).$$

(2) Gierer-Meinhardt system

$$u_t = \Delta u + \gamma(a - bu + \frac{u^2}{v}),$$
$$v_t = d\Delta v + \gamma(u^2 - v).$$

(3) Thomas system

$$u_t = \Delta u + \gamma(a - u - \frac{\rho u v}{1 + u + K u^2}),$$
$$v_t = d\Delta v + \gamma(\alpha(b - v) - \frac{\rho u v}{1 + u + K u^2}).$$

A slightly more general version of (2) is

(2')

$$u_t = \Delta u + \gamma\left(a - u + \frac{u^2}{v(1 + k u^2)}\right),$$
$$v_t = d\Delta v + \gamma(u^2 - v).$$

We turn to the general discussion of the Turing mechanism. We assume that we have a fixed point (u_0, v_0) of the reaction system:

$$f(u_0, v_0) = 0 = g(u_0, v_0).$$

We furthermore assume that this fixed point is linearly stable. This means that for a solution w of the linearized problem

$$w_t = \gamma A w, \quad \text{with } A = \begin{pmatrix} f_u(u_0, v_0) & f_v(u_0, v_0) \\ g_u(u_0, v_0) & g_v(u_0, v_0) \end{pmatrix}, \tag{5.3.2}$$

we have $w \to 0$ for $t \to 0$. Thus, all eigenvalues λ of A must have

$$\mathrm{Re}(\lambda) < 0,$$

as solutions are linear combinations of terms behaving like $e^{\lambda t}$.

The eigenvalues of A are the solutions of

$$\lambda^2 - \gamma(f_u + g_v)\lambda + \gamma^2(f_u g_v - f_v g_u) = 0 \tag{5.3.3}$$

(all derivatives of f and g are evaluated at (u_0, v_0)), hence

$$\lambda_{1,2} = \frac{1}{2}\gamma\left((f_u + g_v) \pm \sqrt{(f_u + g_v)^2 - 4(f_u g_v - f_v g_u)}\right). \tag{5.3.4}$$

We have $\mathrm{Re}(\lambda_1) < 0$ and $\mathrm{Re}(\lambda_2) < 0$ if

$$f_u + g_v < 0, \quad f_u g_v - f_v g_u > 0. \tag{5.3.5}$$

The linearization of the full reaction-diffusion system about (u_0, v_0) is

$$w_t = \begin{pmatrix} 1 & 0 \\ 0 & d \end{pmatrix} \Delta w + \gamma A w. \tag{5.3.6}$$

We let $0 = \lambda_0 < \lambda_1 \le \lambda_2 \le \ldots$ be the eigenvalues of Δ on Ω with Neumann boundary conditions, and y_k be a corresponding orthornormal basis of eigenfunctions, as established in Theorem 9.5.2 below,

$$\Delta y_k + \lambda_k y_k = 0 \quad \text{in } \Omega,$$
$$\frac{\partial y_k}{\partial n} = 0 \quad \text{on } \partial\Omega.$$

When we impose the Dirichlet boundary conditions $u = u_0, v = v_0$ on $\partial\Omega$ in place of Neumann conditions, we should then use the Dirichlet eigenfunctions established in Theorem 9.5.1. We then look for solutions of (5.3.6) of the form

$$w_k e^{\lambda t} = \begin{pmatrix} \alpha y_k \\ \beta y_k \end{pmatrix} e^{\lambda t}$$

with real α, β.

Inserting this into (5.3.6) yields

$$\lambda w_k = - \begin{pmatrix} 1 & 0 \\ 0 & d \end{pmatrix} \lambda_k w_k + \gamma A w_k. \tag{5.3.7}$$

For a nontrivial solution of (5.3.7), λ thus has to be an eigenvalue of

$$\left(\gamma A - \begin{pmatrix} 1 & 0 \\ 0 & d \end{pmatrix} \lambda_k \right).$$

The eigenvalue equation is

$$\begin{aligned} &\lambda^2 + \lambda(\lambda_k(1+d) - \gamma(f_u + g_v)) \\ &+ d\lambda_k{}^2 - \gamma(df_u + g_v)\lambda_k + \gamma^2(f_u g_v - f_v g_u) = 0. \end{aligned} \tag{5.3.8}$$

We denote the solutions by $\lambda(k)_{1,2}$.

(5.3.5) then means that

$$\text{Re } \lambda(0)_{1,2} < 0 \quad (\text{recall } \lambda_0 = 0).$$

We now wish to investigate whether we can have

$$\text{Re } \lambda(k) > 0 \tag{5.3.9}$$

for some higher mode λ_k.

Since by (5.3.5), $\lambda_k > 0, d > 0$, clearly

$$\lambda_k(1+d) - \gamma(f_u + g_v) > 0,$$

we need for (5.3.9) that

$$d\lambda_k{}^2 - \gamma(df_u + g_v)\lambda_k + \gamma^2(f_u g_v - f_v g_u) < 0. \tag{5.3.10}$$

Because of (5.3.5), this can only happen if

$$df_u + g_v > 0.$$

Computing this with the first equation of (5.3.5), we thus need

$$d \neq 1,$$
$$f_u g_v < 0.$$

If we assume

$$f_u > 0, \quad g_v < 0, \tag{5.3.11}$$

then we need

$$d > 1. \tag{5.3.12}$$

This is not enough to get (5.3.10) negative. In order to achieve this for some value of λ_k, we first determine that value μ of λ_k for which the lhs of (5.3.10) is minimized, i. e.

$$\mu = \frac{\gamma}{2d}(df_u + g_v), \tag{5.3.13}$$

and we then need that the lhs of (5.3.10) becomes negative for $\lambda_k = \mu$. This is equivalent to

$$\frac{(df_u + g_v)^2}{4d} > f_u g_v - f_v g_u. \tag{5.3.14}$$

If (5.3.14) holds, then the lhs of (5.3.10) has two values of λ_k where it vanishes, namely

$$\begin{aligned} \mu_\pm &= \frac{\gamma}{2d}\left((df_u + g_v) \pm \sqrt{(df_u + g_v)^2 - 4d(f_u g_v - f_v g_u)}\right) \\ &= \frac{\gamma}{2d}\left((df_u + g_v) \pm \sqrt{(df_u - g_v)^2 + 4df_v g_u}\right) \end{aligned} \tag{5.3.15}$$

and it becomes negative for

$$\mu_- < \lambda_k < \mu_+. \tag{5.3.16}$$

We conclude

Lemma 5.3.1: *Suppose (5.3.14) holds. Then (u_0, v_0) is spatially unstable w. r. t. the mode λ_k, i. e. there exists a solution of (5.3.7) with*

$$Re \, \lambda > 0$$

if λ_k satisfies (5.3.16), where $\mu_\pm$ are given by (5.3.15).

(5.3.14) is satisfied for

$$d > d_c = -\frac{2f_v g_u - f_u g_v}{f_u^2} + \frac{2}{f_u^2}\sqrt{f_v g_u(f_v g_u - f_u g_v)}. \tag{5.3.17}$$

Whether there exists an eigenvalue λ_k of Δ satisfying (5.3.16) depends on the geometry of Ω. In particular, if Ω is small, all nonzero eigenvalues are large (see Corollaries 9.5.2, 9.5.3 for some results in this direction), and so it may happen that for a given Ω, all nonzero eigenvalues are larger than μ_+. In that case, no Turing instability can occur.

We may also view this somewhat differently. Namely, given Ω, we have the smallest nonzero eigenvalue λ_1. Recalling that μ_+ in (5.3.15) depends on the parameter γ, we may choose $\gamma > 0$ so small that

$$\mu_+ < \lambda_1.$$

Then, again, (5.3.16) cannot be solved, and no Turing instability can occur. In other words, for a Turing instability, we need a certain minimal domain size for a given reaction strength, or a certain minimal reaction strength for a given domain size.

If the condition (5.3.16) is satisfied for some eigenvalue λ_k, it is also of geometric significance for which value of k this happens. Namely, by Courant's nodal domain theorem (see the remark at the end of §9.5), the nodal set $\{y_k = 0\}$ of the eigenfunction y_k divides Ω into at most $(k+1)$ regions. On any of these regions, y_k then has a fixed sign, i. e. is either positive or negative on that entire region. Since y_k is the unstable mode, this controls the number of oscillations of the developing instability.

We summarize

Theorem 5.3.1: *Suppose that at a solution (u_0, v_0) of*

$$f(u_0, v_0) = 0 = g(u_0, v_0),$$

we have

$$f_u + g_v < 0, \quad f_u g_v - f_v g_u > 0.$$

Then (u_0, v_0) is linearly stable for the reaction system

$$u_t = \gamma f(u, v),$$
$$v_t = \gamma g(u, v).$$

Suppose that $d > 1$ satisfies

$$d f_u + g_v > 0,$$

$$(d f_u + g_v)^2 - 4d(f_u g_v - f_v g_u) > 0.$$

Then (u_0, v_0) as a solution of the reaction-diffusion system

$$u_t = \Delta u + \gamma f(u, v),$$
$$v_t = d\Delta v + \gamma g(u, v)$$

is linearly unstable against spatial oscillations with eigenvalue λ_k whenever λ_k satisfies (5.3.16).

Since we assume that Ω is bounded, the eigenvalues λ_k of Δ on Ω are discrete, and so it also depends on the geometry of Ω whether such an eigenvalue in the range determined by (5.3.16) exists. The number k controls the frequency of oscillations of the instability about (u_0, v_0), and thus determines the shape of the resulting spatial pattern.

Thus, in the situation described in Theorem 5.3.1, the equilibrium state (u_0, v_0) is unstable, and in the vicinity of it, perturbations grow at a rate $e^{\mathrm{Re}\lambda}$, where λ solves (5.3.8).

Typically, one assumes, however, that the dynamics is confined within a bounded region in $(\mathbb{R}^+)^2$. This means that appropriate assumptions on f and g for $u = 0$ or $v = 0$, or for u and v large ensure that solutions starting in the positive quadrant can neither become zero nor unbounded. It is essentially a consequence of the maximum principle that if this holds for the reaction system, then it also holds for the reaction-diffusion system, see the discussion in §5.1 and §sec4a2.

Thus, even though (u_0, v_0) is locally unstable, small perturbations grow exponentially, this growth has to terminate eventually, and one expects that the corresponding solution of the reaction-diffusion system settles at a spatially inhomogeneous steady state. This is the idea of the Turing mechanism. This has not yet been demonstrated in full rigour and generality. So far, the existence of spatially heterogeneous solutions has only been shown by singular perturbation analysis near the critical parameter d_c in (5.3.17). Thus, from the global and non-linear perspective adopted in this book, the topic has not yet received a complete and satisfactory mathematical treatment.

We want to apply Theorem 5.3.1 to the example (1) above. In that case we have

$$u_0 = a + b,$$
$$v_0 = \frac{b}{(a+b)^2}, \qquad \text{(of course, } a, b > 0\text{)}$$

and at (u_0, v_0) then

$$f_u = \frac{b-a}{a+b},$$
$$f_v = (a+b)^2,$$
$$g_u = -\frac{2b}{a+b},$$
$$g_v = -(a+b)^2,$$
$$f_u g_v - f_v g_u = (a+b)^2 > 0.$$

Since we need that f_u and g_v have opposite signs (in order to get $df_u + g_v > 0$ later on), we require

$$b > a.$$

$f_u + g_v < 0$ then imples

$$0 < b - a < (a + b)^3, \qquad (5.3.18)$$

while $df_u + g_v > 0$ implies

$$d(b - a) > (a + b)^3. \qquad (5.3.19)$$

Finally, $(df_u + g_v)^2 - 4d(f_u g_v - f_v g_u) > 0$ requires

$$\left(d(b - a) - (a + b)^3\right)^2 > 4d(a + b)^4. \qquad (5.3.20)$$

The parameters a, b, d satisfying (5.3.18), (5.3.19), (5.3.20) constitute the so-called Turing space for the reaction-diffusion system investigated here.

For many case studies of the Turing mechanism in biological pattern formation, we recommend [19].

Summary

In this chapter, we have studied reaction-diffusion equations

$$u_t(x, t) - \Delta u(x, t) = F(x, t, u) \quad \text{for } x \in \Omega, t > 0$$

as well as systems of this structure. They are nonlinear because of the u-dependence of F. Solutions of such equations combine aspects of the linear diffusion equation

$$u_t(x, t) - \Delta u(x, t) = 0$$

and of the nonlinear reaction equation

$$u_t(t) = F(t, u),$$

but can also exhibit genuinely new phenomena like travelling waves. The Turing mechanism arises in systems of the form

$$u_t = \Delta u + \gamma f(u, v),$$
$$v_t = d\Delta v + \gamma g(u, v).$$

under appropriate conditions, in particular when an inhibitor v diffuses at a faster rate than an enhancer u, that is, when $d > 1$ and certain conditions on the derivatives f_u, f_v, g_u, g_v are satisfied. A Turing instability means that for such a system, a spatially homogeneous state becomes unstable. Thus, spatially nonconstant patterns will develop. This is obviously a genuinely nonlinear phenomenon.

Exercises

5.1 Consider the nonlinear elliptic equation

$$\Delta u(x) + \sigma u(x) - u^3(x) = 0 \text{ in a domain } \Omega \subset \mathbb{R}^d,$$
$$u(y) = 0 \text{ for } y \in \partial\Omega. \tag{5.3.21}$$

Let λ_1 be the smallest Dirichlet eigenvalue of Ω (cf. Theorem 9.5.1 below). Show that for $\sigma < \lambda_1$, $u \equiv 0$ is the only solution (hint: multiply the equation by u and integrate by parts and use Corollary 9.5.1 below).

5.2 Consider the nonlinear elliptic system

$$d_\alpha \Delta u^\alpha(x) + F^\alpha(x, u) = 0 \quad \text{for } x \in \Omega, \ \alpha = 1, \dots, n, \tag{5.3.22}$$

with homogeneous Neumann boundary conditions

$$\frac{\partial u^\alpha(x)}{\partial \nu} = 0 \text{ for } x \in \partial\Omega, \ \alpha = 1, \dots, n. \tag{5.3.23}$$

Assume that

$$\delta = \lambda_1 \min_{\alpha=1,\dots,n} d_\alpha - L > 0 \tag{5.3.24}$$

as in Theorem 5.2.1. Show that $u \equiv$ const.

5.3 Determine the Turing spaces for the Gierer-Meinhardt and Thomas systems.

5.4 Carry out the analysis of the Turing mechanism for periodic boundary conditions.

6. The Wave Equation and its Connections with the Laplace and Heat Equations

6.1 The One-Dimensional Wave Equation

The wave equation is the PDE

$$\frac{\partial^2}{\partial t^2} u(x,t) - \Delta u(x,t) = 0 \quad \text{for } x \in \Omega \subset \mathbb{R}^d, \, t \in (0,\infty) \text{ or } t \in \mathbb{R}. \quad (6.1.1)$$

As with the heat equation, we consider t as time and x as a spatial variable. For illustration, we first consider the case where the spatial variable x is one-dimensional. We then write the wave equation as

$$u_{tt}(x,t) - u_{xx}(x,t) = 0. \quad (6.1.2)$$

Let $\varphi, \psi \in C^2(\mathbb{R})$. Then

$$u(x,t) = \varphi(x+t) + \psi(x-t) \quad (6.1.3)$$

obviously solves (6.1.2).

This simple fact already leads to the important observation that in contrast to the heat equation, solutions of the wave equation need not be more regular for $t > 0$ than they are at $t = 0$. In particular, they are not necessarily of class C^∞. We shall have more to say about that issue, but right now we first wish to motivate (6.1.3):

$\varphi(x+t)$ solves

$$\varphi_t - \varphi_x = 0, \quad (6.1.4)$$

$\psi(x-t)$ solves

$$\psi_t + \psi_x = 0, \quad (6.1.5)$$

and the wave operator

$$L := \frac{\partial^2}{\partial t^2} - \frac{\partial^2}{\partial x^2} \quad (6.1.6)$$

can be written as

$$L = \left(\frac{\partial}{\partial t} - \frac{\partial}{\partial x} \right) \left(\frac{\partial}{\partial t} + \frac{\partial}{\partial x} \right), \qquad (6.1.7)$$

i.e., as the product of the two operators occurring in (6.1.4), (6.1.5). This suggests the transformation of variables

$$\xi = x + t, \quad \eta = x - t. \qquad (6.1.8)$$

The wave equation (6.1.2) then becomes

$$u_{\xi\eta}(\xi, \eta) = 0, \qquad (6.1.9)$$

and for a solution, u_ξ has to be independent of η, i.e.,

$$u_\xi = \varphi'(\xi) \quad \text{(where "$'$" denotes a derivative as usual)},$$

and consequently,

$$u = \int \varphi'(\xi) + \psi(\eta) = \varphi(\xi) + \psi(\eta). \qquad (6.1.10)$$

Thus, (6.1.3) actually is the most general solution of the wave equation (6.1.2).

Since this solution contains two arbitrary functions, we may prescribe two data at $t = 0$, namely, initial values and initial derivatives, again in contrast to the heat equation, where only initial values could be prescribed. From the initial conditions

$$u(x, 0) = f(x),$$
$$u_t(x, 0) = g(x), \qquad (6.1.11)$$

we obtain

$$\varphi(x) + \psi(x) = f(x),$$
$$\varphi'(x) - \psi'(x) = g(x), \qquad (6.1.12)$$

and thus

$$\varphi(x) = \frac{f(x)}{2} + \frac{1}{2} \int_0^x g(y)dy + c,$$
$$\psi(x) = \frac{f(x)}{2} - \frac{1}{2} \int_0^x g(y)dy - c \qquad (6.1.13)$$

with some constant c. Hence we have the following theorem:

Theorem 6.1.1: *The solution of the initial value problem*

$$u_{tt}(x, t) - u_{xx}(x, t) = 0 \quad \text{for } x \in \mathbb{R}, \ t > 0,$$
$$u(x, 0) = f(x),$$
$$u_t(x, 0) = g(x),$$

is given by

$$u(x,t) = \varphi(x+t) + \psi(x-t)$$
$$= \frac{1}{2}\left\{f(x+t) + f(x-t)\right\} + \frac{1}{2}\int_{x-t}^{x+t} g(y)dy. \tag{6.1.14}$$

(For u to be of class C^2, we need to require $f \in C^2$, $g \in C^1$.)   □

The representation formula (6.1.14) emphasizes another difference between the wave and the heat equations. For the latter, we had found an infinite propagation speed, in the sense that changing the initial values in some local region affected the solution for arbitrary small $t > 0$ in its entire domain of definition. The solution u of the wave equation from formula (6.1.14), however, is determined at (x,t) already by the values of f and g in the interval $[x-t, x+t]$. The value $u(x,t)$ thus is not affected by the choice of f and g outside that interval. Conversely, the initial values at the point $(y,0)$ on the x-axis influence the value of $u(x,t)$ only in the cone

$$y - t \le x \le y + t.$$

Since the rays bounding that region have slope 1, the propagation speed for perturbations of the initial values for the wave equation thus is 1.

In order to compare the wave equation with the Laplace and the heat equations, as in Section 4.1, we now consider some open $\Omega \subset \mathbb{R}^d$ and try to solve the wave equation on

$$\Omega_T = \Omega \times (0,T) \quad (T > 0)$$

by separating variables, i.e., writing the solution u of

$$\begin{aligned} u_{tt}(x,t) &= \Delta_x u(x,t) \quad &\text{on } \Omega_T, \\ u(x,t) &= 0 \quad &\text{for } x \in \partial\Omega, \end{aligned} \tag{6.1.15}$$

as

$$u(x,t) = v(x)w(t) \tag{6.1.16}$$

as in (4.1.2). This yields, as in Section 4.1,

$$\frac{w_{tt}(t)}{w(t)} = \frac{\Delta v(x)}{v(x)}, \tag{6.1.17}$$

and since the left-hand side is a function of t, and the right-hand side one of x, each of them is constant, and we obtain

$$\Delta v(x) = -\lambda v(x), \tag{6.1.18}$$
$$w_{tt}(t) = -\lambda w(t), \tag{6.1.19}$$

for some constant $\lambda \geq 0$.

As in Section 4.1, v is thus an eigenfunction of the Laplace operator on Ω with Dirichlet boundary conditions, to be studied in more detail in Section 9.5 below. From (6.1.19), since $\lambda \geq 0$, w is then of the form

$$w(t) = \alpha \cos \sqrt{\lambda}\, t + \beta \sin \sqrt{\lambda}\, t. \tag{6.1.20}$$

As in Section 4.1, referring to the expansions demonstrated in Section 9.5, we let $0 < \lambda_1 \leq \lambda_2 \leq \lambda_3 \ldots$ denote the sequence of Dirichlet eigenvalues of Δ on Ω, and $v_1, v_2, \ldots$ the corresponding orthonormal eigenfunctions, and we represent a solution of our wave equation (6.1.15) as

$$u(x,t) = \sum_{n \in \mathbb{N}} \left(\alpha_n \cos \sqrt{\lambda_n}\, t + \beta_n \sin \sqrt{\lambda_n}\, t \right) v_n(x). \tag{6.1.21}$$

In particular, for $t = 0$, we have

$$u(x,0) = \sum_{n \in \mathbb{N}} \alpha_n v_n(x), \tag{6.1.22}$$

and so the coefficients α_n are determined by the initial values $u(x,0)$. Likewise,

$$u_t(x,0) = \sum_{n \in \mathbb{N}} \beta_n \sqrt{\lambda_n}\, v_n(x), \tag{6.1.23}$$

and so the coefficients β_n are determined by the initial derivatives $u_t(x,0)$ (the convergence of the series in (6.1.23) is addressed in Theorem 9.5.1 below). So, in contrast to the heat equation, for the wave equation we may supplement the Dirichlet data on $\partial\Omega$ by two additional data at $t = 0$, namely, initial values and initial time derivatives.

From the representation formula (6.1.21), we also see, again in contrast to the heat equation, that solutions of the wave equation do not decay exponentially in time, but rather that the modes oscillate like trigonometric functions. In fact, there is a conservation principle here; namely, the so-called energy

$$E(t) := \frac{1}{2} \int_\Omega \left\{ u_t(x,t)^2 + \sum_{i=1}^{d} u_{x^i}(x,t)^2 \right\} dx \tag{6.1.24}$$

is given by

$$\begin{aligned}
E(t) &= \frac{1}{2} \int_\Omega \left\{ \left(\sum_n \left(-\alpha_n \sqrt{\lambda_n} \sin \sqrt{\lambda_n}\, t + \beta_n \sqrt{\lambda_n} \cos \sqrt{\lambda_n}\, t \right) v_n(x) \right)^2 \right. \\
&\quad \left. + \sum_{i=1}^{d} \left(\sum_n \left(\alpha_n \cos \sqrt{\lambda_n}\, t + \beta_n \sin \sqrt{\lambda_n}\, t \right) \frac{\partial}{\partial x_i} v_n(x) \right)^2 \right\} dx \\
&= \frac{1}{2} \sum_n \lambda_n (\alpha_n^2 + \beta_n^2),
\end{aligned} \tag{6.1.25}$$

since

$$\int_\Omega v_n(x)v_m(x)dx = \begin{cases} 1 & \text{for } n = m, \\ 0 & \text{otherwise,} \end{cases}$$

and

$$\sum_{i=1}^d \int_\Omega \frac{\partial}{\partial x_i}v_n(x)\frac{\partial}{\partial x_i}v_n(x) = \begin{cases} \lambda_n & \text{for } n = m, \\ 0 & \text{otherwise} \end{cases}$$

(see Theorem 9.5.1). Equation (6.1.25) implies that E does not depend on t, and we conclude that the energy for a solution u of (6.1.15), represented by (6.1.21), is conserved in time. This issue will be taken up from a somewhat different perspective in Section 6.3.

6.2 The Mean Value Method: Solving the Wave Equation Through the Darboux Equation

Let $v \in C^0(\mathbb{R}^d)$, $x \in \mathbb{R}^d$, $r > 0$. As in Section 1.2, we consider the spatial mean

$$S(v,x,r) = \frac{1}{d\omega_d r^{d-1}} \int_{\partial B(x,r)} v(y)do(y). \tag{6.2.1}$$

For $r > 0$, we put $S(v,x,-r) := S(v,x,r)$, and $S(v,x,r)$ thus is an even function of $r \in \mathbb{R}$. Since $\frac{\partial}{\partial r}S(v,x,r)|_{r=0} = 0$, the extended function remains sufficiently many times differentiable.

Theorem 6.2.1 (Darboux equation): *For $v \in C^2(\mathbb{R}^d)$,*

$$\left(\frac{\partial}{\partial r^2} + \frac{d-1}{r}\frac{\partial}{\partial r}\right) S(v,x,r) = \Delta_x S(v,x,r). \tag{6.2.2}$$

Proof: We have

$$S(v,x,r) = \frac{1}{d\omega_d} \int_{|\xi|=1} v(x + r\xi)\, do(\xi),$$

and hence

$$\frac{\partial}{\partial r}S(v,x,r) = \frac{1}{d\omega_d} \int_{|\xi|=1} \sum_{i=1}^d \frac{\partial v}{\partial x^i}(x + r\xi)\xi^i\, do(\xi)$$

$$= \frac{1}{d\omega_d r^{d-1}} \int_{\partial B(x,r)} \frac{\partial}{\partial \nu}v(y)\, do(y),$$

where ν is the exterior normal of $B(x,r)$

$$= \frac{1}{d\omega_d r^{d-1}} \int_{B(x,r)} \Delta v(z)\, dz$$

by the Gauss integral theorem.

This implies

$$\frac{\partial^2}{\partial r^2} S(v, x, r) = -\frac{d-1}{d\omega_d r^d} \int_{B(x,r)} \Delta v(z) dz + \frac{1}{d\omega_d r^{d-1}} \int_{\partial B(x,r)} \Delta v(y) \, do(y)$$

$$= -\frac{d-1}{r} \frac{\partial}{\partial r} S(v, x, r) + \frac{1}{d\omega_d r^{d-1}} \Delta_x \int_{\partial B(x,r)} v(y) \, do(y),$$

$$(6.2.4)$$

because

$$\Delta_x \int_{\partial B(x,r)} v(y) \, do(y) = \Delta_x \int_{\partial B(x_0,r)} v(x - x_0 + y) \, do(y)$$

$$= \int_{\partial B(x_0,r)} \Delta_x v(x - x_0 + y) \, do(y)$$

$$= \int_{\partial B(x,r)} \Delta v(y) \, do(y).$$

Equation (6.2.4) is equivalent to (6.2.2).   □

Corollary 6.2.1: *Let* $u(x, t)$ *be a solution of the initial value problem for the wave equation*

$$u_{tt}(x, t) - \Delta(x, t) = 0 \quad \text{for } x \in \mathbb{R}^d, \ t > 0,$$
$$u(x, 0) = f(x), \tag{6.2.5}$$
$$u_t(x, 0) = g(x).$$

We define the spatial mean

$$M(u, x, r, t) := \frac{1}{d\omega_d r^{d-1}} \int_{\partial B(x,r)} u(y, t) \, do(y). \tag{6.2.6}$$

We then have

$$\frac{\partial^2}{\partial t^2} M(u, x, r, t) = \left(\frac{\partial^2}{\partial r^2} + \frac{d-1}{r} \frac{\partial}{\partial r} \right) M(u, x, r, t). \tag{6.2.7}$$

Proof: By the first line of (6.2.4),

$$\left(\frac{\partial^2}{\partial r^2} + \frac{d-1}{r} \frac{\partial}{\partial r} \right) M(u, x, r, t) = \frac{1}{d\omega_d r^{d-1}} \int_{\partial B(x,r)} \Delta_y u(y, t) \, do(y)$$

$$= \frac{1}{d\omega_d r^{d-1}} \int_{\partial B(x,r)} \frac{\partial^2}{\partial t^2} u(y, t) \, do(y),$$

since u solves the wave equation, and this in turn equals

$$\frac{\partial^2}{\partial t^2} M(u, x, r, t).$$

□

For abbreviation, we put

$$w(r, t) := M(u, x, r, t). \tag{6.2.8}$$

Thus w solves the differential equation

$$w_{tt} = w_{rr} + \frac{d-1}{r} w_r \tag{6.2.9}$$

with initial data

$$\begin{aligned} w(r, 0) &= S(f, x, r), \\ w_t(r, 0) &= S(g, x, r). \end{aligned} \tag{6.2.10}$$

If the space dimension d equals 3, for a solution u of (6.2.9), $v := rw$ then solves the one-dimensional wave equation

$$v_{tt} = v_{rr} \tag{6.2.11}$$

with initial data

$$\begin{aligned} v(r, 0) &= rS(f, x, r), \\ v_t(r, 0) &= rS(g, x, r). \end{aligned} \tag{6.2.12}$$

By Theorem 6.1.1, this implies

$$\begin{aligned} rM(u, x, r, t) = \; &\frac{1}{2} \{(r+t)S(f, x, r+t) + (r-t)S(f, x, r-t)\} \\ &+ \frac{1}{2} \int_{r-t}^{r+t} \rho S(g, x, \rho) d\rho. \end{aligned} \tag{6.2.13}$$

Since $S(f, x, r)$ and $S(g, x, r)$ are even functions of r, we obtain

$$\begin{aligned} M(u, x, r, t) = \; &\frac{1}{2r} \{(t+r)S(f, x, r+t) - (t-r)S(f, x, t-r)\} \\ &+ \frac{1}{2r} \int_{t-r}^{t+r} \rho S(g, x, \rho) d\rho. \end{aligned} \tag{6.2.14}$$

We want to let r tend to 0 in this formula. By continuity of u,

$$M(u, x, 0, t) = u(x, t), \tag{6.2.15}$$

and we obtain

$$u(x, t) = tS(g, x, t) + \frac{\partial}{\partial t}(tS(f, x, t)). \tag{6.2.16}$$

By our preceding considerations, every solution of class C^2 of the initial value problem (6.2.5) for the wave equation must be represented in this way, and we thus obtain the following result:

Theorem 6.2.2: *The unique solution of the initial value problem for the wave equation in 3 space dimensions,*

$$u_{tt}(x,t) - \Delta u(x,t) = 0 \quad \text{for } x \in \mathbb{R}^3, \ t > 0,$$
$$u(x,0) = f(x), \tag{6.2.17}$$
$$u_t(x,0) = g(x),$$

for given $f \in C^3(\mathbb{R}^3)$, $g \in C^2(\mathbb{R}^3)$, can be represented as

$$u(x,t) = \frac{1}{4\pi t^2} \int_{\partial B(x,t)} \left(tg(y) + f(y) + \sum_{i=1}^{3} f_{y^i}(y)(y^i - x^i) \right) do(y). \tag{6.2.18}$$

Proof: First of all, (6.2.16) yields

$$u(x,t) = \frac{1}{4\pi t} \int_{\partial B(x,t)} g(y) do(y) + \frac{\partial}{\partial t} \left(\frac{1}{4\pi t} \int_{\partial B(x,t)} f(y) do(y) \right). \tag{6.2.19}$$

In order to carry out the differentiation in the integral, we need to transform the mean value of f back to the unit sphere, i.e.,

$$\frac{1}{4\pi t} \int_{\partial B(x,t)} f(y) do(y) = \frac{t}{4\pi} \int_{|z|=1} f(x + tz) do(z).$$

The Darboux equation implies that u from (6.2.19) solves the wave equation, and the correct initial data result from the relations

$$S(w, x, 0) = w(x), \quad \frac{\partial}{\partial r} S(w, x, r)|_{r=0} = 0$$

satisfied by every continuous w.   □

An important observation resulting from (6.2.18) is that for space dimensions 3 (and higher), a solution of the wave equation can be less regular than its initial values. Namely, if $u(x,0) \in C^k$, $u_t(x,0) \in C^{k-1}$, this implies $u(x,t) \in C^{k-1}$, $u_t(x,t) \in C^{k-2}$ for positive t.

Moreover, as in the case $d = 1$, we may determine the regions of influence of the initial data. It is quite remarkable that the value of u at (x,t) depends on the initial data only on the sphere $\partial B(x,t)$, but not on the data in the interior of the ball $B(x,t)$. This is the so-called Huygens principle. This principle, however, holds only in odd dimensions greater than 1, but not in even dimensions. We want to explain this for the case $d = 2$. Obviously, a solution of the wave equation for $d = 2$ can be considered as a solution for $d = 3$ that happens to be independent of the third spatial coordinate x^3.

We thus put $x^3 = 0$ in (6.2.19) and integrate on the sphere $\partial B(x,t) = \{y \in \mathbb{R}^3 : (y^1 - x^1)^2 + (y^2 - x^2)^2 + (y^3)^2 = t^2\}$ with surface element

$$do(y) = \frac{t}{|y^3|} dy^1 dy^2.$$

Since the points (y^1, y^2, y^3) and $(y^1, y^2, -y^3)$ yield the same contributions, we obtain

$$u(x^1, x^2, t) = \frac{1}{2\pi} \int_{B(x,t)} \frac{g(y)}{\sqrt{t^2 - |x - y|^2}} dy$$
$$+ \frac{\partial}{\partial t} \left(\frac{1}{2\pi} \int_{B(x,t)} \frac{f(y)}{\sqrt{t^2 - |x - y|^2}} dy \right),$$

where $x = (x^1, x^2)$, $y = (y^1, y^2)$, and the ball $B(x, t)$ now is the two-dimensional one.

The values of u at (x, t) now depend on the values on the whole disk $B(x, t)$ and not only on its boundary $\partial B(x, t)$.

A reference for Sections 6.1 and 6.2 is F. John [10].

6.3 The Energy Inequality and the Relation with the Heat Equation

Let u be a solution of the wave equation

$$u_{tt}(x, t) - \Delta u(x, t) = 0 \quad \text{for } x \in \mathbb{R}^d, \ t > 0. \tag{6.3.1}$$

We define the energy norm of u as follows:

$$E(t) := \frac{1}{2} \int_{\mathbb{R}^d} \left\{ u_t(x, t)^2 + \sum_{i=1}^{d} u_{x^i}(x, t)^2 \right\} dx. \tag{6.3.2}$$

We have

$$\frac{dE}{dt} = \int_{\mathbb{R}^d} \left\{ u_t u_{tt} + \sum_{i=1}^{d} u_{x^i} u_{x^i t} \right\} dx$$
$$= \int_{\mathbb{R}^d} \left\{ u_t(u_{tt} - \Delta u) + \sum_{i=1}^{d} (u_t u_{x^i})_{x^i} \right\} dx \tag{6.3.3}$$
$$= 0$$

if $u(x, t) = 0$ for sufficiently large $|x|$ (where that may depend on t, so that this computation may be applied to solutions of (6.3.1) with compactly supported initial values).

In this manner, it is easy to show the following result about the region of dependency of a solution of (6.3.1), partially generalizing the corresponding results of Section 6.2 to arbitrary dimensions:

Theorem 6.3.1: *Let u be a solution of (6.3.1) with*

$$u(x,0) = f(x), \quad u_t(x,0) = 0, \tag{6.3.4}$$

and let $K := \operatorname{supp} f \left(:= \overline{\{x \in \mathbb{R}^d : f(x) \neq 0\}} \right)$ be compact. Then

$$u(x,t) = 0 \quad \text{for } \operatorname{dist}(x,K) > t. \tag{6.3.5}$$

Proof: We show that $f(y) = 0$ for all $y \in B(x,T)$ implies $u(x,T) \geq 0$, which is equivalent to our assertion. We put

$$\overline{E}(t) := \frac{1}{2} \int_{B(x,T-t)} \left\{ u_t^2 + \sum_{i=1}^d u_{y^i}^2 \right\} dy \tag{6.3.6}$$

and obtain as in (6.3.3) (cf. (1.1.1))

$$\frac{d\overline{E}}{dt} = \int_{B(x,T-t)} \left\{ u_t u_{tt} + \sum u_{y^i} u_{y^i t} \right\} dy - \frac{1}{2} \int_{\partial B(x,T-t)} \left\{ u_t^2 + \sum u_{y^i}^2 \right\} do(y)$$

$$= \int_{\partial B(x,T-t)} \left\{ u_t \frac{\partial u}{\partial \nu} - \frac{1}{2} \left(u_t^2 + \sum u_{y^i}^2 \right) \right\} do(y).$$

By the Schwarz inequality, the integrand is nonpositive, and we conclude that

$$\frac{d\overline{E}}{dt} \leq 0 \quad \text{for } t > 0.$$

Since by assumption $\overline{E}(0) = 0$ and $\overline{E}$ is nonnegative, necessarily

$$\overline{E}(t) = 0 \quad \text{for all } t \leq T,$$

and hence

$$u(y,t) = 0 \quad \text{for } |x - y| \leq T - t,$$

so that

$$u(x,T) = 0$$

as desired.   □

Theorem 6.3.2: *As in Theorem 6.3.1, let u be a solution of the wave equation with initial values*

$$u(x,0) = f(x) \quad \text{with compact support}$$

and

$$u_t(x, 0) = 0.$$

Then

$$v(x, t) := \int_{-\infty}^{\infty} \frac{e^{-\frac{s^2}{4t}}}{\sqrt{4\pi t}} u(x, s) ds$$

yields a solution of the heat equation

$$v_t(x, t) - \Delta v(x, t) = 0 \quad \text{for } x \in \mathbb{R}^d, t > 0$$

with initial values

$$v(x, 0) = f(x).$$

Proof: That u solves the heat equation is seen by differentiating under the integral

$$\frac{\partial}{\partial t} v(x, t) = \int_{-\infty}^{\infty} \frac{\partial}{\partial t} \left(\frac{e^{-\frac{s^2}{4t}}}{\sqrt{4\pi t}} \right) u(x, s) ds$$

$$= \int_{-\infty}^{\infty} \frac{\partial^2}{\partial s^2} \left(\frac{e^{-\frac{s^2}{4t}}}{\sqrt{4\pi t}} \right) u(x, s) ds$$

(since the kernel solves the heat equation)

$$= \int_{-\infty}^{\infty} \frac{e^{-\frac{s^2}{4t}}}{\sqrt{4\pi t}} \frac{\partial^2}{\partial s^2} u(x, s) ds$$

$$= \int_{-\infty}^{\infty} \frac{e^{-\frac{s^2}{4t}}}{\sqrt{4\pi t}} \Delta_x u(x, s) ds$$

(since u solves the wave equation)

$$= \Delta v(x, t),$$

where we omit the detailed justification of interchanging differentiation and integration here. Then $v(x, 0) = u(x, 0) = f(x)$ follows as in Section 4.1.   □

Summary

In the present chapter we have studied the wave equation

$$\frac{\partial^2}{\partial t^2} u(x, t) - \Delta u(x, t) = 0 \quad \text{for } x \in \mathbb{R}^d, \ t > 0$$

with initial data

$$u(x, 0) = f(x),$$

$$\frac{\partial}{\partial t} u(x, 0) = g(x).$$

In contrast to the heat equation, there is no gain of regularity compared to the initial data, and in fact, for $d > 1$, there may even occur a loss of regularity.

As was the case with the Laplace equation, mean value constructions are important for the wave equation, and they permit us to reduce the wave equation for $d > 1$ to the Darboux equation for the mean values, which is hyperbolic as well but involves only one spatial coordinate.

The propagation speed for the wave equation is finite, in contrast to the heat equation. The effect of perturbations sets in sharply, and in odd dimensions greater than 1, it also terminates sharply (Huygens principle).

The energy

$$E(t) = \int_{\mathbb{R}^d} \left(|u_t(x,t)|^2 + |\nabla_x u(x,t)|^2 \right) dx$$

is constant in time.

By a certain time averaging, a solution of the wave equation yields a solution of the heat equation.

Exercises

6.1 We consider the wave equation in one space dimension,

$$u_{tt} - u_{xx} = 0 \quad \text{for } 0 < x < \pi, t > 0,$$

with initial data

$$u(x,0) = \sum_{n=1}^{\infty} \alpha_n \sin nx, \quad u_t(x,0) = \sum_{n=1}^{\infty} \beta_n \sin nx$$

and boundary values

$$u(0,t) = u(\pi,t) = 0 \quad \text{for all } t > 0.$$

Represent the solution as a Fourier series

$$u(x,t) = \sum_{n=1}^{\infty} \gamma_n(t) \sin nx$$

and compute the coefficients $\gamma_n(t)$.

6.2 Consider the equation

$$u_t + c u_x = 0$$

for some function $u(x,t)$, $x, t \in \mathbb{R}$, where c is constant. Show that u is constant along any line

$$x - ct = \text{const} = \xi,$$

and thus the general solution of this equation is given as

$$u(x, t) = f(\xi) = f(x - ct)$$

where the initial values are $u(x, 0) = f(x)$. Does this differential equation satisfy the Huygens principle?

6.3 We consider the general quasilinear PDE for a function $u(x, y)$ of two variables,

$$au_{xx} + 2bu_{xy} + cu_{yy} = d,$$

where a, b, c, d are allowed to depend on x, y, u, u_x, and u_y. We consider the curve $\gamma(s) = (\varphi(s), \psi(s))$ in the xy-plane, where we wish to prescribe the function u and its first derivatives:

$$u = f(s), \quad u_x = g(s), \quad u_y = h(s) \quad \text{for } x = \varphi(s), y = \psi(s).$$

Show that for this to be possible, we need the relation

$$f'(s) = g(s)\varphi'(s) + h(s)\psi'(s).$$

For the values of u_{xx}, u_{xy}, u_{yy} along γ, compute the equations

$$\varphi' u_{xx} + \psi' u_{xy} = g',$$
$$\varphi' u_{xy} + \psi' u_{yy} = h'.$$

Conclude that the values of u_{xx}, u_{xy}, and u_{yy} along γ are uniquely determined by the differential equations and the data f, g, h (satisfying the above compatibility conditions), unless

$$a\psi'^2 - 2b\varphi'\psi' + c\varphi'^2 = 0$$

along γ. If this latter equation holds, γ is called a characteristic curve for the solution u of our PDE $au_{xx} + 2bu_{xy} + cu_{yy} = d$. (Since a, b, c, d may depend on u and u_x, u_y, in general it depends not only on the equation, but also on the solution, which curves are characteristic.) How is this existence of characteristic curves related to the classification into elliptic, hyperbolic, and parabolic PDEs discussed in the introduction? What are the characteristic curves of the wave equation $u_{tt} - u_{xx} = 0$?

7. The Heat Equation, Semigroups, and Brownian Motion

7.1 Semigroups

We first want to reinterpret some of our results about the heat equation. For that purpose, we again consider the heat kernel of $\mathbb{R}^d$, which we now denote by $p(x, y, t)$,

$$p(x, y, t) = \frac{1}{(4\pi t)^{\frac{d}{2}}} e^{-\frac{|x-y|^2}{4t}}. \tag{7.1.1}$$

For a continuous and bounded function $f : \mathbb{R}^d \to \mathbb{R}$, by Lemma 4.2.1

$$u(x, t) = \int_{\mathbb{R}^d} p(x, y, t) f(y) dy \tag{7.1.2}$$

then solves the heat equation

$$\Delta u(x, t) - u_t(x, t) = 0. \tag{7.1.3}$$

For $t > 0$, and letting C_b^0 denote the class of bounded continuous functions, we define the operator

$$P_t : C_b^0(\mathbb{R}^d) \to C_b^0(\mathbb{R}^d)$$

via

$$(P_t f)(x) = u(x, t), \tag{7.1.4}$$

with u from (7.1.2). By Lemma 4.2.2

$$P_0 f := \lim_{t \to 0} P_t f = f; \tag{7.1.5}$$

i.e., P_0 is the identity operator. The crucial point is that we have for any $t_1, t_2 \geq 0$,

$$P_{t_1+t_2} = P_{t_2} \circ P_{t_1}. \tag{7.1.6}$$

Written out, this means that for all $f \in C_b^0(\mathbb{R}^d)$,

$$\int_{\mathbb{R}^d} \frac{1}{(4\pi(t_1+t_2))^{\frac{d}{2}}} e^{-\frac{|x-y|^2}{4(t_1+t_2)}} f(y)\, dy$$

$$= \int_{\mathbb{R}^d} \frac{1}{(4\pi t_2)^{\frac{d}{2}}} e^{-\frac{|x-z|^2}{4t_2}} \int_{\mathbb{R}^d} \frac{1}{(4\pi t_1)^{\frac{d}{2}}} e^{-\frac{|z-y|^2}{4t_1}} f(y)\, dy\, dz. \quad (7.1.7)$$

This follows from the formula

$$\frac{1}{(4\pi(t_1+t_2))^{\frac{d}{2}}} e^{-\frac{|x-y|^2}{4(t_1+t_2)}} = \frac{1}{(4\pi t_2)^{\frac{d}{2}}} \frac{1}{(4\pi t_1)^{\frac{d}{2}}} \int_{\mathbb{R}^d} e^{-\frac{|x-z|^2}{4t_2}} e^{-\frac{|z-y|^2}{4t_1}}\, dz,$$

$$(7.1.8)$$

which can be verified by direct computation (cf. also Exercise 4.3).

There exists, however, a deeper and more abstract reason for (7.1.6): $P_{t_1+t_2}f(x)$ is the solution at time $t_1 + t_2$ of the heat equation with initial values f. At time t_1, this solution has the value $P_{t_1}f(x)$. On the other hand, $P_{t_2}(P_{t_1}f)(x)$ is the solution at time t_2 of the heat equation with initial values $P_{t_1}f$. Since by Theorem 4.1.2, the solution of the heat equation is unique within the class of bounded functions, and the heat equation is invariant under time translations, it must lead to the same result starting at time 0 with initial values $P_{t_1}f$ and considering the solution at time t_2, or starting at time t_1 with value $P_{t_1}f$ and considering the solution at time $t_1 + t_2$, since the time difference is the same in both cases. This reasoning is also valid for the initial value problem because solutions here are unique as well, by Corollary 4.1.1. We have the following result:

Theorem 7.1.1: *Let $\Omega \subset \mathbb{R}^d$ be bounded and of class C^2, and let $g : \partial\Omega \to \mathbb{R}$ be continuous. For any $f \in C_b^0(\Omega)$, we let*

$$P_{\Omega,g,t}f(x)$$

be the solution of the initial value problem

$$\begin{aligned} \Delta u - u_t = 0 &\quad \text{in } \Omega \times (0,\infty), \\ u(x,t) = g(x) &\quad \text{for } x \in \partial\Omega, \\ u(x,0) = f(x) &\quad \text{for } x \in \Omega. \end{aligned} \quad (7.1.9)$$

We then have

$$P_{\Omega,g,0}f = \lim_{t\searrow 0} P_{\Omega,g,t}f = f \quad \text{for all } f \in C^0(\Omega), \quad (7.1.10)$$

$$P_{\Omega,g,t_1+t_2} = P_{\Omega,g,t_2} \circ P_{\Omega,g,t_1}. \quad (7.1.11)$$

□

Corollary 7.1.1: *Under the assumptions of Theorem 7.1.1, we have for all $t_0 \geq 0$ and for all $f \in C_b^0(\Omega)$,*

$$P_{\Omega,g,t_0}f = \lim_{t\searrow t_0} P_{\Omega,g,t}f.$$

□

We wish to cover the phenomenon just exhibited by a general definition:

Definition 7.1.1: *Let B be a Banach space, and for $t > 0$, let $T_t : B \to B$ be continuous linear operators with*

(i) $T_0 = \mathrm{Id}$;
(ii) $T_{t_1+t_2} = T_{t_2} \circ T_{t_1}$ for all $t_1, t_2 \geq 0$;
(iii) $\lim_{t \to t_0} T_t v = T_{t_0} v$ for all $t_0 \geq 0$ and all $v \in B$.

Then the family $\{T_t\}_{t \geq 0}$ is called a continuous semigroup (of operators).

A different and simpler example of a semigroup is the following: Let B be the Banach space of bounded, uniformly continuous functions on $[0, \infty)$. For $t \geq 0$, we put

$$T_t f(x) := f(x + t). \tag{7.1.12}$$

Then all conditions of Definition 7.1.1 are satisfied. Both semigroups (for the heat semigroup, this follows from the maximum principle) satisfy the following definition:

Definition 7.1.2: *A continuous semigroup $\{T_t\}_{t \geq 0}$ of continuous linear operators of a Banach space B with norm $\| \cdot \|$ is called contracting if for all $v \in B$ and all $t \geq 0$,*

$$\|T_t v\| \leq \|v\|. \tag{7.1.13}$$

(Here, continuity of the semigroup means continuous dependence of the operators T_t on t.)

7.2 Infinitesimal Generators of Semigroups

If the initial values $f(x) = u(x, 0)$ of a solution u of the heat equation

$$u_t(x, t) - \Delta u(x, t) = 0 \tag{7.2.1}$$

are of class C^2, we expect that

$$\lim_{t \searrow 0} \frac{u(x, t) - u(x, 0)}{t} = u_t(x, 0) = \Delta u(x, 0) = \Delta f(x), \tag{7.2.2}$$

or with the notation

$$u(x, t) = P_t f(u)$$

of the previous section,

$$\lim_{t \searrow 0} \frac{1}{t}(P_t - \mathrm{Id})f = \Delta f. \tag{7.2.3}$$

We want to discuss this in more abstract terms and verify the following definition:

Definition 7.2.1: *Let* $\{T_t\}_{t \geq 0}$ *be a continuous semigroup on a Banach space* B. *We put*

$$D(A) := \left\{ v \in B : \lim_{t \searrow 0} \frac{1}{t}(T_t - \mathrm{Id})v \ exists \right\} \subset B \qquad (7.2.4)$$

and call the linear operator

$$A : D(A) \to B,$$

defined as

$$Av := \lim_{t \searrow 0} \frac{1}{t}(T_t - \mathrm{Id})v, \qquad (7.2.5)$$

the infinitesimal generator of the semigroup $\{T_t\}$.

Then $D(A)$ is nonempty, since it contains 0.

Lemma 7.2.1: *For all* $v \in D(A)$ *and all* $t \geq 0$, *we have*

$$T_t Av = A T_t v. \qquad (7.2.6)$$

Thus A *commutes with all the* T_t.

Proof: For $v \in D(A)$, we have

$$T_t Av = T_t \lim_{\tau \searrow 0} \frac{1}{\tau}(T_\tau - \mathrm{Id})v$$

$$= \lim_{\tau \searrow 0} \frac{1}{\tau}(T_t T_\tau - T_t)v \ (\text{since } T_t \text{ is continuous and linear})$$

$$= \lim_{\tau \searrow 0} \frac{1}{\tau}(T_\tau T_t - T_t)v \ (\text{by the semigroup property})$$

$$= \lim_{\tau \searrow 0} \frac{1}{\tau}(T_\tau - \mathrm{Id})T_t v$$

$$= A T_t v.$$

$\square$

In particular, if $v \in D(A)$, then so is $T_t v$. In that sense, there is no loss of regularity of $T_t v$ when compared with v $(= T_0 v)$.

In the sequel, we shall employ the notation

$$J_\lambda v := \int_0^\infty \lambda e^{-\lambda s} T_s v \, ds \quad \text{for } \lambda > 0 \qquad (7.2.7)$$

for a contracting semigroup $\{T_t\}$. The integral here is a Riemann integral for functions with values in some Banach space. The standard definition of the Riemann integral as a limit of step functions easily generalizes to the

Banach-space-valued case. The convergence of the improper integral follows from the estimate

$$\lim_{K,M\to\infty} \left\| \int_K^M \lambda e^{-\lambda s} T_s v\, ds \right\| \le \lim_{K,M\to\infty} \int_K^M \lambda e^{-\lambda s} \|T_s v\|\, ds$$

$$\le \lim_{K,M\to\infty} \|v\| \int_K^M \lambda e^{-\lambda s}\, ds$$

$$= 0,$$

which holds because of the contraction property and the completeness of B. Since

$$\int_0^\infty \lambda e^{-\lambda s}\, ds = \int_0^\infty -\frac{d}{ds}\left(e^{-\lambda s}\right) ds = 1, \qquad (7.2.8)$$

$J_\lambda v$ is a weighted mean of the semigroup $\{T_t\}$ applied to v. Since

$$\|J_\lambda v\| \le \int_0^\infty \lambda e^{-\lambda s} \|T_s v\|\, ds$$

$$\le \|v\| \int_0^\infty \lambda e^{-\lambda s}\, ds \qquad (7.2.9)$$

$$\text{by the contraction property}$$

$$\le \|v\|$$

by (7.2.8), $J_\lambda : B \to B$ is a bounded linear operator with norm $\|J_\lambda\| \le 1$.

Lemma 7.2.2: *For all $v \in B$, we have*

$$\lim_{\lambda\to\infty} J_\lambda v = v. \qquad (7.2.10)$$

Proof: By (7.2.8),

$$J_\lambda v - v = \int_0^\infty \lambda e^{-\lambda s}(T_s v - v)\, ds.$$

For $\delta > 0$, let

$$I_\lambda^1 := \left\| \int_0^\delta \lambda e^{-\lambda s}(T_s v - v)\, ds \right\|, \qquad I_\lambda^2 := \left\| \int_\delta^\infty \lambda e^{-\lambda s}(T_s v - v)\, ds \right\|.$$

Now let $\varepsilon > 0$ be given. Since $T_s v$ is continuous in s, there exists $\delta > 0$ such that

$$\|T_s v - v\| < \frac{\varepsilon}{2} \quad \text{for } 0 \le s \le \delta$$

and thus also

$$I_\lambda^1 \leq \frac{\varepsilon}{2} \int_0^\delta \lambda e^{-\lambda s} ds < \frac{\varepsilon}{2}$$

by (7.2.8). For each $\delta > 0$, there also exists $\lambda_0 \in \mathbb{R}$ such that for all $\lambda \geq \lambda_0$,

$$I_\lambda^2 \leq \int_\delta^\infty \lambda e^{-\lambda s} \left(\|T_s v\| + \|v\| \right) ds$$

$$\leq 2 \|v\| \int_\delta^\infty \lambda e^{-\lambda s} ds \text{ (by the contraction property)}$$

$$< \frac{\varepsilon}{2}.$$

This easily implies (7.2.10).   □

Theorem 7.2.1: Let $\{T_t\}_{t\geq 0}$ be a contracting semigroup with infinitesimal generator A. Then $D(A)$ is dense in B.

Proof: We shall show that for all $\lambda > 0$ and all $v \in B$,

$$J_\lambda v \in D(A). \tag{7.2.11}$$

Since by Lemma 7.2.2,

$$\{J_\lambda v : \lambda > 0, v \in B\}$$

is dense in B, this will imply the assertion. We have

$$\frac{1}{t}(T_t - \mathrm{Id})J_\lambda v = \frac{1}{t}\int_0^\infty \lambda e^{-\lambda s} T_{t+s} v \, ds - \frac{1}{t}\int_0^\infty \lambda e^{-\lambda s} T_s v \, ds$$

since T_t is continuous and linear

$$= \frac{1}{t}\int_t^\infty \lambda e^{\lambda t} e^{-\lambda \sigma} T_\sigma v \, d\sigma - \frac{1}{t}\int_0^\infty \lambda e^{-\lambda s} T_s v \, ds$$

$$= \frac{e^{\lambda t} - 1}{t} \int_t^\infty \lambda e^{-\lambda \sigma} T_\sigma v \, d\sigma - \frac{1}{t}\int_0^t \lambda e^{-\lambda s} T_s v \, ds$$

$$= \frac{e^{\lambda t} - 1}{t} \left(J_\lambda v - \int_0^t \lambda e^{-\lambda \sigma} T_\sigma v \, d\sigma \right) - \frac{1}{t}\int_0^t \lambda e^{-\lambda s} T_s v \, ds.$$

The last term, the integral being continuous in s, for $t \to 0$ tends to $-\lambda T_0 v = -\lambda v$, while the first term in the last line tends to $\lambda J_\lambda v$. This implies

$$A J_\lambda v = \lambda \left(J_\lambda - \mathrm{Id} \right) v \quad \text{for all } v \in B, \tag{7.2.12}$$

which in turn implies (7.2.11).   □

For a contracting semigroup $\{T_t\}_{t\geq 0}$, we now define operators

$$D_t T_t : D(D_t T_t)(\subset B) \to B$$

by

$$D_t T_t v := \lim_{h \to 0} \frac{1}{h} \left(T_{t+h} - T_t \right) v, \tag{7.2.13}$$

where $D(D_t T_t)$ is the subspace of B where this limit exists.

Lemma 7.2.3: $v \in D(A)$ *implies* $v \in D(D_t T_t)$, *and we have*

$$D_t T_t v = A T_t v = T_t A v \quad \text{for } t \geq 0. \tag{7.2.14}$$

Proof: The second equation has already established shown in Lemma 7.2.1. We thus have for $v \in D(A)$,

$$\lim_{h \searrow 0} \frac{1}{h} \left(T_{t+h} - T_t \right) v = A T_t v = T_t A v. \tag{7.2.15}$$

Equation (7.2.15) means that the right derivative of $T_t v$ with respect to t exists for all $v \in D(A)$ and is continuous in t. By a well-known calculus lemma, this then implies that the left derivative exists as well and coincides with the right one, implying differentiability and (7.2.14). (The proof of the calculus lemma goes as follows: Let $f : [0, \infty) \to B$ be continuous, and suppose that for all $t \geq 0$, the right derivative $d^+ f(t) := \lim_{h \searrow 0} \frac{1}{h}(f(t + h) - f(t))$ exists and is continuous. The continuity of $d^+ f$ implies that on every interval $[0, T]$ this limit relation even holds uniformly in t. In order to conclude that f is differentiable with derivative $d^+ f$, one argues that

$$\lim_{h \searrow 0} \left\| \frac{1}{h} \left(f(t) - f(t \quad h) \right) - d^+ f(t) \right\|$$

$$\leq \lim_{h \searrow 0} \left\| \frac{1}{h} (f((t - h) + h) - f(t - h)) - d^+ f(t - h) \right\|$$

$$+ \lim_{h \searrow 0} \left\| d^+ f(t - h) - d^+ f(t) \right\|$$

$$= 0.)$$

$\square$

Theorem 7.2.2: *For $\lambda > 0$, the operator $(\lambda \operatorname{Id} - A) : D(A) \to B$ is invertible (A being the infinitesimal generator of a contracting semigroup), and we have*

$$(\lambda \operatorname{Id} - A)^{-1} = R(\lambda, A) := \frac{1}{\lambda} J_\lambda, \tag{7.2.16}$$

i.e.,

$$(\lambda \operatorname{Id} - A)^{-1} v = R(\lambda, A) v = \int_0^\infty e^{-\lambda s} T_s v \, ds. \tag{7.2.17}$$

Proof: In order that $(\lambda \, \text{Id} - A)$ be invertible, we need to show first that $(\lambda \, \text{Id} - A)$ is injective. So, we need to exclude that there exists $v_0 \in D(A)$, $v_0 \neq 0$, with

$$\lambda v_0 = A v_0. \tag{7.2.18}$$

For such a v_0, we would have by (7.2.14)

$$D_t T_t v_0 = T_t A v_0 = \lambda T_t v_0, \tag{7.2.19}$$

and hence

$$T_t v_0 = e^{\lambda t} v_0. \tag{7.2.20}$$

Since $\lambda > 0$, for $v_0 \neq 0$ this would violate the contraction property

$$\|T_t v_0\| \leq \|v_0\|,$$

however. Therefore, $(\lambda \, \text{Id} - A)$ is invertible for $\lambda > 0$. In order to obtain (7.2.16), we start with (7.2.12), i.e.,

$$A J_\lambda v = \lambda (J_\lambda - \text{Id}) v,$$

and get

$$(\lambda \, \text{Id} - A) J_\lambda v = \lambda v. \tag{7.2.21}$$

Therefore, $(\lambda \, \text{Id} - A)$ maps the image of J_λ bijectively onto B. Since this image is dense in $D(A)$ by (7.2.11), and since $(\lambda \, \text{Id} - A)$ is injective, $(\lambda \, \text{Id} - A)$ then also has to map $D(A)$ bijectively onto B. Thus, $D(A)$ has to coincide with the image of J_λ, and (7.2.21) then implies (7.2.16).   □

Lemma 7.2.4 (resolvent equation): *Under the assumptions of Theorem 7.2.2, we have for $\lambda, \mu > 0$,*

$$R(\lambda, A) - R(\mu, A) = (\mu - \lambda) R(\lambda, A) R(\mu, A). \tag{7.2.22}$$

Proof:

$$\begin{aligned}
R(\lambda, A) &= R(\lambda, A)(\mu \, \text{Id} - A) R(\mu, A) \\
&= R(\lambda, A)((\mu - \lambda) \, \text{Id} + (\lambda \, \text{Id} - A)) R(\mu, A) \\
&= (\mu - \lambda) R(\lambda, A) R(\mu, A) + R(\mu, A).
\end{aligned}$$

□

We now want to compute the infinitesimal generators of the two examples we have considered with the help of the preceding formalism. We begin with the translation semigroup: B here is the Banach space of bounded, uniformly continuous functions on $[0, \infty]$, and $T_t f(x) = f(x+t)$ for $f \in B$, $x, t \geq 0$. We then have

$$(J_\lambda f)(x) = \int_0^\infty \lambda e^{-\lambda s} f(x+s) ds = \int_x^\infty \lambda e^{-\lambda(s-x)} f(s) ds, \qquad (7.2.23)$$

and hence

$$\frac{d}{dx}(J_\lambda f)(x) = -\lambda f(x) + \lambda (J_\lambda f)(x). \qquad (7.2.24)$$

By (7.2.12), the infinitesimal generator satisfies

$$A J_\lambda f(x) = \lambda (J_\lambda f - f)(x), \qquad (7.2.25)$$

and consequently

$$A J_\lambda f = \frac{d}{dx} J_\lambda f. \qquad (7.2.26)$$

At the end of the proof of Theorem 7.2.2, we have seen that the image of J_λ coincides with $D(A)$, and we thus have

$$Ag = \frac{d}{dx} g \quad \text{for all } g \in D(A). \qquad (7.2.27)$$

We now intend to show that $D(A)$ contains precisely those $g \in B$ for which $\frac{d}{dx} g$ belongs to B as well. For such a g, we define $f \in B$ by

$$\frac{d}{dx} g(x) - \lambda g(x) = -\lambda f(x). \qquad (7.2.28)$$

By (7.2.24), we then also have

$$\frac{d}{dx}(J_\lambda f)(x) - \lambda J_\lambda f(x) = -\lambda f(x). \qquad (7.2.29)$$

Thus

$$\varphi(x) := g(x) - J_\lambda f(x)$$

satisfies

$$\frac{d}{dx}\varphi(x) = \lambda \varphi(x), \qquad (7.2.30)$$

whence $\varphi(x) = ce^{\lambda x}$, and since $\varphi \in B$, necessarily $c = 0$, and so $g = J_\lambda f$.

We thus have verified that the infinitesimal generator A is given by (7.2.27), with the domain of definition $D(A)$ containing precisely those $g \in B$ for which $\frac{d}{dx} g \in B$ as well.

We now want to study the heat semigroup according to the same pattern. Let B be the Banach space of bounded, uniformly continuous functions on $\mathbb{R}^d$, and

$$P_t f(x) = \frac{1}{(4\pi t)^{\frac{d}{2}}} \int e^{-\frac{|x-y|^2}{4t}} f(y) dy \quad \text{for } t > 0. \tag{7.2.31}$$

We now have

$$J_\lambda f(x) = \int_{\mathbb{R}^d} \int_0^\infty \frac{\lambda}{(4\pi t)^{\frac{d}{2}}} e^{-\lambda t - \frac{|x-y|^2}{4t}} dt f(y) dy. \tag{7.2.32}$$

We compute

$$\Delta J_\lambda f(x) = \int_{\mathbb{R}^d} \int_0^\infty \frac{\lambda}{(4\pi t)^{\frac{d}{2}}} \Delta_x e^{-\lambda t - \frac{|x-y|^2}{4t}} dt f(y) dy$$

$$= \int_{\mathbb{R}^d} \int_0^\infty \lambda e^{-\lambda t} \frac{\partial}{\partial t} \left(\frac{1}{(4\pi t)^{\frac{d}{2}}} e^{-\frac{|x-y|^2}{4t}} \right) dt f(y) dy$$

$$= -\lambda f(x) - \int_{\mathbb{R}^d} \int_0^\infty \frac{\partial}{\partial t} \left(\lambda e^{-\lambda t} \right) \frac{1}{(4\pi t)^{\frac{d}{2}}} e^{-\frac{|x-y|^2}{4t}} dt f(y) dy$$

$$= -\lambda f(x) + \lambda J_\lambda f(x).$$

It follows as before that

$$A J_\lambda f = \Delta J_\lambda f, \tag{7.2.33}$$

and thus

$$Ag = \Delta g \quad \text{for all } g \in D(A). \tag{7.2.34}$$

We now want to show that this time, $D(A)$ contains all those $g \in B$ for which Δg is contained in B as well. For such a g, we define $f \in B$ by

$$\Delta g(x) - \lambda g(x) = -\lambda f(x) \tag{7.2.35}$$

and compare this with

$$\Delta J_\lambda f(x) - \lambda J_\lambda f(x) = -\lambda f(x). \tag{7.2.36}$$

Thus $\varphi := g - J_\lambda f$ is bounded and satisfies

$$\Delta \varphi - \lambda \varphi = 0 \quad \text{for } \lambda > 0. \tag{7.2.37}$$

The next lemma will imply $\varphi \equiv 0$, whence $g = J_\lambda f$ as desired:

Lemma 7.2.5: *Let $\lambda > 0$. There does not exist $\varphi \not\equiv 0$ with*

$$\Delta \varphi(x) = \lambda \varphi(x) \quad \text{for all } x \in \mathbb{R}^d. \tag{7.2.38}$$

Proof: For a solution of (7.2.38), we compute

$$\Delta\varphi^2 = 2\left|\nabla\varphi\right|^2 + 2\varphi\Delta\varphi \quad \left(\text{with } \nabla\varphi = \left(\frac{\partial}{\partial x^1}\varphi, \ldots, \frac{\partial}{\partial x^d}\varphi\right)\right)$$

$$= 2\left|\nabla\varphi\right|^2 + 2\lambda\varphi^2 \quad \text{by (7.2.38).} \tag{7.2.39}$$

Let $x_0 \in \mathbb{R}^d$. We choose C^2-functions η_R for $R \geq 1$ with

$$0 \leq \eta_R(x) \leq 1 \quad \text{for all } x \in \mathbb{R}^d, \tag{7.2.40}$$
$$\eta_R(x) = 0 \quad \text{for } |x - x_0| \geq R + 1, \tag{7.2.41}$$
$$\eta_R(x) = 1 \quad \text{for } |x - x_0| \leq R, \tag{7.2.42}$$
$$|\nabla\eta_R(x)| + |\Delta\eta_R(x)| \leq c_0 \quad \text{with a constant } c_0 \text{ that does} \tag{7.2.43}$$
$$\text{not depend on } x \text{ and } R.$$

We compute

$$\Delta\left(\eta_R^2\varphi^2\right) = \eta_R^2\Delta\varphi^2 + \varphi^2\Delta\eta_R^2 + 8\eta_R\varphi\nabla\eta_R \cdot \nabla\varphi$$
$$\geq 2\eta_R^2\left|\nabla\varphi\right|^2 + 2\lambda\eta_R^2\varphi^2 + \left(\Delta\eta_R^2\right)\varphi^2 - 2\eta_R^2\left|\nabla\varphi\right|^2 - 8\left|\nabla\eta_R\right|^2\varphi^2$$
$$\text{by (7.2.39) and the Schwarz inequality}$$
$$= 2\lambda\eta_R^2\varphi^2 + \left(\Delta\eta_R^2 - 8\left|\nabla\eta_R\right|^2\right)\varphi^2. \tag{7.2.44}$$

Together with (7.2.40)–(7.2.43), this implies

$$0 = \int_{B(x_0,R+1)} \Delta\left(\eta_R^2\varphi^2\right) \geq 2\lambda \int_{B(x_0,R)} \varphi^2 - c_1 \int_{B(x_0,R+1)\backslash B(x_0,R)} \varphi^2, \tag{7.2.45}$$

where the constant c_1 does not depend on $\mathbb{R}$.

By assumption, φ is bounded, so

$$\varphi^2 \leq K. \tag{7.2.46}$$

Thus (7.2.45) implies

$$\int_{B(x_0,R)} \varphi^2 \leq \frac{c_2 K}{\lambda} R^{d-1}, \tag{7.2.47}$$

where the constant c_2 again is independent of R. Equation (7.2.39) implies that φ is subharmonic. The mean value inequality (cf. Theorem 7.2.2) thus implies

$$\varphi^2(x_0) \leq \frac{1}{\omega_d R^d} \int_{B(x_0,R)} \varphi^2 \leq \frac{c_2 K}{\omega_d \lambda R} \quad \text{(by (7.2.47))} \quad \rightarrow 0 \quad \text{for } R \rightarrow \infty. \tag{7.2.48}$$

Thus, $\varphi(x_0) = 0$. Since this holds for all $x_0 \in \mathbb{R}^d$, φ has to vanish identically. $\square$

Lemma 7.2.6: *Let B be a Banach space, $L : B \to B$ a continuous linear operator with $\|L\| \leq 1$. Then for every $t \geq 0$ and each $x \in B$, the series*

$$\exp(tL)x := \sum_{\nu=0}^{\infty} \frac{1}{\nu!}(tL)^{\nu}x$$

converges and defines a continuous semigroup with infinitesimal generator L.

Proof: Because of $\|L\| \leq 1$, we also have

$$\|L^n\| \leq 1 \quad \text{for all } n \in \mathbb{N}. \tag{7.2.49}$$

Thus

$$\left\| \sum_{\nu=m}^{n} \frac{1}{\nu!}(tL)^{\nu}x \right\| \leq \sum_{\nu=m}^{n} \frac{1}{\nu!}t^{\nu}\|L^{\nu}x\| \leq \|x\| \sum_{\nu=m}^{n} \frac{t^{\nu}}{\nu!}. \tag{7.2.50}$$

By the Cauchy property of the real-valued exponential series, the last expression becomes arbitrarily small for sufficiently large m, n, and thus our Banach-space-valued exponential series satisfies the Cauchy property as well, and therefore it converges, since B is complete. The limit $\exp(tL)$ is bounded, because by (7.2.50)

$$\left\| \sum_{\nu=0}^{n} \frac{1}{\nu!}(tL)^{\nu}x \right\| \leq e^t \|x\|$$

and thus also

$$\|\exp(tL)x\| \leq e^t \|x\| . \tag{7.2.51}$$

As for the real exponential series, we have

$$\sum_{\nu=0}^{\infty} \frac{(t+s)^{\nu}}{\nu!}L^{\nu}x = \left(\sum_{\mu=0}^{\infty} \frac{t^{\mu}}{\mu!}L^{\mu} \right) \left(\sum_{\sigma=0}^{\infty} \frac{s^{\sigma}}{\sigma!}L^{\sigma} \right) x, \tag{7.2.52}$$

i.e.,

$$\exp((t+s)L) = \exp tL \circ \exp sL, \tag{7.2.53}$$

whence the semigroup property. Furthermore ,

$$\left\| \frac{1}{h}\left(\exp(hL) - \mathrm{Id} \right)x - Lx \right\| \leq \sum_{\nu=2}^{\infty} \frac{h^{\nu-1}}{\nu!}\|L^{\nu}x\| \leq \|x\| \sum_{\nu=2}^{\infty} \frac{h^{\nu-1}}{\nu!}.$$

Since the last expression tends to 0 as $h \to 0$, h is the infinitesimal generator of the semigroup $\{\exp(tL)\}_{t \geq 0}$. $\square$

In the same manner as (7.2.53), one proves (cf. (7.2.52)) the following lemma:

Lemma 7.2.7: *Let $L, M : B \to B$ be continuous linear operators satisfying the assumptions of Lemma 7.2.6, and suppose*

$$LM = ML. \tag{7.2.54}$$

Then

$$\exp(t(M + L)) = \exp(tM) \circ \exp(tL). \tag{7.2.55}$$

$\square$

Theorem 7.2.3 (Hille–Yosida): *Let $A : D(A) \to B$ be a linear operator whose domain of definition $D(A)$ is dense in the Banach space B. Suppose that the resolvent $R(n, A) = (n\,\mathrm{Id} - A)^{-1}$ exists for all $n \in \mathbb{N}$, and that*

$$\left\| \left(\mathrm{Id} - \frac{1}{n} A \right)^{-1} \right\| \leq 1 \quad \text{for all } n \in \mathbb{N}. \tag{7.2.56}$$

Then A generates a unique contracting semigroup.

Proof: As before, we put

$$J_n := \left(\mathrm{Id} - \frac{1}{n} A \right)^{-1} \quad \text{for } n \in \mathbb{N} \text{ (cf. Theorem 7.2.2)}.$$

The proof will consist of several steps:

(1) We claim

$$\lim_{n \to \infty} J_n x = x \quad \text{for all } x \in B, \tag{7.2.57}$$

and

$$J_n x \in D(A) \quad \text{for all } x \in B. \tag{7.2.58}$$

Namely, for $x \in D(A)$, we first have

$$A J_n x = J_n A x = J_n (A - n\,\mathrm{Id})x + n J_n x = n(J_n - \mathrm{Id})x, \tag{7.2.59}$$

and since by assumption $\|J_n A x\| \leq \|Ax\|$, it follows that

$$J_n x - x = \frac{1}{n} J_n A x \to 0 \quad \text{for } n \to \infty.$$

As $D(A)$ is dense in B and the operators J_n are equicontinuous by our assumptions, (7.2.57) follows. (7.2.59) then also implies (7.2.58).

(2) By Lemma 7.2.6, the semigroup

$$\{\exp(sJ_n)\}_{s \geq 0}$$

exists, because of (7.2.56). Putting $s = tn$, we obtain the semigroup

$$\{\exp(tnJ_n)\}_{t \geq 0}$$

and likewise the semigroup

$$T_t^{(n)} := \exp(tAJ_n) = \exp(tn(J_n - \mathrm{Id})) \quad (t \geq 0)$$

(cf. (7.2.59)). By Lemma 7.2.7, we then have

$$T_t^{(n)} = \exp(-tn)\exp(tnJ_n). \tag{7.2.60}$$

Since by (7.2.56)

$$\|\exp(tnJ_n)x\| \leq \sum_{\nu=0}^{\infty} \frac{(nt)^\nu}{\nu!} \|J_n^\nu x\| \leq \exp(nt)\|x\|,$$

it follows that

$$\left\| T_t^{(n)} \right\| \leq 1, \tag{7.2.61}$$

and thus in particular, the operators are equicontinuous in $t \geq 0$ and $n \in \mathbb{N}$.

(3) For all $m, n \in \mathbb{N}$, we have

$$J_m J_n = J_n J_m. \tag{7.2.62}$$

Since by (7.2.60), J_n commutes with $T_t^{(n)}$, then also J_m commutes with $T_t^{(n)}$ for all $n, m \in \mathbb{N}$, $t \geq 0$. By Lemmas 7.2.3, 7.2.6, we have for $x \in B$,

$$D_t T_t^{(n)} x = AJ_n T_t^{(n)} x = T_t^{(n)} AJ_n x; \tag{7.2.63}$$

hence

$$\begin{aligned}
\left\| T_t^{(n)} x - T_t^{(m)} x \right\| &= \left\| \int_0^t D_s \left(T_{t-s}^{(m)} T_s^{(n)} x \right) ds \right\| \\
&= \left\| \int_0^t T_{t-s}^{(m)} T_s^{(n)} (AJ_n - AJ_m) x \, ds \right\| \\
&\leq t \left\| (AJ_n - AJ_m)x \right\|
\end{aligned} \tag{7.2.64}$$

with (7.2.61). For $x \in D(A)$, we have by (7.2.59)

$$(AJ_n - AJ_m)x = (J_n - J_m)Ax. \tag{7.2.65}$$

Equations (7.2.64), (7.2.65), (7.2.57) imply that for $x \in D(A)$,

$$\left(T_t^{(n)} x \right)_{n \in \mathbb{N}}$$

is a Cauchy sequence, and the Cauchy property holds uniformly on $0 \leq t \leq t_0$, for any t_0. Since the operators $T_t^{(n)}$ are equicontinuous by (7.2.61), and $D(A)$ is dense in B by assumption, then

$$\left(T_t^{(n)} x \right)_{n \in \mathbb{N}}$$

is even a Cauchy sequence for all $x \in B$, again locally uniformly with respect to t. Thus the limit

$$T_t x := \lim_{n \to \infty} T_t^{(n)} x$$

exists locally uniformly in t, and T_t is a continuous linear operator with

$$\|T_t\| \leq 1 \tag{7.2.66}$$

(cf. (7.2.61)).

(4) We claim that $(T_t)_{t \geq 0}$ is a semigroup. Namely, since $\{T_t^{(n)}\}_{t \geq 0}$ is a semigroup for all $n \in \mathbb{N}$, using (7.2.61), we get

$$\begin{aligned}
\|T_{t+s}x - T_t T_s x\| &\leq \left\| T_{t+s}x - T_{t+s}^{(n)}x \right\| + \left\| T_{t+s}^{(n)}x - T_t^{(n)} T_s x \right\| \\
&\quad + \left\| T_t^{(n)} T_s x - T_t T_s x \right\| \\
&\leq \left\| T_{t+s}x - T_{t+s}^{(n)}x \right\| + \left\| T_s^{(n)}x - T_s x \right\| \\
&\quad + \left\| \left(T_t^{(n)} - T_t \right) T_s x \right\|,
\end{aligned}$$

and this tends to 0 for $n \to \infty$.

(5) By (4) and (7.2.66), $\{T_t\}_{t \geq 0}$ is a contracting semigroup. We now want to show that A is the infinitesimal generator of this semigroup. Letting $\bar{A}$ be the infinitesimal generator, we are thus claiming

$$\bar{A} = A. \tag{7.2.67}$$

Let $x \in D(A)$. From (7.2.57) and (7.2.59), we easily obtain

$$T_t A x = \lim_{n \to \infty} T_t^{(n)} A J_n x, \tag{7.2.68}$$

again locally uniformly with respect to t. Thus, for $x \in D(A)$,

$$\lim_{t \searrow 0} \frac{1}{t} \left(T_t x - x \right) = \lim_{t \searrow 0} \frac{1}{t} \lim_{n \to \infty} \left(T_t^{(n)} x - x \right)$$

$$= \lim_{t \searrow 0} \frac{1}{t} \lim_{n \to \infty} \int_0^t T_s^{(n)} A J_n x \, ds \text{ by (7.2.63)}$$

$$= \lim_{t \searrow 0} \frac{1}{t} \int_0^t T_s A x \, ds$$

$$= A x.$$

Thus, for $x \in D(A)$, we also have $x \in D(\bar{A})$, and $A x = \bar{A} x$. All that remains is to show that $D(A) = D(\bar{A})$. By the proof of Theorem 7.2.2, $(n \, \text{Id} - \bar{A})$ maps $D(A)$ bijectively onto B. Since $(n \, \text{Id} - A)$ already maps $D(A)$ bijectively onto B, we must have $D(A) = D(\bar{A})$ as desired.

(6) It remains to show the uniqueness of the semigroup $\{T_t\}_{t \geq 0}$ generated by A. Let $\{\bar{T}_t\}_{t \geq 0}$ be another contracting semigroup generated by A. Since A then commutes with $\bar{T}_t$, so do $A J_n$ and $T_t^{(n)}$. We thus obtain as in (7.2.64) for $x \in D(A)$,

$$\left\| T_t^{(n)} x - \bar{T}_t x \right\| = \left\| \int_0^t D_s \left(\bar{T}_{t-s} T_s^{(n)} x \right) ds \right\|$$

$$= \left\| \int_0^t \left(-\bar{T}_{t-s} T_s^{(n)} (A - A J_n) x \right) ds \right\|.$$

Then (7.2.57) implies

$$\bar{T}_t x = \lim_{n \to \infty} T_t^{(n)}$$

for all $x \in D(A)$ and then as usual also for all $x \in B$; hence $\bar{T}_t = T_t$.

$\square$

We now wish to show that the two examples that we have been considering satisfy the assumptions of the Hille–Yosida theorem. Again, we start with the translation semigroup and continue to employ the previous notation. We had identified

$$A = \frac{d}{dx} \tag{7.2.69}$$

as the infinitesimal generator, and we want to show that A satisfies condition (7.2.56). Thus, assume

$$\left(\text{Id} - \frac{1}{n} \frac{d}{dx} \right)^{-1} f = g, \tag{7.2.70}$$

and we have to show that

$$\sup_{x \geq 0} |g(x)| \leq \sup_{x \geq 0} |f(x)|. \tag{7.2.71}$$

Equation (7.2.70) is equivalent to

$$f(x) = g(x) - \frac{1}{n}g'(x). \qquad (7.2.72)$$

We first consider the case where g assumes its supremum at some $x_0 \in [0, \infty)$. We then have

$$g'(x_0) \leq 0 \quad (= 0, \text{ if } x_0 > 0).$$

From this,

$$\sup_x g(x) = g(x_0) \leq g(x_0) - \frac{1}{n}g'(x_0) = f(x_0) \leq \sup_x f(x). \qquad (7.2.73)$$

If g does not assume its supremum, we can at least find a sequence $(x_\nu)_{\nu \in \mathbb{N}} \subset [0, \infty)$ with

$$g(x_\nu) \to \sup_x g(x). \qquad (7.2.74)$$

We claim that for every $\varepsilon_0 > 0$ there exists $\nu_0 \in \mathbb{N}$ such that for all $\nu \geq \nu_0$,

$$g'(x_\nu) < \varepsilon_0. \qquad (7.2.75)$$

Namely, if we had

$$g'(x_\nu) \geq \varepsilon_0 \qquad (7.2.76)$$

for some ε_0 and almost all ν, by the uniform continuity of g' that follows from (7.2.72) because $f, g \in B$, there would also exist $\delta > 0$ such that

$$g'(x) \geq \frac{\varepsilon_0}{2} \quad \text{if } |x - x_\nu| \leq \delta$$

for all ν with (7.2.76). Thus we would have

$$g(x_\nu + \delta) = g(x_\nu) + \int_0^\delta g'(x_\nu + t)dt \geq g(x_\nu) + \frac{\varepsilon_0 \delta}{2}. \qquad (7.2.77)$$

On the other hand, by (7.2.74), we may assume

$$g(x_\nu) \geq \sup_x g(x) - \frac{\varepsilon_0 \delta}{4},$$

which in conjunction with (7.2.77) yields the contradiction

$$g(x_\nu + \delta) > \sup g(x).$$

Consequently, (7.2.75) must hold. As in (7.2.73), we now obtain for each $\varepsilon > 0$

$$\sup_x g(x) = \lim_{\nu \to \infty} g(x_\nu) \le \lim_{\nu \to \infty} \left(g(x_\nu) - \frac{1}{n} g'(x_\nu) \right) + \frac{\varepsilon}{n}$$
$$= \lim_{\nu \to \infty} f(x_\nu) + \frac{\varepsilon}{n} \le \sup_x f(x) + \frac{\varepsilon}{n}.$$

The case of an infimum is treated analogously, and (7.2.70) follows.

We now want to carry out the corresponding analysis for the heat semi-group, again using the notation already established. In this case, the infinitesimal generator is the Laplace operator,

$$A = \Delta. \tag{7.2.78}$$

We again consider the equation

$$\left(\mathrm{Id} - \frac{1}{n} \Delta \right)^{-1} f = g, \tag{7.2.79}$$

or equivalently,

$$f(x) = g(x) - \frac{1}{n} \Delta g(x), \tag{7.2.80}$$

and we again want to verify (7.2.56), i.e.,

$$\sup_{x \in \mathbb{R}^d} |g(x)| \le \sup_{x \in \mathbb{R}^d} |f(x)|. \tag{7.2.81}$$

Again, we first consider the case where g achieves its supremum at some $x_0 \in \mathbb{R}^d$. Then

$$\Delta g(x_0) \le 0,$$

and consequently,

$$\sup_x g(x) = g(x_0) \le g(x_0) - \frac{1}{n} \Delta g(x_0) = f(x_0) \le \sup_x f(x). \tag{7.2.82}$$

If g does not assume its supremum, we select some $x_0 \in \mathbb{R}^d$, and for every $\eta > 0$, we consider the function

$$g_\eta(x) := g(x) - \eta |x - x_0|^2 .$$

Since

$$\lim_{|x| \to \infty} g_\eta(x) = -\infty,$$

g_η assumes its supremum at some $x_\eta \in \mathbb{R}^d$. Then

$$\Delta g_\eta(x_\eta) \le 0,$$

i.e.,

$$\Delta g(x_\eta) \le 2d\eta.$$

For $y \in \mathbb{R}^d$, we obtain

$$g(y) \le g(x_\eta) + \eta \, |y - x_0|^2$$

$$\le g(x_\eta) - \frac{1}{n} \Delta g(x_\eta) + \eta \left(\frac{2d}{n} + |y - x_0|^2 \right)$$

$$= f(x_\eta) + \eta \left(\frac{2d}{n} + |y - x_0|^2 \right)$$

$$\le \sup_{x \in \mathbb{R}^d} f(x) + \eta \left(\frac{2d}{n} + |y - x_0|^2 \right).$$

Since $\eta > 0$ can be chosen arbitrarily small, we thus get for every $y \in \mathbb{R}^d$

$$g(y) \le \sup_{x \in \mathbb{R}^d} f(x),$$

i.e., (7.2.81) if we treat the infimum analogously.

It is no longer so easy to verify directly that (7.2.80) is solvable with respect to g for given f. By our previous considerations, however, we already know that Δ generates a contracting semigroup, namely, the heat semigroup, and the solvability of (7.2.80) therefore follows from Theorem 7.2.2. Of course, we could have deduced (7.2.56) in the same way, since it is easy to see that (7.2.56) is also necessary for generating a contracting semigroup. The direct proof given here, however, was simple and instructive enough to be presented.

7.3 Brownian Motion

We consider a particle that moves around in some set S, for simplicity assumed to be a measurable subset of $\mathbb{R}^d$, obeying the following rules: The probability that the particle that is at the point x at time t happens to be in the set $E \subset S$ for $s \ge t$ is denoted by $P(t, x; s, E)$. In particular,

$$P(t, x; s, S) = 1,$$
$$P(t, x; s, \emptyset) = 0.$$

This probability should not depend on the positions of the particles at any times less than t. Thus, the particle has no memory, or, as one also says, the process has the Markov property. This means that for $t < \tau \le s$, the Chapman–Kolmogorov equation

$$P(t, x; s, E) = \int_S P(\tau, y; s, E) P(t, x; \tau, y) dy \qquad (7.3.1)$$

holds. Here, $P(t, x; \tau, y)$ has to be considered as a probability density, i.e., $P(t, x; \tau, y) \geq 0$ and $\int_S P(t, x; \tau, y) dy = 1$ for all x, t, τ. We want to assume that the process is homogeneous in time, meaning that $P(t, x; s, E)$ depends only on $(s - t)$. We thus have

$$P(t, x; s, E) = P(0, x; s - t, E) =: P(s - t, x, E),$$

and (7.3.1) becomes

$$P(t + \tau, x, E) := \int_S P(\tau, y, E) P(t, x, y) dy. \qquad (7.3.2)$$

We express this property through the following definition:

Definition 7.3.1: *Let $\mathcal{B}$ a σ-additive set of subsets of S with $S \in \mathcal{B}$. For $t > 0$, $x \in S$, and $E \in \mathcal{B}$, let $P(t, x, E)$ be defined satisfying*

(i) $P(t, x, E) \geq 0$, $P(t, x, S) = 1$.
(ii) $P(t, x, E)$ *is σ-additive with respect to $E \in \mathcal{B}$ for all t, x.*
(iii) $P(t, x, E)$ *is $\mathcal{B}$-measurable with respect to x for all t, E.*
(iv) $P(t + \tau, x, E) = \int_S P(\tau, y, E) P(t, x, y) dy$ *(Chapman–Kolmogorov equation) for all $t, \tau > 0$, x, E.*

Then $P(t, x, E)$ is called a Markov process on $(S, \mathcal{B})$.

Let $L^\infty(S)$ be the space of bounded functions on S. For $f \in L^\infty(S), t > 0$, we put

$$(T_t f)(x) := \int_S P(t, x, y) f(y) dy. \qquad (7.3.3)$$

The Chapman–Kolmogorov equation implies the semigroup property

$$T_{t+s} = T_t \circ T_s \quad \text{for } t, s > 0. \qquad (7.3.4)$$

Since by (i), $P(t, x, y) \geq 0$ and

$$\int_S P(t, x, y) dy = 1, \qquad (7.3.5)$$

it follows that

$$\sup_{x \in S} |T_t f(x)| \leq \sup_{x \in S} |f(x)|, \qquad (7.3.6)$$

i.e., the contraction property.

In order that T_t map continuous functions to continuous functions and that $\{T_t\}_{t \geq 0}$ define a continuous semigroup, we need additional assumptions. For simplicity, we consider only the case $S = \mathbb{R}^d$.

Definition 7.3.2: *The Markov process $P(t, x, E)$ is called spatially homogeneous if for all translations $i : \mathbb{R}^d \to \mathbb{R}^d$,*

$$P(t, i(x), i(E)) = P(t, x, E). \qquad (7.3.7)$$

A spatially homogeneous Markov process is called a Brownian motion if for all $\varrho > 0$ and all $x \in \mathbb{R}^d$,

$$\lim_{t \searrow 0} \frac{1}{t} \int_{|x-y| > \varrho} P(t, x, y) dy = 0. \qquad (7.3.8)$$

Theorem 7.3.1: *Let B be the Banach space of bounded and uniformly continuous functions on $\mathbb{R}^d$, equipped with the supremum norm. Let $P(t, x, E)$ be a Brownian motion. We put*

$$(T_t f)(x) := \int_{\mathbb{R}^d} P(t, x, y) f(y) dy \quad for\ t > 0,$$
$$T_0 f = f.$$

Then $\{T_t\}_{t \geq 0}$ constitutes a contracting semigroup on B.

Proof: As already explained, $P(t, x, E) \geq 0$, $P(t, x, \mathbb{R}^d) = 1$ implies the contraction property

$$\sup_{x \in \mathbb{R}^d} |(T_t f)(x)| \leq \sup_{x \in \mathbb{R}^d} |f(x)| \quad \text{for all } f \in B, t \geq 0, \qquad (7.3.9)$$

and the semigroup property follows from the Chapman–Kolmogorov equation. Let i be a translation of Euclidean space. We put

$$if(x) := f(ix)$$

and obtain

$$iT_t f(x) = T_t f(ix) = \int_{\mathbb{R}^d} P(t, ix, y) f(y) dy$$
$$= \int_{\mathbb{R}^d} P(t, ix, iy) f(iy) dy,$$
$$\text{since } d(iy) = dy \text{ for a translation,}$$
$$= \int_{\mathbb{R}^d} P(t, x, y) f(iy) dy,$$
$$\text{since the process is spatially homogeneous,}$$
$$= T_t if(x),$$

i.e.,

$$iT_t = T_t i. \qquad (7.3.10)$$

For $x, y \in \mathbb{R}^d$, we may find a translation $i : \mathbb{R}^d \to \mathbb{R}^d$ with

$$ix = y.$$

We then have

$$|(T_t f)(x) - (T_t f)(y)| = |(T_t f)(x) - (iT_t f)(x)| = |T_t(f - if)(x)|.$$

Since f is uniformly continuous, this implies that $T_t f$ is uniformly continuous as well; namely,

$$|T_t(f - if)(x)| = \left| \int P(t, x, z)(f(z) - f(iz))dz \right| \leq \sup_z |f(z) - f(iz)|,$$

and if $|x - y| < \delta$, then also $|z - iz| < \delta$ for all $z \in \mathbb{R}^d$, and δ may be chosen such that this expression becomes smaller than any given $\varepsilon > 0$. Note that this estimate does not depend on t.

It remains to show continuity with respect to t. Let $t \geq s$. For $f \in B$, we consider

$$|T_t f(x) - T_s f(x)| = |T_\tau g(x) - g(x)| \quad \text{for } \tau := t - s, g := T_s f$$

$$= \left| \int_{\mathbb{R}^d} P(\tau, x, y)(g(y) - g(x))dy \right|$$

$$\text{because of } \int_{\mathbb{R}^d} P(t, x, y)dy = 1$$

$$\leq \left| \int_{|x-y| \leq \varrho} P(\tau, x, y)(g(y) - g(x))dy \right|$$

$$+ \left| \int_{|x-y| > \varrho} P(\tau, x, y)(g(y) - g(x))dy \right|$$

$$\leq \left| \int_{|x-y| \leq \varrho} P(\tau, x, y)(g(y) - g(x))dy \right|$$

$$+ 2 \sup_{z \in \mathbb{R}^d} |f(z)| \int_{|x-y| > \varrho} P(\tau, x, y)dy$$

by (7.3.9). Since we have checked already that $g = T_s f$ satisfies the same continuity estimates as f, for given $\varepsilon > 0$ we may choose $\varrho > 0$ so small that the first term on the right-hand side becomes smaller than $\varepsilon/2$. For that value of ϱ we may then choose τ so small that the second term becomes smaller than $\varepsilon/2$ as well. Note that because of the spatial homogeneity, τ can be chosen independently of x and y. This shows that $\{T_t\}_{t \geq 0}$ is a continuous semigroup, and the proof of Theorem 7.3.1 is complete.   □

An example of Brownian motion is given by the heat kernel

$$P(t, x, y) = \frac{1}{(4\pi t)^{\frac{d}{2}}} e^{-\frac{|x-y|^2}{4t}}. \tag{7.3.11}$$

We shall now see that this already is the typical case of a Brownian motion.

Theorem 7.3.2: *Let $P(t, x, E)$ be a Brownian motion that is invariant under all isometries of Euclidean space, i.e.,*

$$P(t, i(x), i(E)) = P(t, x, E) \tag{7.3.12}$$

for all Euclidean isometries i. Then the infinitesimal generator of the contracting semigroup defined by this process is

$$A = c\Delta, \tag{7.3.13}$$

$c = \text{const} > 0$, $\Delta =$ *Laplace operator, and this semigroup then coincides with the heat semigroup up to reparametrization, according to the uniqueness result of Theorem 7.2.3. More precisely, we have*

$$P(t, x, y) = \frac{1}{(4\pi ct)^{\frac{d}{2}}} e^{-\frac{|x-y|^2}{4ct}}. \tag{7.3.14}$$

Proof: (1) Let B again be the Banach space of bounded, uniformly continuous functions on $\mathbb{R}^d$, equipped with the supremum norm. By Theorem 7.3.1, our semigroup operates on B. By Theorem 7.2.1, the domain of definition $D(A)$ of the infinitesimal operator A is dense in B.

(2) We claim that $D(A) \cap C^\infty(\mathbb{R}^d)$ is still dense in B. To verify that, as in Section 2.1 we consider mollifications with a smooth kernel, i.e., for $f \in D(A)$,

$$f_r(x) = \frac{1}{r^d} \int_{\mathbb{R}^d} \varrho\left(\frac{|x-y|}{r}\right) f(y) dy \quad \text{as in (1.2.6)}$$

$$= \int_{\mathbb{R}^d} \rho(|z|) f(x - rz) dz. \tag{7.3.15}$$

Since we are assuming translation invariance, if the function $f(x)$ is contained in $D(A)$, so is $(i_{rz}f)(x) = f(x - rz)$ for all $r > 0$, $z \in \mathbb{R}^d$ in $D(A)$, and the defining criterion, namely,

$$\lim_{t \to 0} \frac{1}{t} \left(\int_{\mathbb{R}^d} P(t, x, y) f(y - rz) - f(x - rz) \right) = 0,$$

holds uniformly in r, z. Approximating the preceding integral by step functions of the form $\sum_\nu c_\nu f(x - rz_\nu)$ (where we have only finitely many summands, since g has compact support), we see that since f does, f_r also satisfies $\lim_{t \to 0} \frac{1}{t} \left(\int_{\mathbb{R}^d} P(t, x, y) f_r(y) \, dy - f_r(x) \right) = 0$, hence is contained in $D(A)$. Since f_r is contained in $C^\infty(\mathbb{R}^d)$ for $r > 0$, and converges to f uniformly as $r \to 0$, the claim follows.

(3) We claim that there exists a function $\varphi \in D(A) \cap C^\infty(\mathbb{R}^d)$ with

$$x^j x^k \frac{\partial^2 \varphi}{\partial x^j \partial x^k}(0) \geq \sum_{j=1}^{d} (x^j)^2 \quad \text{for all } x \in \mathbb{R}^d. \qquad (7.3.16)$$

For that purpose, we select $\psi \in B$ with

$$\frac{\partial^2 \psi}{\partial x^j \partial x^k}(0) = 2\delta_{jk} \quad \left(\delta_{jk} = \begin{cases} 1 & \text{for } j = k \\ 0 & \text{otherwise} \end{cases} \right),$$

and from (2), we find a sequence $(f^{(\nu)})_{\nu \in \mathbb{N}} \subset D(A) \cap C^\infty(\mathbb{R}^d)$, converging uniformly to ψ. Then

$$\frac{\partial^2}{\partial x^j \partial x^k} f_r^{(\nu)}(0) = \frac{1}{r^d} \int \frac{\partial^2}{\partial x^j \partial x^k} \varrho \left(\frac{|y-x|}{r} \right) \Bigg|_{x=0} f^{(\nu)}(y)\, dy$$

$$\to \frac{1}{r^d} \int \frac{\partial^2}{\partial x^j \partial x^k} \varrho \left(\frac{|y-x|}{r} \right) \Bigg|_{x=0} \psi(y)\, dy \quad \text{for } \nu \to \infty$$

$$= \frac{1}{r^d} \int \rho \left(\frac{|y-x|}{r} \right) \frac{\partial^2}{\partial x^j \partial x^k} \psi(y)\, dy$$

replacing the derivative with respect to x by one with respect to y and integrating by parts

$$\to \frac{\partial^2}{\partial x^j \partial x^k} \psi(0) \quad \text{for } r \to 0$$

$$= 2\delta_{jk}.$$

We may thus put $\varphi = f_r^{(\nu)}$ for suitable $\nu \in \mathbb{N}$, $r > 0$, in order to achieve (7.3.16). By Euclidean invariance, for every $x_0 \in \mathbb{R}^d$, there then exists a function in $D(A) \cap C^\infty(\mathbb{R}^d)$, again denoted by φ for simplicity, with

$$(x^j - x_0^j)(x^k - x_0^k) \frac{\partial^2 \varphi}{\partial x^j \partial x^k}(x_0) \geq \sum (x^j - x_0^j)^2 \quad \text{for all } x \in \mathbb{R}^d. \qquad (7.3.17)$$

(4) For all $x_0 \in \mathbb{R}^d$, $j = 1, \ldots, d$, $r > 0$, $t > 0$,

$$\int_{|x-x_0| \leq r} (x^j - x_0^j) P(t, x_0, x)\, dx = 0, \quad x_0 = (x_0^1, \ldots, x_0^d); \qquad (7.3.18)$$

namely, let

$$i : \mathbb{R}^d \to \mathbb{R}^d$$

be the Euclidean isometry defined by

$$\begin{aligned} i(x^j - x_0^j) &= -(x^j - x_0^j), \\ i(x^k - x_0^k) &= x^k - x_0^k \quad \text{for } k \neq j \end{aligned} \qquad (7.3.19)$$

(reflection across the hyperplane through x_0 that is orthogonal to the jth coordinate axis). We then have

$$\int_{|x-x_0|\leq r} (x^j - x_0^j)P(t, x_0, x)dx = \int_{|x-x_0|\leq r} i(x^j - x_0^j)P(t, ix_0, ix)dx$$

$$= -\int_{|x-x_0|\leq r} (x^j - x_0^j)P(t, x_0, x)dx$$

because of (7.3.19) and the assumed invariance of P, and this indeed implies (7.3.18).

Similarly, the invariance of P under rotations of $\mathbb{R}^d$ yields

$$\int_{|x-x_0|\leq r} (x^j - x_0^j)^2 P(t, x_0, x)dx = \int_{|x-x_0|\leq r} (x^k - x_0^k)^2 P(t, x_0, x)dx$$

$$\text{for all } x_0 \in \mathbb{R}^d, r > 0, t > 0, j, k = 1, \ldots, d, \quad (7.3.20)$$

and finally as in (7.3.18),

$$\int_{|x_0-x|\leq r} (x^j - x_0^j)(x^k - x_0^k)P(t, x_0, x)dx = 0 \quad \text{for } j \neq k, \quad (7.3.21)$$

if $x_0 \in \mathbb{R}^d$, $r > 0$, $t > 0$, $j, k \in \{1, \ldots, d\}$.

(5) Let $\varphi \in D(A) \cap C^2(\mathbb{R}^d)$. We then obtain the existence of

$$A\varphi(x_0) = \lim_{t\searrow 0} \frac{1}{t} \int_{\mathbb{R}^d} P(t, x_0, x)(\varphi(x) - \varphi(x_0))dx$$

$$= \lim_{t\searrow 0} \frac{1}{t} \int_{|x-x_0|\leq\varepsilon} P(t, x_0, x)(\varphi(x) - \varphi(x_0))dx \quad \text{by (7.3.8)}$$

$$= \lim_{t\searrow 0} \frac{1}{t} \int_{|x-x_0|\leq\varepsilon} \sum_{j=1}^d (x^j - x_0^j)\frac{\partial\varphi}{\partial x^j}(x_0)P(t, x_0, x)dx$$

$$+ \lim_{t\searrow 0} \frac{1}{t} \int_{|x-x_0|\leq\varepsilon} \frac{1}{2}\sum_{j,k}(x^j - x_0^j)(x^k - x_0^k)$$

$$\times \frac{\partial^2\varphi}{\partial x^j \partial x^k}(x_0 + \tau(x - x_0))P(t, x_0, x)dx$$

by Taylor expansion for some $\tau \in [0, 1)$, as $\varphi \in C^2(\mathbb{R}^d)$.

The first term on the right-hand side vanishes by (7.3.18). Thus, the limit for $t \searrow 0$ of the second term exists, and it follows from (7.3.17) and $P(t, x_0, x) \geq 0$ that

$$\limsup_{t\searrow 0} \frac{1}{t} \int_{|x-x_0|\leq\varepsilon} \sum (x^j - x_0^j)^2 P(t, x_0, x)dx < \infty. \quad (7.3.22)$$

By (7.3.8), this limit superior does not depend on $\varepsilon > 0$, and neither does the corresponding limit inferior.

(6) Now let $f \in D(A) \cap C^2(\mathbb{R}^d)$. As in (5), we obtain, by Taylor expanding f at x_0,

$$\frac{1}{t}(T_t f(x_0) - f(x_0))$$

$$= \frac{1}{t} \int_{\mathbb{R}^d} (f(x) - f(x_0)) P(t, x_0, x) dx$$

$$= \frac{1}{t} \int_{|x-x_0|>\varepsilon} (f(x) - f(x_0)) P(t, x_0, x) dx$$

$$+ \frac{1}{t} \int_{|x-x_0|\le\varepsilon} \sum_j (x^j - x_0^j) \frac{\partial f}{\partial x^j}(x_0) P(t, x_0, x) dx$$

$$+ \frac{1}{t} \int_{|x-x_0|\le\varepsilon} \frac{1}{2} \sum_{j,k} (x^j - x_0^j)(x^k - x_0^k) \frac{\partial^2 f}{\partial x^j \partial x^k}(x_0) P(t, x_0, x) dx$$

$$+ \frac{1}{t} \int_{|x-x_0|\le\varepsilon} \sum_{j,k} (x^j - x_0^j)(x^k - x_0^k) \sigma_{ij}(\varepsilon) P(t, x_0, x) dx$$

(where the notation suppresses the x-dependence of the remainder term $\sigma_{ij}(\varepsilon)$, since this converges to 0 for $\varepsilon \to 0$ uniformly in x, since $f \in C^2(\mathbb{R}^d)$)

$$= \frac{1}{t} \int_{|x-x_0|>\varepsilon} (f(x) - f(x_0)) P(t, x_0, x) dx$$

$$+ \frac{1}{t} \int_{|x-x_0|\le\varepsilon} \sum_j (x^j - x_0^j)^2 \frac{\partial^2 f}{(\partial x^j)^2}(x_0) P(t, x_0, x) dx$$

$$+ \frac{1}{t} \int_{|x-x_0|\le\varepsilon} \sum_{j,k} (x^j - x_0^j)(x^k - x_0^k) \sigma_{ij}(\varepsilon) P(t, x_0, x) dx$$

by (7.3.18), (7.3.21).   (7.3.23)

By (7.3.8), the first term on the right-hand side tends to 0 as $t \to 0$ for every $\varepsilon > 0$. Because of (7.3.22) and $\lim_{\varepsilon \to 0} \sigma_{ij}(\varepsilon) = 0$ (since $f \in C^2$), the last term converges to 0 as $\varepsilon \to 0$ for every $t > 0$. Since we have observed at the end of (5), however, that in the second term on the right-hand side, limits can be performed independently of ε, for all $\varepsilon > 0$, we obtain the existence of

$$\lim_{t\searrow 0} \frac{1}{t} \int_{|x-x_0|\le\varepsilon} \sum_j (x^j - x_0^j)^2 \frac{\partial^2 f}{(\partial x^j)^2}(x_0) P(t, x_0, x) dx = Af(x_0), \quad (7.3.24)$$

by performing the limit $t \to 0$ on the right-hand side of (7.3.23).

The argument of (3) shows that for $f \in D(A)$,

$$\frac{\partial^2 f}{(\partial x^j)^2}(x_0)$$

may approximate arbitrary values, and so in particular, we infer the existence of

$$\lim_{t \searrow 0} \frac{1}{t} \int_{|x-x_0| \leq \varepsilon} \sum (x^j - x_0^j)^2 P(t, x_0, x) dx$$

independently of ε. By (7.3.20), for each $j = 1, \ldots, d$,

$$\lim_{t \searrow 0} \frac{1}{t} \int_{|x-x_0| \leq \varepsilon} (x^j - x_0^j)^2 P(t, x_0, x) dx$$

exists and is independent of j and by translation invariance independent of x_0 as well. We thus call this limit c. By (7.3.24), we then have

$$Af(x_0) = c\Delta f(x_0).$$

The rest follows from Theorem 7.2.3. $\square$

Remark: If we assume only spatial homogeneity, i.e., translation invariance, but not invariance under reflections and rotations, the infinitesimal generator still is a second-order differential operator; namely, it is of the form

$$Af(x) = \sum_{j,k=1}^{d} a^{jk}(x) \frac{\partial^2 f}{\partial x^j \partial x^k}(x) + \sum_{j=1}^{d} b^j(x) \frac{\partial f}{\partial x^j}(x)$$

with

$$a^{jk}(x) = \lim_{t \searrow 0} \frac{1}{t} \int_{|y-x| \leq \varepsilon} (y^j - x^j)(y^k - x^k) P(t, x, y) dy,$$

and thus in particular,

$$a^{jk} = a^{kj}, \quad a^{jj} \geq 0 \quad \text{for all } j, k,$$

and

$$b^j(x) = \lim_{t \searrow 0} \frac{1}{t} \int_{|y-x| \leq \varepsilon} (y^j - x^j) P(t, x, y) dy,$$

where the limits again are independent of $\varepsilon > 0$. The proof can be carried out with the same methods as employed for demonstrating Theorem 7.3.2.

A reference for the present chapter is Yosida [23].

Summary

The heat equation satisfies a Markov property in the sense that the solution $u(x, t)$ at time $t_1 + t_2$ with initial values $u(x, 0) = f(x)$ equals the solution at time t_2 with initial values $u(x, t_1)$. Putting

$$(P_t f)(x) := u(x, t),$$

we thus have

$$(P_{t_1+t_2} f)(x) = P_{t_2}(P_{t_1} f)(x);$$

i.e., P_t satisfies the semigroup property

$$P_{t_1+t_2} = P_{t_2} \circ P_{t_1} \quad \text{for } t_1, t_2 \geq 0.$$

Moreover, $\{P_t\}_{t \geq 0}$ is continuous on the space C^0 in the sense that

$$\lim_{t \searrow t_0} P_t = P_{t_0}$$

for all $t_0 \geq 0$ (in particular, this also holds for $t_0 = 0$, with $P_0 = \text{Id}$).
Moreover, P_t is contracting because of the maximum principle, i.e.,

$$\|P_t f\|_{C^0} \leq \|f\|_{C^0} \quad \text{for } t \geq 0, f \in C^0.$$

The infinitesimal generator of the semigroup P_t is the Laplace generator, i.e.,

$$\Delta = \lim_{t \searrow 0} \frac{1}{t}(P_t - \text{Id}).$$

Upon these properties one may found an abstract theory of semigroups in Banach spaces. The Hille–Yosida theorem says that a linear operator $A : D(A) \to B$ whose domain of definition $D(A)$ is dense in the Banach space B and for which $\text{Id} - \frac{1}{n}A$ is invertible for all $n \in \mathbb{N}$ and

$$\left\| \left(\text{Id} - \frac{1}{n}A\right)^{-1} \right\| \leq 1$$

generates a unique contracting semigroup of operators

$$T_t : B \to B \quad (t \geq 0).$$

For a stochastic interpretation, one considers the probability density $P(t, x, y)$ that some particle that during the random walk happened to be at the point x at a certain time can be found at y at a time that is larger by the amount t. This constitutes a Markov process inasmuch as this probability density depends only on the time difference, but not on the individual values of the times involved. In particular, $P(t, x, y)$ does not depend on where the particle had been before reaching x (random walk without memory). Such a random walk on the set S satisfies the Chapman–Kolmogorov equation

$$P(t_1 + t_2, x, y) = \int_S P(t_1, x, z) P(t_2, z, y) dz$$

and thus constitutes a semigroup.

If such a process on $\mathbb{R}^d$ is spatially homogeneous and satisfies

$$\lim_{t \searrow 0} \frac{1}{t} \int_{|x-y|>\rho} P(t,x,y)\, dy = 0$$

for all $\rho > 0$ and $x \in \mathbb{R}^d$, it is called a Brownian motion. One shows that up to a scaling factor, such a Brownian motion has to be given by the heat semigroup, i.e.,

$$P(t,x,y) = \frac{1}{(4\pi ct)^{d/2}} e^{-\frac{|x-y|^2}{4ct}}.$$

Exercises

7.1 Let $f \in C^0(\mathbb{R}^d)$ be bounded, $u(x,t)$ a solution of the heat equation

$$u_t(x,t) = \Delta u(x,t) \quad \text{for } x \in \mathbb{R}^d, t > 0,$$
$$u(x,0) = f(x).$$

Show that the derivatives of u satisfy

$$|\frac{\partial}{\partial x^j} u(x,t)| \leq \text{const } \sup |f| \cdot t^{-1/2}.$$

(Hint: Use the representation formula (4.2.3) from Section 4.2.)

7.2 As in Section 7.2, we consider a continuous semigroup

$$\exp(tA) : B \to B \quad (t \geq 0), B \text{ a Banach space.}$$

Let B_1 be another Banach space, and for $t > 0$ suppose

$$\exp(tA) : B_1 \to B$$

is defined, and we have for $0 < t \leq 1$ and for all $\varphi \in B_1$,

$$\|\exp(tA)\varphi\|_B \leq \text{const } t^{-\alpha}\|\varphi\|_{B_1} \quad \text{for some } \alpha < 1.$$

Finally, let

$$\Phi : B \to B_1$$

be Lipschitz continuous.

Show that for every $f \in B$ there exists $T > 0$ with the property that the evolution equation

$$\frac{\partial v}{\partial t} = Av + \Phi(v(t)) \quad \text{for } t > 0,$$
$$v(0) = f,$$

has a unique, continuous solution $v : [0, T] \to B$.
(Hint: Convert the problem into the integral equation

$$v(t) = \exp(tA)f + \int_0^t \exp((t - s)A)\Phi(v(s))ds$$

and use the Banach fixed-point theorem (as in the standard proof of the Picard–Lindelöf theorem for ODEs) to obtain a solution of that integral equation.)

7.3 Apply the results of Exercises 6.1, 6.2 to the initial value problem for the following semilinear parabolic PDE:

$$\frac{\partial u(x, t)}{\partial t} = \Delta u(x, t) + F(t, x, u(x), Du(x)) \quad \text{for } x \in \mathbb{R}^d, t > 0,$$

$$u(x, 0) = f(x),$$

for compactly supported $f \in C^0(\mathbb{R}^d)$. We assume that F is smooth with respect to all its arguments.

7.4 Demonstrate the assertion in the remark at the end of Section 7.3.

8. The Dirichlet Principle.
Variational Methods for the Solution of PDEs
(Existence Techniques III)

8.1 Dirichlet's Principle

We consider the Dirichlet problem for harmonic functions once more:
We want to find a solution $u : \Omega \to \mathbb{R}$, $\Omega \in \mathbb{R}^d$ a domain, of

$$\Delta u = 0 \quad \text{in } \Omega,$$
$$u = f \quad \text{on } \partial\Omega, \tag{8.1.1}$$

with given f.

Dirichlet's principle is based on the following observation: Let $u \in C^2(\Omega)$ be a function with $u = f$ on $\partial\Omega$ and

$$\int_\Omega |\nabla u(x)|^2 \, dx = \min \left\{ \int_\Omega |\nabla v(x)|^2 \, dx : v : \Omega \to \mathbb{R} \text{ with } v = f \text{ on } \partial\Omega \right\}. \tag{8.1.2}$$

We now claim that u then solves (8.1.1). To show this, let

$$\eta \in C_0^\infty(\Omega).^{[1]}$$

According to (8.1.2), the function

$$\alpha(t) := \int_\Omega |\nabla(u + t\eta)(x)|^2 \, dx$$

possesses a minimum at $t = 0$, because $u + t\eta = f$ on $\partial\Omega$, since η vanishes on $\partial\Omega$. Expanding this expression, we obtain

$$\alpha(t) = \int_\Omega |\nabla u(x)|^2 \, dx + 2t \int_\Omega \nabla u(x) \cdot \nabla \eta(x) dx + t^2 \int_\Omega |\nabla \eta(x)|^2 \, dx. \tag{8.1.3}$$

In particular, α is differentiable with respect to t, and the minimality at $t = 0$ implies

$$\dot{\alpha}(0) = 0. \tag{8.1.4}$$

---

[1] $C_0^\infty(A) := \{\varphi \in C^\infty(A) : \text{the closure of } \{x : \varphi(x) \neq 0\} \text{ is compact and contained in } A\}$.

By (8.1.3) this implies

$$\int_\Omega \nabla u(x) \cdot \nabla \eta(x) dx = 0, \tag{8.1.5}$$

and this holds for all $\eta \in C_0^\infty(\Omega)$.

Integrating (8.1.5) by parts, we obtain

$$\int_\Omega \Delta u(x)\eta(x) dx = 0 \quad \text{for all } \eta \in C_0^\infty(\Omega). \tag{8.1.6}$$

We now recall the following well-known and elementary fact:

Lemma 8.1.1: *Suppose* $g \in C^0(\Omega)$ *satisfies*

$$\int_\Omega g(x)\eta(x) dx = 0 \quad \text{for all } \eta \in C_0^\infty(\Omega).$$

Then $g \equiv 0$ *in* Ω.   □

Applying Lemma 8.1.1 to (8.1.6) (which is possible, since $\Delta u \in C^0(\Omega)$ by our assumption $u \in C^2(\Omega)$), we indeed obtain

$$\Delta u(x) = 0 \quad \text{in } \Omega,$$

as claimed.

This observation suggests that we try to minimize the so-called Dirichlet integral

$$D(u) := \int_\Omega |\nabla u(x)|^2 \, dx \tag{8.1.7}$$

in the class of all functions $u : \Omega \to \mathbb{R}$ with $u = f$ on $\partial\Omega$. This is Dirichlet's principle.

It is by no means evident, however, that the Dirichlet integral assumes its infimum within the considered class of functions. This constitutes the essential difficulty of Dirichlet's principle. In any case, so far we have not specified which class of functions $u : \Omega \to \mathbb{R}$ (with the given boundary values) we allow for competition; the possibilities include functions of class C^∞, which would be natural, since we have shown already in Chapter 1 that any solution of (8.1.1) automatically is of regularity class C^∞; functions of class C^2, which would be natural, since then the differential equation $\Delta u(x) = 0$ would have a meaning; and functions of class C^1 because then at least (assuming Ω bounded and f sufficiently regular, e.g., $f \in C^1$) the Dirichlet integral $D(u)$ would be finite. Posing the question somewhat differently, should we try to minimize $D(U)$ in a space of functions that is as large as possible, in order to increase the chance that a minimizing sequence possesses a limit in that space that then would be a natural candidate for a minimizer, or should we rather select a smaller space in order to facilitate the verification that a tentative solution is a minimizer?

In order to analyze this question, we consider a minimzing sequence $(u_n)_{n \in \mathbb{N}}$ for D, i.e.,

$$\lim_{n \to \infty} D(u_n) = \inf \{D(v) : v : \Omega \to \mathbb{R}, v = f \text{ on } \partial\Omega\} =: \kappa, \qquad (8.1.8)$$

where, of course, we assume $u_n = f$ on $\partial\Omega$ for all u_n. To find properties of such a minimizing sequence, we shall employ the following simple lemma:

Lemma 8.1.2: *Dirichlet's integral is convex, i.e.,*

$$D(tu + (1-t)v) \leq tD(u) + (1-t)D(v) \qquad (8.1.9)$$

for all u, v and all $t \in [0, 1]$.

Proof:

$$D(tu + (1-t)v) = \int_\Omega |t\nabla u + (1-t)\nabla v|^2$$
$$\leq \int_\Omega \left\{ t |\nabla u|^2 + (1-t) |\nabla v|^2 \right\}$$
$$\text{because of the convexity of } w \mapsto |w|^2$$
$$= tD(u) + (1-t)Dv.$$

$\square$

Now let $(u_n)_{n \in \mathbb{N}}$ be a minimizing sequence. Then

$$D(u_n - u_m) = \int_\Omega |\nabla(u_n - u_m)|^2$$
$$= 2 \int_\Omega |\nabla u_n|^2 + 2 \int_\Omega |\nabla u_m|^2 - 4 \int_\Omega \left| \nabla \left(\frac{u_n + u_m}{2} \right) \right|^2$$
$$= 2D(u_n) + 2D(u_m) - 4D \left(\frac{u_n + u_m}{2} \right). \qquad (8.1.10)$$

We now have

$$\kappa \leq D \left(\frac{u_n + u_m}{2} \right) \text{ by definition of } \kappa \ ((8.1.8))$$
$$\leq \frac{1}{2}D(u_n) + \frac{1}{2}D(u_m) \text{ by Lemma 8.1.2}$$
$$\to \kappa \quad \text{for } n, m \to \infty, \qquad (8.1.11)$$

since (u_n) is a minimizing sequence. This implies that the right-hand side of (8.1.10) converges to 0 for $n, m \to \infty$, and so then does the left-hand side.

This means that $(\nabla u_n)_{n \in \mathbb{N}}$ is a Cauchy sequence with respect to the topology of the space $L^2(\Omega)$. (Since ∇u_n has d components, i.e., is vector-valued, this says that $\frac{\partial u_n}{\partial x^i}$ is a Cauchy seqeunce in $L^2(\Omega)$ for $i = 1, \ldots, d$.) Since $L^2(\Omega)$ is a Hilbert space, hence complete, ∇u_n thus converges to some $w \in L^2(\Omega)$. The question now is whether w can be represented as the gradient ∇u of some function $u : \Omega \to \mathbb{R}$. At the moment, however, we know only that $w \in L^2(\Omega)$, and so it is not clear what regularity properties u should possess. In any case, this consideration suggests that we seek a minimum of D in the space of those functions whose gradient is in $L^2(\Omega)$. In a subsequent step we would then have to analyze the regularity proprties of such a minimizer u. For that step, the starting point would be relation (8.1.5), i.e.,

$$\int_\Omega \nabla u(x) \cdot \nabla \eta(x) dx = 0 \quad \text{for all } \eta \in C_0^\infty(\Omega), \qquad (8.1.12)$$

which continues to hold in the context presently considered. By Corollary 1.2.1 this already implies $u \in C^\infty(\Omega)$. In the next chapter, however, we shall investigate this problem in greater generality.

Dividing the problem into two steps as just sketched, namely, first proving the existence of a minimizer and afterwards establishing its regularity, proves to be a fruitful approach indeed, as we shall find in the sequel. For that purpose, we first need to investigate the space of functions just considered in more detail. This is the task of the next section.

8.2 The Sobolev Space $W^{1,2}$

Definition 8.2.1: Let $\Omega \subset \mathbb{R}^d$ be open and $u \in L^1_{\text{loc}}(\Omega)$. A function $v \in L^1_{\text{loc}}(\Omega)$ is called weak derivative of u in the direction x^i ($x = (x^1, \ldots, x^d) \in \mathbb{R}^d$) if

$$\int_\Omega \phi v = -\int_\Omega u \frac{\partial \phi}{\partial x^i} dx \qquad (8.2.1)$$

for all $\phi \in C_0^1(\Omega)$.[2] We write $v = D_i u$.

A function u is called weakly differentiable if it possesses a weak derivative in the direction x^i for all $i \in \{1, \ldots, d\}$.

It is obvious that each $u \in C^1(\Omega)$ is weakly differentiable, and the weak derivatives are simply given by the ordinary derivatives. Equation (8.2.1) is then the formula for integrating by parts. Thus, the idea behind the definition of weak derivatives is to use the integration by parts formula as an abstract axiom.

---

[2] $C_0^k(\Omega) := \{f \in C^k(\Omega) : \text{the closure of } \{x : f(x) \neq 0\} \text{ is a compact subset of } \Omega\}$ ($k = 1, 2, \ldots$).

Lemma 8.2.1: *Let $u \in L^1_{loc}(\Omega)$, and suppose $v = D_i u$ exists. If $\operatorname{dist}(x, \partial\Omega)$*
$> h$, we have

$$D_i(u_h(x)) = (D_i u)_h(x).$$

Proof: By differentiating under the integral, we obtain

$$D_i(u_h(x)) = \frac{1}{h^d} \int \frac{\partial}{\partial x^i} \varrho\left(\frac{x-y}{h}\right) u(y) dy$$

$$= \frac{-1}{h^d} \int \frac{\partial}{\partial y^i} \varrho\left(\frac{x-y}{h}\right) u(y) dy$$

$$= \frac{1}{h^d} \int \varrho\left(\frac{x-y}{h}\right) D_i u(y) dy \text{ by } (8.2.1)$$

$$= (D_i u)_h(x).$$

$\square$

Lemmas A.3 and 8.2.1 and formula (8.2.1) imply the following theorem:

Theorem 8.2.1: *Let $u, v \in L^2(\Omega)$. Then*

$$v = D_i u$$

precisely if there exists a sequence $(u_n) \subset C^\infty(\Omega)$ with

$$u_n \to u, \quad \frac{\partial}{\partial x^i} u_n \to v \quad \text{in } L^2(\Omega') \quad \text{for any } \Omega' \subset\subset \Omega.$$

$\square$

Definition 8.2.2: *The Sobolev space $W^{1,2}(\Omega)$ is defined as the space of those $u \in L^2(\Omega)$ that possess a weak derivative of class $L^2(\Omega)$ for each direction x^i $(i = 1, \ldots, d)$.*
   In $W^{1,2}(\Omega)$ we define a scalar product

$$(u, v)_{W^{1,2}(\Omega)} := \int_\Omega uv + \sum_{i=1}^d \int_\Omega D_i u \cdot D_i v$$

and a norm

$$\|u\|_{W^{1,2}(\Omega)} := (u, u)^{\frac{1}{2}}_{W^{1,2}(\Omega)}.$$

We also define $H^{1,2}(\Omega)$ as the closure of $C^\infty(\Omega) \cap W^{1,2}(\Omega)$ with respect to the $W^{1,2}$-norm, and $H^{1,2}_0(\Omega)$ as the closure of $C^\infty_0(\Omega)$ with respect to this norm.

Corollary 8.2.1: *$W^{1,2}(\Omega)$ is complete with respect to $\|\cdot\|_{W^{1,2}}$, and is hence a Hilbert space. $W^{1,2}(\Omega) = H^{1,2}(\Omega)$.*

Proof: Let $(u_n)_{n\in\mathbb{N}}$ be a Cauchy sequence in $W^{1,2}(\Omega)$. Then $(u_n)_{n\in\mathbb{N}}$, $(D_i u_n)_{n\in\mathbb{N}}$ $(i = 1,\ldots,d)$ are Cauchy sequences in $L^2(\Omega)$. Since $L^2(\Omega)$ is complete, there exist $u, v^i \in L^2(\Omega)$ with

$$u_n \to u, \quad D_i u_n \to v^i \quad \text{in } L^2(\Omega) \quad (i = 1,\ldots,d).$$

For $\phi \in C_0^1(\Omega)$, we have

$$\int D_i u_n \cdot \phi = -\int u_n D_i \phi,$$

and the left-hand side converges to $\int v^i \cdot \phi$, the right-hand side to $-\int u \cdot D_i \phi$. Therefore, $D_i u = v^i$, and thus $u \in W^{1,2}(\Omega)$. This shows completeness.

In order to prove the equality $H^{1,2}(\Omega) = W^{1,2}(\Omega)$, we need to verify that the space $C^\infty(\Omega) \cap W^{1,2}(\Omega)$ is dense in $W^{1,2}(\Omega)$. For $n \in \mathbb{N}$, we put

$$\Omega_n := \left\{ x \in \Omega : \|x\| < n, \operatorname{dist}(x, \partial\Omega) > \frac{1}{n} \right\},$$

with $\Omega_0 := \Omega_{-1} := \emptyset$. Thus,

$$\Omega_n \subset\subset \Omega_{n+1} \quad \text{and} \quad \bigcup_{n\in\mathbb{N}} \Omega_n = \Omega.$$

We let $\{\varphi_j\}_{j\in\mathbb{N}}$ be a partition of unity subordinate to the cover

$$\left\{ \Omega_{n+1} \setminus \bar\Omega_{n-1} \right\}$$

of Ω. Let $u \in W^{1,2}(\Omega)$. By Theorem 8.2.1, for every $\varepsilon > 0$, we may find a positive number h_n for any $n \in \mathbb{N}$ such that

$$h_n \leq \operatorname{dist}(\Omega_n, \partial\Omega_{n+1}),$$

$$\|(\varphi_n u)_{h_n} - \varphi_n u\|_{W^{1,2}(\Omega)} < \frac{\varepsilon}{2^n}.$$

Since the φ_n constitute a partition of unity, on any $\Omega' \subset\subset \Omega$, at most finitely many of the smooth functions $(\varphi_n u)_{h_n}$ are non-zero. Consequently,

$$\tilde{u} := \sum_n (\varphi_n u)_{h_n} \in C^\infty(\Omega).$$

We have

$$\|u - \tilde{u}\|_{W^{1,2}(\Omega)} \leq \sum_n \|(\varphi_n u)_{h_n} - \varphi_n u\| < \varepsilon,$$

and we see that every $u \in W^{1,2}(\Omega)$ can be approximated by C^∞-functions.

□

Corollary 8.2.1 answers one of the questions raised in Section 8.1, namely whether the function w considered there can be represented as the gradient of an L^2-function.

Examples: $\Omega = (-1, 1) \subset \mathbb{R}$.

(i) $u(x) := |x|$

In that case, $u \in W^{1,2}((-1, 1))$, and

$$Du(x) = \begin{cases} 1 & \text{for } 0 < x < 1, \\ -1 & \text{for } -1 < x < 0, \end{cases}$$

because for every $\phi \in C_0^1((-1, 1))$,

$$\int_{-1}^0 -\phi(x)dx + \int_0^1 \phi(x)dx = -\int_{-1}^1 \phi'(x) \cdot |x| \, dx.$$

(ii)

$$u(x) := \begin{cases} 1 & \text{for } 0 \leq x < 1, \\ 0 & \text{for } -1 < x < 0, \end{cases}$$

is not weakly differentiable, for if it were, necessarily $Du(x) = 0$ for $x \neq 0$; hence as an L_{loc}^1 function $Du \equiv 0$, but we do not have, for every $\phi \in C_0^1((-1, 1))$,

$$0 = \int_{-1}^1 \phi(x) \cdot 0 \, dx = -\int_{-1}^1 \phi'(x)u(x)dx = -\int_0^1 \phi'(x)dx = \phi(0).$$

Remark: Any $u \in L_{\text{loc}}^1(\Omega)$ defines a distribution (cf. Section 1.1) l_u by

$$l_u[\varphi] := \int_\Omega u(x)\varphi(x)dx \quad \text{for } \varphi \in C_0^\infty(\Omega).$$

Every distribution l possesses distributional derivatives $D_i l$, $i = 1, \ldots, d$, defined by

$$D_i l[\varphi] := -l\left[\frac{\partial \varphi}{\partial x^i}\right].$$

If $v = D_i u \in L_{\text{loc}}^1(\Omega)$ is the weak derivative of u, then

$$D_i l_u = l_v,$$

because

$$l_v[\varphi] = \int_\Omega D_i u(x)\varphi(x)dx = -\int_\Omega u(x)\frac{\partial \varphi}{\partial x^i}(x)dx = D_i l_u[\varphi]$$

for all $\varphi \in C_0^\infty(\Omega)$.

Whereas the distributional derivative $D_i l_u$ always exists, the weak derivative need not exist. Thus, in general, the distributional derivative is not of the form l_v for some $v \in L^1_{loc}(\Omega)$, i.e., not represented by a locally integrable function. This is what happens in Example 2. Here, $Dl_u = \delta_0$, the delta distribution at 0, because

$$Dl_u[\varphi] = -l_u[\varphi'] = -\int_{-1}^{1} \varphi'(x)dx = -\int_{0}^{1} \varphi'(x)dx = \varphi(0).$$

The delta distribution cannot be represented by some locally integrable function v, because, as one easily verifies, there is no function $v \in L^1_{loc}((-1,1))$ with

$$\int_{-1}^{1} v(x)\varphi(x)dx = \varphi(0) \quad \text{for all } \varphi \in C_0^\infty(\Omega).$$

This explains why u from Example 2 is not weakly differentiable.

We now prove a replacement lemma exhibiting a characteristic property of Sobolev functions:

Lemma 8.2.2: *Let* $\Omega_0 \subset\subset \Omega$, $g \in W^{1,2}(\Omega)$, $u \in W^{1,2}(\Omega_0)$, $u - g \in H_0^{1,2}(\Omega_0)$. *Then*

$$v(x) := \begin{cases} u(x) & \text{for } x \in \Omega_0, \\ g(x) & \text{for } x \in \Omega \setminus \Omega_0, \end{cases}$$

is contained in $W^{1,2}(\Omega)$, *and*

$$D_i v(x) = \begin{cases} D_i u(x) & \text{for } x \in \Omega_0, \\ D_i g(x) & \text{for } x \in \Omega \setminus \Omega_0. \end{cases}$$

Proof: By Corollary 8.2.1, there exist $g_n \in C^\infty(\Omega)$, $u_n \in C^\infty(\Omega_0)$ with

$$g_n \to g \quad \text{in } W^{1,2}(\Omega),$$
$$u_n \to u \quad \text{in } W^{1,2}(\Omega_0),$$
$$u_n - g_n = 0 \quad \text{on } \partial\Omega_0. \tag{8.2.2}$$

We put

$$w_n^i(x) := \begin{cases} D_i u_n(x) & \text{for } x \in \Omega_0, \\ D_i g_n(x) & \text{for } x \in \Omega \setminus \Omega_0, \end{cases}$$

$$v_n(x) := \begin{cases} u_n(x) & \text{for } x \in \Omega_0, \\ g_n(x) & \text{for } x \in \Omega \setminus \Omega_0, \end{cases}$$

$$w^i(x) := \begin{cases} D_i u(x) & \text{for } x \in \Omega_0, \\ D_i g(x) & \text{for } x \in \Omega \setminus \Omega_0. \end{cases}$$

We then have for $\varphi \in C_0^1(\Omega)$,

$$\int_\Omega \varphi w_n^i = \int_{\Omega_0} \varphi w_n^i + \int_{\Omega \setminus \Omega_0} \varphi w_n^i = \int_{\Omega_0} \varphi D_i u_n + \int_{\Omega \setminus \Omega_0} \varphi D_i g_n$$

$$= - \int_{\Omega_0} u_n D_i \varphi - \int_{\Omega \setminus \Omega_0} g_n D_i \varphi$$

since the two boundary terms resulting from integrating the two integrals by parts have opposite signs and thus cancel because of $g_n = u_n$ on $\partial \Omega_0$

$$= - \int_\Omega v_n D_i \varphi$$

by (8.2.2). Now for $n \to \infty$,

$$\int_\Omega \varphi w_n^i \to \int_{\Omega_0} \varphi D_i u + \int_{\Omega \setminus \Omega_0} \varphi D_i g,$$

$$\int_\Omega v_n D_i \varphi \to \int_\Omega v D_i \varphi,$$

and the claim follows.   □

The next lemma is a chain rule for Sobolev functions:

Lemma 8.2.3: *For $u \in W^{1,2}(\Omega)$, $f \in C^1(\mathbb{R})$, suppose*

$$\sup_{y \in \mathbb{R}} |f'(y)| < \infty.$$

Then $f \circ u \in W^{1,2}(\Omega)$, and the weak derivative satisfies $D(f \circ u) = f'(u) Du$.

Proof: Let $u_n \in C^\infty(\Omega)$, $u_n \to u$ in $W^{1,2}(\Omega)$ for $n \to \infty$. Then

$$\int_\Omega |f(u_n) - f(u)|^2 \, dx \le \sup |f'|^2 \int_\Omega |u_n - u|^2 \, dx \to 0$$

and

$$\int_\Omega |f'(u_n) Du_n - f'(u) Du|^2 \, dx \le 2 \sup |f'|^2 \int_\Omega |Du_n - Du|^2 \, dx$$

$$+ 2 \int_\Omega |f'(u_n) - f'(u)|^2 |Du|^2 \, dx.$$

By a well-known result about L^2-functions, after selection of a subsequence, u_n converges to u pointwise almost everywhere in Ω.[3] Since f' is continuous, $f'(u_n)$ then also converges pointwise almost everywhere to $f'(u)$, and since

---

[3] See J. Jost, *Postmodern Analysis*, p. 240 [12].

f' is also bounded, the last integral converges to 0 for $n \to \infty$ by Lebesgue's theorem on dominated convergence.

Thus

$$f(u_n) \to f(u) \quad \text{in } L^2(\Omega)$$

and

$$D(f(u_n)) = f'(u_n)Du_n \to f'(u)Du \quad \text{in } L^2(\Omega),$$

and hence $f \circ u \in W^{1,2}(\Omega)$ and $D(f \circ u) = f'(u)Du$. $\qquad\square$

Corollary 8.2.2: *If $u \in W^{1,2}(\Omega)$, then also $|u| \in W^{1,2}(\Omega)$, and $D|u| = \text{sign } u \cdot Du$.*

Proof: We consider $f_\varepsilon(u) := (u^2 + \varepsilon^2)^{\frac{1}{2}} - \varepsilon$, apply Lemma 8.2.3, and let $\varepsilon \to 0$, using once more Lebesgue's theorem on dominated convergence to justify the limit as before. $\qquad\square$

We next prove the Poincaré inequality (see also Corollary 9.5.1 below).

Theorem 8.2.2: *For $u \in H_0^{1,2}(\Omega)$, we have*

$$\|u\|_{L^2(\Omega)} \leq \left(\frac{|\Omega|}{\omega_d}\right)^{\frac{1}{d}} \|Du\|_{L^2(\Omega)}, \tag{8.2.3}$$

where $|\Omega|$ denotes the (Lebesgue) measure of Ω, and ω_d is the measure of the unit ball in $\mathbb{R}^d$. In particular, for any $u \in H_0^{1,2}(\Omega)$, its $W^{1,2}$-norm is controlled by the L^2-norm of Du:

$$\|u\|_{W^{1,2}(\Omega)} \leq \left(1 + \left(\frac{|\Omega|}{\omega_d}\right)^{\frac{1}{d}}\right) \|Du\|_{L^2(\Omega)}.$$

Proof: Suppose first $u \in C_0^1(\Omega)$; we put $u(x) = 0$ for $x \in \mathbb{R}^d \setminus \Omega$. For $\omega \in \mathbb{R}^d$ with $|\omega| = 1$, by the fundamental theorem of calculus we obtain by integrating along the ray $\{r\omega : 0 \leq r < \infty\}$ that

$$u(x) = -\int_0^\infty \frac{\partial}{\partial r} u(x + r\omega)dr.$$

Integrating with respect to ω then yields, as in the proof of Theorem 1.2.1,

$$\begin{aligned}
u(x) &= -\frac{1}{d\omega_d} \int_0^\infty \int_{|\omega|=1} \frac{\partial}{\partial r} u(x + r\omega)\, d\omega dr \\
&= -\frac{1}{d\omega_d} \int_0^\infty \int_{\partial B(x,r)} \frac{1}{r^{d-1}} \frac{\partial u}{\partial \nu}(z)d\sigma(z)dr \\
&= -\frac{1}{d\omega_d} \int_\Omega \frac{1}{|x-y|^{d-1}} \sum_{i=1}^d \frac{\partial}{\partial y^i} u(y) \frac{x^i - y^i}{|x-y|} dy,
\end{aligned} \tag{8.2.4}$$

and thus with the Schwarz inequality,

$$|u(x)| \leq \frac{1}{d\omega_d} \int_\Omega \frac{1}{|x-y|^{d-1}} \cdot |Du(y)| \, dy. \tag{8.2.5}$$

We now need a lemma:

Lemma 8.2.4: *For $f \in L^1(\Omega)$, $0 < \mu \leq 1$, let*

$$(V_\mu f)(x) := \int_\Omega |x-y|^{d(\mu-1)} f(y) dy.$$

Then

$$\|V_\mu f\|_{L^2(\Omega)} \leq \frac{1}{\mu} \omega_d^{1-\mu} |\Omega|^\mu \|f\|_{L^2(\Omega)}.$$

Proof: $B(x,R) := \{y \in \mathbb{R}^d : |x-y| \leq R\}$. Let R be chosen such that $|\Omega| = |B(x,R)| = \omega_d R^d$. Since in that case

$$|\Omega \setminus (\Omega \cap B(x,R))| = |B(x,R) \setminus (\Omega \cap B(x,R))|$$

and

$$|x-y|^{d(\mu-1)} \leq R^{d(\mu-1)} \quad \text{for } |x-y| \geq R,$$
$$|x-y|^{d(\mu-1)} \geq R^{d(\mu-1)} \quad \text{for } |x-y| \leq R,$$

it follows that

$$\int_\Omega |x-y|^{d(\mu-1)} \, dy \leq \int_{B(x,R)} |x-y|^{d(\mu-1)} \, dy = \frac{1}{\mu}\omega_d R^{d\mu} = \frac{1}{\mu}\omega_d^{1-\mu} |\Omega|^\mu. \tag{8.2.6}$$

We now write

$$|x-y|^{d(\mu-1)} |f(y)| = \left(|x-y|^{\frac{d}{2}(\mu-1)}\right)\left(|x-y|^{\frac{d}{2}(\mu-1)} |f(y)|\right)$$

and obtain, applying the Cauchy Schwarz inequality,

$$|(V_\mu f)(x)| \leq \int_\Omega |x-y|^{d(\mu-1)} |f(y)| \, dy$$
$$\leq \left(\int_\Omega |x-y|^{d(\mu-1)} \, dy\right)^{\frac{1}{2}} \left(\int_\Omega |x-y|^{d(\mu-1)} |f(y)|^2 \, dy\right)^{\frac{1}{2}},$$

and hence

$$\int_\Omega |V_\mu f(x)|^2 \, dx \leq \frac{1}{\mu}\omega_d^{1-\mu}|\Omega|^\mu \int_\Omega \int_\Omega |x-y|^{d(\mu-1)}|f(y)|^2 \, dy \, dx$$

by estimating the first integral of the preceding inequality with (8.2.6)

$$\leq \left(\frac{1}{\mu}\omega_d^{1-\mu}|\Omega|^\mu\right)^2 \int_\Omega |f(y)|^2 dy$$

by interchanging the integrations with respect to x and y and applying (8.2.6) once more, whence the claim.   □

We may now complete the *proof of Theorem 8.2.2*: Applying Lemma 8.2.4 with $\mu = \frac{1}{d}$ and $f = |Du|$ to the right-hand side of (8.2.5), we obtain (8.2.3) for $u \in C_0^1(\Omega)$. Since by definition of $H_0^{1,2}(\Omega)$, it contains $C_0^1(\Omega)$ as a dense subspace, we may approximate u in the $H^{1,2}$-norm by some sequence $(u_n)_{n \in \mathbb{N}} \subset C_0^1(\Omega)$. Thus, u_n converges to u in L^2, and Du_n to u. Thus, the inequality (8.2.3) that has been proved for u_n extends to u.   □

Remark: The assumption that u is contained in $H_0^{1,2}(\Omega)$, and not only in $H^{1,2}(\Omega)$, is necessary for Theorem 8.2.2, since otherwise the nonzero constants would constitute counterexamples. However, the assumption $u \in H_0^{1,2}(\Omega)$ may be replaced by other assumptions that exclude nonzero constants, for example by $\int_\Omega u(x)dx = 0$.

For our treatment of eigenvalues of the Laplace operator in Section 9.5, the fundamental tool will be the compactness theorem of Rellich:

Theorem 8.2.3: *Let $\Omega \in \mathbb{R}^d$ be open and bounded. Then $H_0^{1,2}(\Omega)$ is compactly embedded in $L^2(\Omega)$; i.e., any sequence $(u_n)_{n \in \mathbb{N}} \subset H_0^{1,2}(\Omega)$ with*

$$\|u_n\|_{W^{1,2}(\Omega)} \leq c_0 \tag{8.2.7}$$

contains a subsequence that converges in $L^2(\Omega)$.

Proof: The strategy is to find functions $w_{n,\varepsilon} \in C^1(\Omega)$, for every $\varepsilon > 0$, with

$$\|u_n - w_{n,\varepsilon}\|_{W^{1,2}(\Omega)} < \frac{\varepsilon}{2} \tag{8.2.8}$$

and

$$\|w_{n,\varepsilon}\|_{W^{1,2}(\Omega)} \leq c_1 \tag{8.2.9}$$

(the constant c_1 will depend on ε, but not on n). By the Aszela–Ascoli theorem, $(w_{n,\varepsilon})_{n \in \mathbb{N}}$ then contains a subsequence that converges uniformly, hence also in L^2. Since this holds for every $\varepsilon > 0$, one may appeal to a general theorem about compact subsets of metric spaces to conclude that the closure of $(u_n)_{n \in \mathbb{N}}$ is compact in $L^2(\Omega)$ and thus contains a convergent subsequence. That theorem[4] states that a subset of a metric space is compact precisely if it is complete and totally bounded, i.e., if for any $\varepsilon > 0$, it is contained in the union of a finite number of balls of radius ε.

Applying this result to the (closure of the) sequence $(w_{n,\varepsilon})_{n \in \mathbb{N}}$, we infer that there exist finitely many z_ν, $\nu = 1, \ldots, N$, in $L^2(\Omega)$ such that for every $n \in \mathbb{N}$,

---

[4] see, e.g., J. Jost, *Postmodern Analysis*, Springer, 1998, Theorem 7.38.

$$\|w_{n,\varepsilon} - z_\nu\|_{L^2(\Omega)} < \frac{\varepsilon}{2} \quad \text{for some } \nu \in \{1, \ldots, N\}. \tag{8.2.10}$$

Hence, from (8.2.8) and (8.2.10), for every $n \in \mathbb{N}$,

$$\|u_n - z_\nu\|_{L^2(\Omega)} < \varepsilon \quad \text{for some } \nu.$$

Since this holds for every $\varepsilon > 0$, the sequence $(u_n)_{n \in \mathbb{N}}$ is totally bounded, and so its closure is compact in $L^2(\Omega)$, and we get the desired convergent subsequence in $L^2(\Omega)$.

It remains to construct the $w_{n,\varepsilon}$. First of all, by definition of $H_0^{1,2}(\Omega)$, there exists $w_n \in C_0^1(\Omega)$ with

$$\|u_n - w_n\|_{W^{1,2}(\Omega)} < \frac{\varepsilon}{4}. \tag{8.2.11}$$

By (8.2.7), then also

$$\|w_n\|_{W^{1,2}(\Omega)} \le c_0' \quad \text{for some constant } c_0'. \tag{8.2.12}$$

We then define $w_{n,\varepsilon}$ as the mollification of w_n with a parameter $h = h(\varepsilon)$ to be determined subsequently:

$$w_{n,\varepsilon}(x) = \frac{1}{h^d} \int_\Omega \varrho\left(\frac{x-y}{h}\right) w_n(y) dy.$$

The crucial step now is to control the L^2-norm of the difference $w_n - w_{n,\varepsilon}$ with the help of the $W^{1,2}$-bound on the original u_n. This goes as follows:

$$\int_\Omega |w_n(x) - w_{n,\varepsilon}(x)|^2 dx = \int_\Omega \left(\int_{|y| \le 1} \varrho(y)(w_n(x) - w_n(x - hy)) dy \right)^2 dx$$

$$\le \int_\Omega \left(\int_{|y| \le 1} \varrho(y) \int_0^{h|y|} \left| \frac{\partial}{\partial r} w_n(x - r\omega) \right| dr\, dy \right)^2 dx \quad \text{with } \omega = \frac{y}{|y|}$$

$$= \int_\Omega \left(\int_{|y| \le 1} \varrho(y)^{\frac{1}{2}} \varrho(y)^{\frac{1}{2}} \int_0^{h|y|} \left| \frac{\partial}{\partial r} w_n(x - r\omega) \right| dr\, dy \right)^2 dx$$

$$\le \left(\int_{|y| \le 1} \varrho(y) dy \right) \left(\int_{|y| \le 1} \varrho(y) h^2 |y|^2 \int |Dw_n(x)|^2 dx\, dy \right)$$

by Hölder's inequality ((A.4) of the Appendix) and Fubini's theorem. Since $\int_{|y| \le 1} \varrho(y) dy = 1$, we obtain the estimate

$$\|w_n - w_{n,\varepsilon}\|_{L^2(\Omega)} \le h \|Dw_n\|_{L^2(\Omega)}.$$

Because of (8.2.12), we may then choose h such that

$$\|w_n - w_{n,\varepsilon}\|_{L^2(\Omega)} < \frac{\varepsilon}{4}. \tag{8.2.13}$$

Then (8.2.11) and (8.2.13) yield the desired estimate (8.2.8). $\square$

8.3 Weak Solutions of the Poisson Equation

As before, let Ω be an open and bounded subset of $\mathbb{R}^d$, $g \in H^{1,2}(\Omega)$. With the concepts introduced in the previous section, we now consider the following version of the Dirichlet principle. We seek a solution of

$$\Delta u = 0 \quad \text{in } \Omega,$$

$$u = g \quad \text{for } \partial\Omega \quad \left(\text{meaning } u - g \in H_0^{1,2}(\Omega) \right),$$

by minimizing the Dirichlet integral

$$\int_\Omega |Dv|^2 \quad (\text{here, } Dv = (D_1 v, \ldots, D_d v))$$

among all $v \in H^{1,2}(\Omega)$ with $v - g \in H_0^{1,2}(\Omega)$. We want to convince ourselves that this approach indeed works. Let

$$\kappa := \inf \left\{ \int_\Omega |Dv|^2 : v \in H^{1,2}(\Omega), v - g \in H_0^{1,2}(\Omega) \right\},$$

and let $(u_n)_{n\in\mathbb{N}}$ be a minimizing sequence, meaning that $u_n - g \in H_0^{1,2}(\Omega)$, and

$$\int_\Omega |Du_n|^2 \to \kappa.$$

We have already argued in Section 8.1 that for a minimizing sequence $(u_n)_{n\in\mathbb{N}}$, the sequence of (weak) derivatives (Du_n) is a Cauchy sequence in $L^2(\Omega)$. Theorem 8.2.2 implies

$$\|u_n - u_m\|_{L^2(\Omega)} \leq \text{const} \, \|Du_n - Du_m\|_{L^2(\Omega)} .$$

Thus, (u_n) also is a Cauchy sequence in $L^2(\Omega)$. We conclude that $(u_n)_{n\in\mathbb{N}}$ converges in $W^{1,2}(\Omega)$ to some u. This u satisfies

$$\int_\Omega |Du|^2 = \kappa$$

as well as

$$u - g \in H_0^{1,2}(\Omega),$$

because $H_0^{1,2}(\Omega)$ is a closed subspace of $W^{1,2}(\Omega)$. Furthermore, for every $v \in H_0^{1,2}(\Omega)$, $t \in \mathbb{R}$, putting $Du \cdot Dv := \sum_{i=1}^d D_i u \cdot D_i v$, we have

$$\kappa \leq \int_\Omega |D(u + tv)|^2 = \int_\Omega |Du|^2 + 2t \int_\Omega Du \cdot Dv + t^2 \int_\Omega |Dv|^2 ,$$

and differentiating with respect to t at $t = 0$ yields

$$0 = \frac{d}{dt} \int_\Omega |D(u + tv)|^2 |_{t=0} = 2 \int_\Omega Du \cdot Dv \quad \text{for all } v \in H_0^{1,2}(\Omega).$$

Definition 8.3.1: *A function $u \in H^{1,2}(\Omega)$ is called weakly harmonic, or a weak solution of the Laplace equation, if*

$$\int_\Omega Du \cdot Dv = 0 \quad \text{for all } v \in H_0^{1,2}(\Omega). \tag{8.3.1}$$

Any harmonic function obviously satisfies (8.3.1). In order to obtain a harmonic function from the Dirichlet principle one has to show that, conversely, any solution of (8.3.1) is twice continuously differentiable, hence harmonic. In the present case, this follows directly from Corollary 1.2.1:

Corollary 8.3.1: *Any weakly harmonic function is smooth and harmonic. In particular, applying the Dirichlet principle yields harmonic functions. More precisely, for any open and bounded Ω in $\mathbb{R}^d$, $g \in H^{1,2}(\Omega)$, there exists a function $u \in H^{1,2}(\Omega) \cap C^\infty(\Omega)$ with*

$$\Delta u = 0 \quad in \ \Omega$$

and

$$u - g \in H_0^{1,2}(\Omega).$$

The proof of Corollary 8.3.1 depends on the rotational invariance of the Laplace operator and therefore cannot be generalized. For that reason, in the sequel, we want to develop a more general approach to regularity theory. Before turning to that theory, however, we wish to slightly extend the situation just considered.

Definition 8.3.2: *Let $f \in L^2(\Omega)$. A function $u \in H^{1,2}(\Omega)$ is called a weak solution of the Poisson equation $\Delta u = f$ if for all $v \in H_0^{1,2}(\Omega)$,*

$$\int_\Omega Du \cdot Dv + \int_\Omega fv = 0. \tag{8.3.2}$$

Remark: For given boundary values g (meaning $u - g \in H_0^{1,2}(\Omega)$), a solution can be obtained by minimizing

$$\frac{1}{2} \int_\Omega |Dw|^2 + \int_\Omega fw$$

inside the class of all $w \in H^{1,2}(\Omega)$ with $w - g \in H_0^{1,2}(\Omega)$. Note that this expression is bounded from below by the Poincaré inequality (Theorem 8.2.2), because we are assuming fixed boundary values g.

Lemma 8.3.1 (stability lemma): *Let $u_{i=1,2}$ be a weak solution of $\Delta u_i = f_i$ with $u_1 - u_2 \in H_0^{1,2}(\Omega)$. Then*

$$\|u_1 - u_2\|_{W^{1,2}(\Omega)} \leq \text{const } \|f_1 - f_2\|_{L^2(\Omega)}.$$

In particular, a weak solution of $\Delta u = f$, $u - g \in H_0^{1,2}(\Omega)$ is uniquely determined.

Proof: We have

$$\int_\Omega D(u_1 - u_2)Dv = -\int_\Omega (f_1 - f_2)v \quad \text{for all } v \in H_0^{1,2}(\Omega),$$

and thus in particular,

$$
\begin{aligned}
\int_\Omega D(u_1 - u_2)D(u_1 - u_2) &= -\int_\Omega (f_1 - f_2)(u_1 - u_2) \\
&\leq \|f_1 - f_2\|_{L^2(\Omega)} \|u_1 - u_2\|_{L^2(\Omega)} \\
&\leq \text{const } \|f_1 - f_2\|_{L^2(\Omega)} \|Du_1 - Du_2\|_{L^2(\Omega)}
\end{aligned}
$$

by Theorem 8.2.2, and hence

$$\|Du_1 - Du_2\|_{L^2(\Omega)} \leq \text{const } \|f_1 - f_2\|_{L^2(\Omega)}.$$

The claim follows by applying Theorem 8.2.2 once more.   □

We have thus obtained the existence and uniqueness of weak solutions of the Poisson equation in a very simple manner. The task of regularity theory then consists in showing that (for sufficiently well behaved f) a weak solution is of class C^2 and thus also a classical solution of $\Delta u = f$.

We shall present three different methods, namely the so-called L^2-theory, the theory of strong solutions, and the C^α-theory. The L^2-theory will be developed in Chapter 9, the theory of strong solutions in Chapter 10, and the C^α-theory in Chapter 11.

8.4 Quadratic Variational Problems

We may ask whether the Dirichlet principle can be generalized to obtain solutions of other PDEs. In general, of course, a minimizer u of some variational problem has to satisfy the corresponding Euler–Lagrange equations, first in the weak sense, and if u is regular, also in the classical sense. In the general case, however, regularity theory encounters obstacles, and weak solutions of Euler–Lagrange equations need not always be regular. We therefore restrict ourselves to quadratic variational problems and consider

$$
\begin{aligned}
I(u) := \int_\Omega \Bigg\{ &\sum_{i,j=1}^d a^{ij}(x)D_i u(x)D_j u(x) \\
&+ 2\sum_{j=1}^d b^j(x)D_j u(x)u(x) + c(x)u(x)^2 \Bigg\} \, dx.
\end{aligned}
\tag{8.4.1}
$$

We require the symmetry condition $a^{ij} = a^{ji}$ for all i, j. In addition, the coefficients $a^{ij}(x)$, $b^j(x)$, $c(x)$ should all be bounded. Then $I(u)$ is defined for $u \in H^{1,2}(\Omega)$. As before, we compute, for $\varphi \in H_0^{1,2}(\Omega)$,

$$I(u + t\varphi) = I(u)$$
$$+ 2t \int_\Omega \left\{ \sum_{i,j} a^{ij} D_i u D_j \varphi + \sum_j b^j u D_j \varphi + \left(\sum_j b^j D_j u + cu \right) \varphi \right\} dx$$
$$+ t^2 I(\varphi). \quad (8.4.2)$$

A minimizer u thus satisfies, as before,

$$\frac{d}{dt} I(u + t\varphi)|_{t=0} = 0 \quad \text{for all } \varphi \in H_0^{1,2}(\Omega); \quad (8.4.3)$$

hence

$$\int_\Omega \left\{ \sum_j \left(\sum_i a^{ij} D_i u + b^j u \right) D_j \varphi + \left(\sum_j b^j D_j u + cu \right) \varphi \right\} dx = 0 \quad (8.4.4)$$

for all $\varphi \in H_0^{1,2}(\Omega)$.

If $u \in C^2(\Omega)$ and $a^{ij}, b^j \in C^1(\Omega)$, then (8.4.4) implies the differential equation

$$\sum_{j=1}^d \frac{\partial}{\partial x^j} \left(\sum_{i=1}^d a^{ij}(x) \frac{\partial u}{\partial x^i} + b^j(x) u \right) - \sum_{j=1}^d b^j(x) \frac{\partial u}{\partial x^j} - c(x) u = 0. \quad (8.4.5)$$

As the Euler–Lagrange equation of a quadratic variational integral, we thus obtain a linear PDE of second order. This equation is elliptic when we assume that the matrix $(a^{ij}(x))_{i,j=1,\ldots,d}$ is positive definite at every $x \in \Omega$.

In the next chapter we should see that weak solutions of (8.4.5) (i.e., solutions of (8.4.4)) are regular, provided that appropriate assumptions for the coefficients a^{ij}, b^j, c hold. The direct method of the calculus of variations, as this generalization of the Dirichlet principle is called, consists in finding a weak solution of (8.4.5) by minimizing $I(u)$, and then demonstrating its regularity. We finally wish to study the transformation behavior of the Dirichlet integral and the Laplace operator with respect to changes of the independent variables. We shall also need that transformation rule for our investigation of boundary regularity in the next chapter.

Thus let

$$\xi \to x(\xi)$$

be a diffeomorphism from Ω' to Ω. We put

$$g_{ij} := \sum_{\alpha=1}^{d} \frac{\partial x^\alpha}{\partial \xi^i} \frac{\partial x^\alpha}{\partial \xi^j}, \tag{8.4.6}$$

$$g^{ij} := \sum_{\alpha=1}^{d} \frac{\partial \xi^i}{\partial x^\alpha} \frac{\partial \xi^j}{\partial x^\alpha}, \tag{8.4.7}$$

i.e.,

$$\sum_{k=1}^{d} g_{ki} g^{kj} = \delta_{ij} = \begin{cases} 1 & \text{for } i = j, \\ 0 & \text{for } i \neq j, \end{cases}$$

and

$$g := \det \left(g_{ij} \right)_{i,j=1,\ldots,d}. \tag{8.4.8}$$

We then have, for $u(\xi(x))$,

$$\sum_{\alpha=1}^{d} \left(\frac{\partial u}{\partial x^\alpha} \right)^2 = \sum_{\alpha=1}^{d} \sum_{i,j=1}^{d} \frac{\partial u}{\partial \xi^i} \frac{\partial \xi^i}{\partial x^\alpha} \frac{\partial u}{\partial \xi^j} \frac{\partial \xi^j}{\partial x^\alpha} = \sum_{i,j=1}^{d} g^{ij} \frac{\partial u}{\partial \xi^i} \frac{\partial u}{\partial \xi^j}. \tag{8.4.9}$$

The Dirichlet integral thus transforms via

$$\int_\Omega \sum_{\alpha=1}^{d} \left(\frac{\partial u}{\partial x^\alpha} \right)^2 dx = \int_{\Omega'} \sum_{i,j=1}^{d} g^{ij} \frac{\partial u}{\partial \xi^i} \frac{\partial u}{\partial \xi^j} \sqrt{g} d\xi. \tag{8.4.10}$$

By (8.4.5), the Euler–Lagrange equation for the integral on the right-hand side is

$$\frac{1}{\sqrt{g}} \sum_{j=1}^{d} \left(\frac{\partial}{\partial \xi^j} \left(\sqrt{g} \sum_{i=1}^{d} g^{ij} \frac{\partial u}{\partial \xi^i} \right) \right) = 0, \tag{8.4.11}$$

where we have added the normalization factor $1/\sqrt{g}$. This means that under our substitution $x = x(\xi)$ of the independent variables, the Laplace equation, i.e., the Euler–Lagrange equation for the Dirichlet integral, is transformed into (8.4.11).

Likewise, (8.4.5) is transformed into

$$\frac{1}{\sqrt{g}} \sum_{j=1}^{d} \frac{\partial}{\partial \xi^j} \left(\sqrt{g} \left(\sum_{i,\alpha,\beta=1}^{d} a^{\alpha\beta}(x) \frac{\partial \xi^i}{\partial x^\alpha} \frac{\partial \xi^j}{\partial x^\beta} \frac{\partial}{\partial \xi^i} u + \sum_\alpha b^\alpha(x) \frac{\partial \xi^j}{\partial x^\alpha} u \right) \right)$$
$$- \sum_{j,\alpha} b^\alpha(x) \frac{\partial \xi^j}{\partial x^\alpha} \frac{\partial u}{\partial \xi^j} - c(x) u = 0, \tag{8.4.12}$$

where $x = x(\xi)$ has to be inserted, of course.

8.5 Abstract Hilbert Space Formulation of the Variational Problem. The Finite Element Method

The present section presents an abstract version of the approach described in Section 8.3 together with a method for constructing an approximate solution.

We again set out from from some model problem, the Poisson equation with homogeneous boundary data

$$\Delta u = f \quad \text{in } \Omega,$$
$$u = 0 \quad \text{on } \partial\Omega. \tag{8.5.1}$$

In Definition 8.3.2 we introduced a weak version of that problem, namely the problem of finding a solution u in the Hilbert space $H_0^{1,2}(\Omega)$ of

$$\int_\Omega Du\, D\varphi + \int_\Omega f\varphi = 0 \quad \text{for all } \varphi \in H_0^{1,2}(\Omega). \tag{8.5.2}$$

This problem can be generalized as an abstract Hilbert space problem that we now wish to describe:

Definition 8.5.1: *Let $(H, (\cdot, \cdot))$ be a Hilbert space with associated norm $\|\cdot\|$, $A : H \times H \to \mathbb{R}$ a continuous symmetric bilinear form. Here, continuity means that there exists a constant C such that for all $u, v \in H$,*

$$A(u, v) \leq C \|u\| \|v\|.$$

Symmetry means that for all $u, v \in H$,

$$A(u, v) = A(v, u).$$

The form A is called elliptic, or coercive, if there exists a positive λ such that for all $v \in H$,

$$A(v, v) \geq \lambda \|v\|^2. \tag{8.5.3}$$

In our example, $H = H_0^{1,2}(\Omega)$, and

$$A(u, v) = \frac{1}{2} \int_\Omega Du \cdot Dv. \tag{8.5.4}$$

Symmetry is obvious here, continuity follows from Hölder's inequality, and ellipticity results from

$$\frac{1}{2} \int Du \cdot Du = \frac{1}{2} \|Du\|_{L^2(\Omega)}^2$$

and the Poincaré inequality (Theorem 8.2.2), which implies for $u \in H_0^{1,2}(\Omega)$,

$$\|u\|_{H_0^{1,2}(\Omega)} \leq \text{const } \|Du\|_{L^2(\Omega)}.$$

Moreover, for $f \in L^2(\Omega)$,

$$L : H_0^{1,2}(\Omega) \to \mathbb{R}, \qquad v \mapsto \int_\Omega fv,$$

yields a continuous linear map on $H_0^{1,2}(\Omega)$ (even on $L^2(\Omega)$).
Namely,

$$\|L\| := \sup_{v \neq 0} \frac{|Lv|}{\|v\|_{W^{1,2}(\Omega)}} \leq \|f\|_{L^2(\Omega)},$$

for by Hölder's inequality,

$$\int_\Omega fv \leq \|f\|_{L^2(\Omega)} \|v\|_{L^2(\Omega)} \leq \|f\|_{L^2(\Omega)} \|v\|_{W^{1,2}(\Omega)}.$$

Of course, the purpose of Definition 8.5.1 is to isolate certain abstract assumptions that allow us to treat not only the Dirichlet integral, but also more general variational problems as considered in Section 8.4. However, we do need to impose certain restrictions, in particular for satisfying the ellipticity condition. We consider

$$A(u, v) := \frac{1}{2} \int_\Omega \left\{ \sum_{i,j=1}^d a^{ij}(x) D_i u(x) D_j v(x) + c(x) u(x) v(x) \right\} dx,$$

with $u, v \in H = H_0^{1,2}(\Omega)$, where we assume:

(A) Symmetry:
$$a^{ij}(x) = a^{ji}(x) \quad \text{for all } i, j, \text{ and } x \in \Omega.$$

(B) Ellipticity: There exists $\lambda > 0$ with

$$\sum_{i,j=1}^d a^{ij}(x) \xi_i \xi_j \geq \lambda |\xi|^2 \quad \text{for all } x \in \Omega, \xi \in \mathbb{R}^d.$$

(C) Boundedness: There exists $\Lambda < \infty$ with

$$|c(x)|, |a^{ij}| \leq \Lambda \quad \text{for all } i, j, \text{ and } x \in \Omega.$$

(D) Nonnegativity:
$$c(x) \geq 0 \quad \text{for all } x \in \Omega.$$

The ellipticity condition (B) and the nonnegativity (D) imply that

$$A(v, v) \geq \frac{1}{2} \lambda \int_\Omega Dv \cdot Dv \quad \text{for all } v \in H_0^{1,2}(\Omega),$$

and using the Poincaré inequality, we obtain

$$A(v, v) \geq \frac{\lambda}{2} \|v\|_{H^{1,2}(\Omega)} \quad \text{for all } v \in H_0^{1,2}(\Omega);$$

i.e., A is elliptic in the sense of Definition 8.5.1. The continuity of A of course follows from the boundedness condition (C), and the symmetry is condition (A).

Theorem 8.5.1: *Let $(H, (\cdot, \cdot))$ be a Hilbert space with norm $\|\cdot\|$, $V \subset H$ convex and closed, $A : H \times H \to \mathbb{R}$ a continuous symmetric elliptic bilinear form, $L : H \to \mathbb{R}$ a continuous linear map. Then*

$$J(v) := A(v, v) + L(v)$$

has precisely one minimizer u in V.

Remark: The solution u depends not only on A and L, but also on V, for it solves the problem

$$J(u) = \inf_{v \in V} J(v).$$

Proof: By ellipticity of A, J is bounded from below; namely,

$$J(v) \geq \lambda \|v\|^2 - \|L\| \, \|v\| \geq -\frac{\|L\|^2}{4\lambda}.$$

We put

$$\kappa := \inf_{v \in V} J(v).$$

Now let $(u_n)_{n \in \mathbb{N}} \subset V$ be a minimizing sequence, i.e.,

$$\lim_{n \to \infty} J(u_n) = \kappa. \tag{8.5.5}$$

We claim that $(u_n)_{n \in \mathbb{N}}$ is a Cauchy sequence, from which we then deduce, since V is closed, the existence of a limit

$$u = \lim_{n \to \infty} u_n \in V.$$

The Cauchy property is verified as follows: By definition of κ,

$$\kappa \leq J\left(\frac{u_n + u_m}{2}\right) = \frac{1}{2} J(u_n) + \frac{1}{2} J(u_m) - \frac{1}{4} A(u_n - u_m, u_n - u_m).$$

(Here, we have used that if u_n and u_m are in V, so is $\frac{u_n + u_m}{2}$, because V is convex.)

Since $J(u_n)$ and $J(u_m)$ by (8.4.5) for $n, m \to \infty$ both converge to κ, we deduce that

$$A(u_n - u_m, u_n - u_m)$$

converges to 0 for $n, m \to \infty$. Ellipticity then implies that $\|u_n - u_m\|$ converges to 0 as well, and hence the Cauchy property.

Since J is continuous, the limit u satisfies

$$J(u) = \lim_{n \to \infty} J(u_n) = \inf_{v \in V} J(v)$$

by the choice of the sequence $(u_n)_{n \in \mathbb{N}}$.

The preceding proof yields uniqueness of u, too. It is instructive, however, to see this once more as a consequence of the convexity of J: Thus, let u_1, u_2 be two minimizers, i.e.,

$$J(u_1) = J(u_2) = \kappa = \inf_{v \in V} J(v).$$

Since together with u_1 and u_2, $\frac{u_1 + u_2}{2}$ is also contained in the convex set V, we have

$$\kappa \leq J(\frac{u_1 + u_2}{2}) = \frac{1}{2}J(u_1) + \frac{1}{2}J(u_2) - \frac{1}{4}A(u_1 - u_2, u_1 - u_2)$$

$$= \kappa - \frac{1}{4}A(u_1 - u_2, u_1 - u_2),$$

and thus $A(u_1 - u_2, u_1 - u_2) = 0$, which by ellipticity of A implies $u_1 = u_2$.

$\square$

Remark: Theorem 8.5.1 remains true without the symmetry assumption for A. This is the content of the Lax–Milgram theorem, proved in Appendix A.

This remark allows us also to treat variational integrands that in addition to the symmetric terms

$$\sum_{i,j=1}^{d} a^{ij}(x) D_i D_j v(x) \quad (a^{ij} = a^{ji})$$

and $c(x)u(x)v(x)$ also contain terms of the form $2 \sum_{j=1}^{d} b^j(x) D_j u(x) v(x)$ as in (8.4.1). Of course, we need to impose conditions on the function $b^j(x)$ so as to guarantee boundedness and nonnegativity (the latter requires bounds on $|b^j(x)|$ depending on λ and a lower bound for $|c(x)|$). We leave the details to the reader.

Corollary 8.5.1: *The other assumptions of the previous theorem remaining in force, now let V be a closed linear (hence convex) subspace of H. Then there exists precisely one $u \in V$ that solves*

$$2A(u, \varphi) + L(\varphi) = 0 \quad \text{for all } \varphi \in V. \tag{8.5.6}$$

Proof: The point u is a critical point (e.g., a minimum) of the functional

$$J(v) = A(v, v) + L(v)$$

in V precisely if

$$2A(v, \varphi) + L(\varphi) = 0 \quad \text{for all } \varphi \in V.$$

Namely, that u is a critical point means here that

$$\frac{d}{dt} J(u + t\varphi)_{|t=0} = 0 \quad \text{for all } \varphi \in V.$$

This, however, is equivalent to

$$0 = \frac{d}{dt}(A(u + t\varphi, u + t\varphi) + L(u + t\varphi))_{|t=0} = 2A(u, \varphi) + L(\varphi).$$

Conversely, if that holds, then

$$J(u + t\varphi) = J(u) + t(2A(u, \varphi) + L(\varphi)) + t^2 A(\varphi, \varphi) \geq J(u)$$

for all $\varphi \in V$, and u thus is a minimizer. The existence and uniqueness of a minimizer established in the theorem thus yields the corollary.   □

For our example $A(v, v) = \frac{1}{2} \int Dv \cdot Dv, L(v) = \int fv$ with $f \in L^2(\Omega)$, Corollary 8.5.1 thus yields the existence of some $u \in H_0^{1,2}(\Omega)$ satisfying

$$\int_\Omega Du \cdot D\varphi + \int_\Omega f\varphi = 0, \tag{8.5.7}$$

i.e, a weak solution of the Poisson equation in the sense of Definition 8.3.2.

As explained above, the assumptions apply to more general variational problems, and we deduce the following result from Corollary 8.5.1:

Corollary 8.5.2: *Let $\Omega \subset \mathbb{R}^d$ be open and bounded, and let the functions $a^{ij}(x)$ $(i, j = 1, \ldots, d)$ and $c(x)$ satisfy the above assumptions (A)–(D). Let $f \in L^2(\Omega)$. Then there exists a unique $u \in H_0^{1,2}(\Omega)$ satisfying*

$$\int_\Omega \left\{ \sum_{i,j=1}^d a^{ij}(x) D_i u(x) D_j \varphi(x) + c(x) u(x) \varphi(x) \right\} dx$$

$$= \int_\Omega f(x) \varphi(x) dx \quad \text{for all } \varphi \in H_0^{1,2}(\Omega).$$

Thus, we obtain a weak solution of

$$-\sum_{i,j=1}^d \frac{\partial}{\partial x^i} \left(a^{ij}(x) \frac{\partial}{\partial x^j} u(x) \right) + c(x) u(x) = f(x)$$

with $u = 0$ on $\partial\Omega$. Of course, so far, this equation does not yet make sense, since we do not know yet whether our weak solution u is regular, i.e., of class $C^2(\Omega)$. This issue, however, will be addresssed in the next chapter.

We now want to compare the solution of our variational problem $J(v) \to$ min in H with the one obtained in the subspace V of H.

Lemma 8.5.1: *Let $A : H \times H \to \mathbb{R}$ be a continuous, symmetric, elliptic, bilinear form in the sense of Definition 8.5.1, and let $L : H \to \mathbb{R}$ be linear and continuous. We consider once more the problem*

$$J(v) := A(v,v) + L(v) \to \min. \tag{8.5.8}$$

Let u be the solution in H, u_V the solution in the closed linear subspace V. Then

$$\|u - u_V\| \leq \frac{C}{\lambda} \inf_{v \in V} \|u - v\| \tag{8.5.9}$$

with the constants C and λ from Definition 8.5.1.

Proof: By Corollary 8.5.1,

$$2A(u,\varphi) + L(\varphi) = 0 \quad \text{for all } \varphi \in H,$$
$$2A(u_V,\varphi) + L(\varphi_V) = 0 \quad \text{for all } \varphi \in V,$$

hence also

$$2A(u - u_V, \varphi) = 0 \quad \text{for all } \varphi \in V. \tag{8.5.10}$$

For $v \in V$, we thus obtain

$$\|u - u_V\|^2 \leq \frac{1}{\lambda} A(u - u_V, u - u_V) \text{ by ellipticity of } A$$
$$= \frac{1}{\lambda} A(u - u_V, u - v) + \frac{1}{\lambda} A(u - u_V, v - u_V)$$
$$= \frac{1}{\lambda} A(u - u_V, u - v) \text{ from (8.5.10) with } \varphi = v - u_V \in V$$
$$\leq \frac{C}{\lambda} \|u - u_V\| \|u - v\|,$$

and since the inequality holds for arbitrary $v \in V$, (8.5.9) follows. $\square$

This lemma is the basis for an important numerical method for the approximative solution of variational problems. Since numerically only finite-dimensional problems can be solved, it is necessary to approximate infinite-dimensional problems by finite-dimensional ones. Thus, $J(v) \to$ min cannot be solved in an infinite-dimensional Hilbert space like $H = H_0^{1,2}(\Omega)$, but one needs to replace H by some finite-dimensional subspace V of H that on the

one hand can easily be handled numerically and on the other hand possesses
good approximation properties. These requirements are satisfied well by the
finite element spaces. Here, the region Ω is subdivided into polyhedra that
are as uniform as possible, e.g., triangles or squares in the 2-dimensional case
(if the boundary of Ω is curved, of course, it can only be approximated by
such a polyhedral subdivision). The finite elements then are simply piecewise
polynomials of a given degree. This means that the restriction of such a finite
element ψ onto each polyhedron occurring in the subdivision is a polynomial. In addition, one usually requires that across the boundaries between the
polyhedra, ψ be continuous or even satisfy certain specified differentiability
properties. The simplest such finite elements are piecewise linear functions
on triangles, where the continuity requirement is satisfied by choosing the
coefficients on neighboring triangles approximately. The theory of numerical mathematics then derives several approximation theorems of the type
sketched above. This is not particulary difficult and rather elementary, but
somewhat lengthy and therefore not pursued here. We rather refer to the
corresponding textbooks like Strang–Fix [20] or Braess [2].

The quality of the approximation of course depends not only on the degree of the polynomials, but also on the scale of the subdivision employed.
Typically, it makes sense to work with a fixed polynomial degree, for example admitting only piecewise linear or quadratic elements, and make the
subdivision finer and finer.

As presented here, the method of finite elements depends on the fact
that according to some abstract theorem, one is assured of the existence
(and uniqueness) of a solution of the variational problem under investigation
and that one can approximate that solution by elements of cleverly chosen
subspaces. Even though that will not be necessary for the theoretical analysis
of the method, for reasons of mathematical consistency it might be preferable
to avoid the abstract existence result and to convert the finite-dimensional
approximations into a constructive existence proof instead. This is what we
now wish to do.

Theorem 8.5.2: *Let $A : H \times H \to \mathbb{R}$ be a continuous, symmetric, elliptic,
bilinear form on the Hilbert space $(H, (\cdot, \cdot))$ with norm $\|\cdot\|$, and let $L : H \to \mathbb{R}$
be linear and continuous. We consider the variational problem*

$$J(v) = A(v, v) + L(v) \to \min.$$

*Let $(V_n)_{n \in \mathbb{N}} \subset H$ be an increasing (i.e., $V_n \subset V_{n+1}$ for all n) sequence of
closed linear subspaces exhausting H in the sense that for all $v \in H$ and
$\delta > 0$, there exist $n \in \mathbb{N}$ and $v_n \in V_n$ with*

$$\|v - v_n\| < \delta.$$

Let u_n be the solution of the problem

$$J(v) \to \min \; in \; V_n$$

obtained in Theorem 8.5.1. Then $(u_n)_{n \in \mathbb{N}}$ converges for $n \to \infty$ towards a solution of

$$J(v) \to \min \ in \ H.$$

Proof: Let

$$\kappa := \inf_{v \in H} J(v).$$

We want to show that

$$\lim_{n \to \infty} J(u_n) = \kappa.$$

In that case, $(u_n)_{n \in \mathbb{N}}$ will be a minimizing sequence for J in H, and thus it will converge to a minimizer of J in H by the proof of Theorem 8.5.1. We shall proceed by contradiction and thus assume that for some $\varepsilon > 0$ and all $n \in \mathbb{N}$,

$$J(u_n) \geq \kappa + \varepsilon \qquad (8.5.11)$$

(since $V_n \subset V_{n+1}$, we have $J(u_{n+1}) \leq J(u_n)$ for all n, by the way).

By definition of κ, there exists some $u_0 \in H$ with

$$J(u_0) < \kappa + \varepsilon/2. \qquad (8.5.12)$$

For every $\delta > 0$, by assumption, there exist some $n \in \mathbb{N}$ and some $v_n \in V_n$ with

$$\|u_0 - v_n\| < \delta.$$

With $w_n := v_n - u_0$, we then have

$$
\begin{aligned}
|J(v_n) - J(u_0)| &\leq |A(v_n, v_n) - A(u_0, u_0)| + |L(v_n) - L(u_0)| \\
&\leq A(w_n, w_n) + 2|A(w_n, u_0)| + \|L\| \, \|w_n\| \\
&\leq C \, \|w_n\|^2 + 2C \, \|w_n\| \, \|u_0\| + \|L\| \, \|w_n\| \\
&< \varepsilon/2
\end{aligned}
$$

for some appropriate choice of δ.

Thus

$$J(v_n) < J(u_0) + \varepsilon/2 < \kappa + \varepsilon \quad \text{by (8.5.12)} < J(u_n) \quad \text{by (8.5.11)},$$

contradicting the minimizing property of u_n.

This contradiction shows that $(u_n)_{n \in \mathbb{N}}$ indeed is a minimizing sequence, implying the convergence to a minimizer as already explained.   □

We thus have a constructive method for the (approximative) solution of our variational problem when we choose all the V_n as suitable finite-dimensional subspaces of H. For each V_n, by Corollary 8.5.1 one needs to solve only a finite linear system, with $\dim V_n$ equations; namely, let $e_1, \ldots, e_N$ be a basis of V_n. Then (8.5.6) is equivalent to the N linear equations for $u_n \in V_n$,

$$2A(u_n, e_j) + L(e_j) = 0 \quad \text{for } j = 1, \ldots, N. \tag{8.5.13}$$

Of course, the more general quadratic variational problems studied in Section 8.4 can also be covered by this method; we leave this as an exercise.

8.6 Convex Variational Problems

In the preceding sections, we have studied quadratic variational problems, and we provided an abstract Hilbert space interpretation of Dirichlet's principle. In this section, we shall find out that what is essential is not the quadratic structure of the integrand, but rather the fact that the integrand satisfies suitable bounds. In addition, we need the key assumption of convexity of the integrand, and hence, as we shall see, also of the variational integral.

For simplicity, we consider only variational integrals of the form

$$I(u) = \int_\Omega f(x, Du(x))dx, \tag{8.6.1}$$

where $Du = (D_1 u, \ldots, D_d u)$ denotes the weak derivatives of $u \in H^{1,2}(\Omega)$, instead of admitting more general integrands of the type

$$f(x, u(x), Du(x)). \tag{8.6.2}$$

The additional dependence on the function u itself, instead of just on its derivatives, does not change the results significantly, but it makes the proofs technically more complicated. In Section 12.3 below, when we address the regularity of minimizers, we shall even drop the dependence on x and consider only integrands of the form

$$f(Du(x)),$$

in order to make the proofs as transparent as possible while still preserving the essential features.

The main result of this section then is the following theorem:

Theorem 8.6.1: *Let $\Omega \subset \mathbb{R}^d$ be open, and consider a function*

$$f : \Omega \times \mathbb{R}^d \to \mathbb{R}$$

satisfying:

(i) $f(\cdot, v)$ is measurable for all $v \in \mathbb{R}^d$.
(ii) $f(x, \cdot)$ is convex for all $x \in \Omega$.
(iii) $f(x, v) \geq -\gamma(x) + \kappa|v|^2$ for almost all $x \in \Omega$, all $v \in \mathbb{R}^d$, with $\gamma \in L^1(\Omega)$,
$\kappa > 0$.

We let $g \in H^{1,2}(\Omega)$, and we consider the variational problem

$$I(u) := \int_\Omega f(x, Du(x))dx \to \min$$

among all $u \in H^{1,2}(\Omega)$ with $u - g \in H_0^{1,2}(\Omega)$ (thus, g are boundary values prescribed in the Sobolev sense).

Then I assumes its infimum; i.e., there exists such a u_0 with

$$I(u_0) = \inf_{u - g \in H_0^{1,2}(\Omega)} I(u).$$

To simplify our further considerations, we first observe that it suffices to consider the case $g = 0$. Namely, otherwise, we consider, for $w = u - g$,

$$\tilde{f}(x, w(x)) := f(x, w(x) + g(x)).$$

The function $\tilde{f}$ satisfies the same structural assumptions that f does; this is clear for (i) and (ii), and for (iii), we observe that

$$\tilde{f}(x, w(x)) \geq -\gamma(x) + \kappa|w(x) + g(x)|^2 \geq -\gamma(x) + \kappa\left(\frac{1}{2}|w(x)|^2 - |g(x)|^2\right),$$

and so $\tilde{f}$ satisfies the analogue of (iii) with

$$\tilde{\gamma}(x) := \gamma(x) + \kappa|g(x)|^2 \in L^1$$

and $\tilde{\kappa} := \frac{1}{2}\kappa$. Thus, for the rest of this section we assume

$$g = 0. \tag{8.6.3}$$

In order to prepare the proof of the Theorem 8.6.1, we shall first derive some properties of the variational integral I. We point out that in the next two lemmas the function v takes its values in $\mathbb{R}^d$, i.e., is vector- instead of scalar-valued, but that will not influence our reasoning at all.

Lemma 8.6.1: *Suppose that f is as in Theorem 8.6.1, but with (ii) weakened to*
(ii') $f(x, \cdot)$ is continuous for all $x \in \Omega$,
and supposing in (iii) only $\kappa \in \mathbb{R}$, but not necessarily $\kappa > 0$.
Then

$$J(v) := \int_\Omega f(x, v(x))dx$$

is a lower semicontinuous functional on $L^2(\Omega; \mathbb{R}^d)$.

Proof: We first observe that if v is in L^2, it is measurable, and since $f(x, v)$ is continuous with respect to v, $f(x, v(x))$ then is measurable by a basic result in Lebesgue integration theory.[5] Now let $(v_n)_{n \in \mathbb{N}}$ converge to v in $L^2(\Omega; \mathbb{R}^d)$. By another basic result in Lebesgue integration theory,[6] after selection of a subsequence, (v_n) also converges to v pointwise almost everywhere. (It is legitimate to select a subsequence here, because the subsequent arguments can be applied to any subsequence of (v_n).) By continuity of f,

$$f(x, v(x)) - \kappa|v(x)|^2 = \lim_{n \to \infty} \left(f(x, v_n(x)) - \kappa|v_n(x)|^2 \right).$$

Since $f(x, v_n(x)) - \kappa|v(x)|^2 \geq -\gamma(x)$, and γ is integrable, we may apply Fatou's lemma[7] to obtain

$$\int_\Omega \left(f(x, v(x)) - \kappa|v(x)|^2 \right) dx \leq \liminf_{n \to \infty} \int_\Omega \left((f(x, v_n(x)) - |v_n(x)|^2 \right) dx,$$

and since (v_n) converges to v in L^2, then also

$$\int_\Omega f(x, v(x)) dx \leq \liminf_{n \to \infty} \int_\Omega f(x, v_n(x)) dx.$$

$\square$

Lemma 8.6.2: *Let f be as in Theorem 8.6.1, without necessarily requiring κ in (iii) to be positive. Then*

$$J(v) = \int_\Omega f(x, v(x)) dx$$

is convex on $L^2(\Omega; \mathbb{R}^d)$.

Proof: Let $v_0, v_1 \in L^2(\Omega, \mathbb{R}^d), 0 \leq t \leq 1$. We have

$$J(tv_0 + (1-t)v_1) = \int f(x, tv_0(x) + (1-t)v_1(x))$$

$$\leq \int (tf(x, v_0(x)) + (1-t)f(x, v_1(x))) \quad \text{by (ii)}$$

$$= tJ(v_0) + (1-t)J(v_1).$$

Thus, J is convex. $\square$

Lemma 8.6.1 and Lemma 8.6.2 imply the following result:

---

[5] See J. Jost, *Postmodern Analysis*, p. 214 [12].
[6] See Lemma A.1 or J. Jost, *Postmodern Analysis*, p. 240 [12].
[7] See J. Jost, *Postmodern Analysis*, p. 202 [12].

Lemma 8.6.3: *Let f be as in Theorem 8.6.1, still not necessarily requiring $\kappa > 0$. With our previous simplification $g = 0$ (8.6.3), the functional*

$$I(u) = \int_\Omega f(x, Du(x))dx$$

is a convex and lower semicontinuous functional on $H_0^{1,2}(\Omega)$. □

With Lemma 8.6.3, Theorem 8.6.1 is a consequence of the following abstract result:

Theorem 8.6.2: *Let H be a Hilbert space, with norm $\|\cdot\|$,*

$$I : H \to \mathbb{R} \cup \{\infty\}$$

be bounded from below, not identically equal to $+\infty$, convex and lower semicontinuous. Then, for every $\lambda > 0$, and $u \in H$,

$$I_\lambda(u) := \inf_{y \in H} \left(I(y) + \lambda \|u - y\|^2 \right) \tag{8.6.4}$$

is realized by a unique $u_\lambda \in H$, i.e.,

$$I_\lambda(u) = I(u_\lambda) + \lambda \|u - u_\lambda\|^2, \tag{8.6.5}$$

and if $(u_\lambda)_{\lambda > 0}$ remains bounded as $\lambda \searrow 0$, then

$$u_0 := \lim_{\lambda \to 0} u_\lambda$$

exists and minimizes I, i.e.,

$$I(u_0) = \inf_{u \in H} I(u).$$

Proof: We first verify the auxiliary statement about the uniqueness and existence of u_λ. We let $(y_n)_{n \in \mathbb{N}}$ be a minimizing sequence for (8.6.4), i.e.,

$$I(y_n) + \lambda \|u - y_n\|^2 \to \inf_{y \in H} \left(I(y) + \lambda \|u - y\|^2 \right).$$

For $m, n \in \mathbb{N}$, we put

$$y_{m,n} := \frac{1}{2}(y_m + y_n).$$

We then have

$$I(y_{m,n}) + \lambda \|u - y_{m,n}\|^2 \le \frac{1}{2} \left(I(y_m) + \lambda \|u - y_m\|^2 \right) \tag{8.6.6}$$

$$+ \frac{1}{2} \left(I(y_n) + \lambda \|u - y_n\|^2 \right) - \frac{\lambda}{4} \|y_m - y_n\|^2$$

by the convexity of I and the general Hilbert space identity

$$\left\| x - \frac{1}{2}(y_1 + y_2) \right\|^2 = \frac{1}{2}\left(\|x - y_1\|^2 + \|x - y_2\|^2 \right) - \frac{1}{4}\|y_1 - y_2\|^2 \quad (8.6.7)$$

for any $x, y_1, y_2 \in H$, which is easily derived from expressing the norm squares as scalar products and expanding these scalar products.

Now, by definition of $I_\lambda(u)$, the left-hand side of (8.6.6) has to be $\geq I_\lambda(u)$, whereas for $k = m$ and n, $I(y_k) + \lambda \|u - y_k\|^2$ converges to $I_\lambda(u)$, by choice of the sequence (y_k), for $k \to \infty$. This implies that

$$\|y_m - y_n\|^2 \to 0$$

for $m, n \to \infty$. Thus, $(y_n)_{n \in \mathbb{N}}$ is a Cauchy sequence, and it converges to a unique limit u_λ. Since $\|\cdot\|^2$ is continuous, and I is lower semicontinuous, u_λ realizes the infimum in (8.6.4); i.e., (8.6.5) holds.

If (u_λ) then remains bounded for $\lambda \to 0$, this minimizing property implies that

$$\lim_{\lambda \to 0} I(u_\lambda) = \inf_{y \in H} I(y). \quad (8.6.8)$$

Thus, for any sequence $\lambda_n \to 0$, (u_{λ_n}) is a minimizing sequence for I.

We now let $0 < \lambda_1 < \lambda_2$. From the definition of u_{λ_1},

$$I(u_{\lambda_2}) + \lambda_1 \|u - u_{\lambda_2}\|^2 \geq I(u_{\lambda_1}) + \lambda_1 \|u - u_{\lambda_1}\|^2,$$

and so

$$I(u_{\lambda_2}) + \lambda_2 \|u - u_{\lambda_2}\|^2 \geq I(u_{\lambda_1}) + \lambda_2 \|u - u_{\lambda_1}\|^2$$
$$+ (\lambda_1 - \lambda_2)\left(\|u - u_{\lambda_1}\|^2 - \|u - u_{\lambda_2}\|^2 \right).$$

Since u_{λ_2} minimizes $I(y) + \lambda_2 \|u - y\|^2$, we conclude from this and $\lambda_1 < \lambda_2$ that

$$\|u - u_{\lambda_1}\|^2 \geq \|u - u_{\lambda_2}\|^2.$$

This means that

$$\|u - u_\lambda\|^2$$

is a decreasing function of λ, or in other words, it increases as $\lambda \searrow 0$. Since this expression is also bounded by assumption, it has to converge as $\lambda \searrow 0$. In particular, for any $\varepsilon > 0$, we may find $\lambda_0 > 0$ such that for $0 < \lambda_1, \lambda_2 < \lambda_0$,

$$\left| \|u - u_{\lambda_1}\|^2 - \|u - u_{\lambda_2}\|^2 \right| < \frac{\varepsilon}{2}. \quad (8.6.9)$$

We put

$$u_{1,2} := \frac{1}{2} \left(u_{\lambda_1} + u_{\lambda_2} \right).$$

If we assume, without loss of generality, $I(u_{\lambda_1}) \geq I(u_{\lambda_2})$, the convexity of I implies

$$I(u_{1,2}) \leq I(u_{\lambda_1}). \qquad (8.6.10)$$

We then have

$$I(u_{1,2}) + \lambda_1 \|u - u_{1,2}\|^2$$
$$\leq I(u_{\lambda_1}) + \lambda_1 \left(\frac{1}{2} \|u - u_{\lambda_1}\| + \frac{1}{2} \|u - u_{\lambda_2}\|^2 - \frac{1}{4} \|u_{\lambda_1} - u_{\lambda_2}\|^2 \right)$$
$$\text{by (8.6.10) and (8.6.7)}$$
$$< I(u_{\lambda_1}) + \lambda_1 \left(\|u - u_{\lambda_1}\|^2 + \frac{\varepsilon}{4} - \frac{1}{4} \|u_{\lambda_1} - u_{\lambda_2}\|^2 \right) \quad \text{by (8.6.9)}.$$

Since u_{λ_1} minimizes $I(y) + \lambda_1 \|u - y\|^2$, we conclude that

$$\|u_{\lambda_1} - u_{\lambda_2}\|^2 < \varepsilon.$$

So, we have shown the Cauchy property of u_λ for $\lambda \searrow 0$, and therefore, we obtain the existence of

$$u_0 = \lim_{\lambda \to 0} u_\lambda.$$

By (8.6.8) and the lower semicontinuity of I, we see that

$$I(u_0) = \inf_{y \in H} I(y).$$

Thus, we have shown the existence of a minimizer of I. This concludes the proof of Theorem 8.6.2, as well as that of Theorem 8.6.1. $\qquad \square$

While we shall see in Chapter 9 that the minimizers of the quadratic variational problems studied in the preceding sections of this chapter are smooth, we have to wait until Chapter 12 until we can derive a regularity theorem for minimizers of a class of variational integrals that satisfy similar structural conditions as in Theorem 8.6.1. Let us anticipate here Theorem 12.3.1 below:

Let $f : \mathbb{R}^d \to \mathbb{R}$ be of class C^∞ and satisfy:

(i) There exists a constant $K < \infty$ with

$$\left| \frac{\partial f}{\partial v_i}(v) \right| \leq K|v| \quad \text{for } i = 1, \ldots, d \quad (v = (v^1, \ldots, v^d) \in \mathbb{R}^d).$$

(ii) There exist constants $\lambda > 0$, $\Lambda < \infty$ with

$$\lambda|\xi|^2 \le \sum_{i,j=1}^{d} \frac{\partial^2 f(v)}{\partial v_i v_j} \xi_i \xi_j \le \Lambda|\xi|^2 \quad \text{for all } \xi \in \mathbb{R}^d.$$

Let $\Omega \subset \mathbb{R}^d$ be open and bounded. Let $u_0 \in W^{1,2}(\Omega)$ minimize

$$I(u) := \int_\Omega f(Du(x))dx$$

among all $u \in W^{1,2}(\Omega)$ with $u - u_0 \in H_0^{1,2}(\Omega)$. Then

$$u_0 \in C^\infty(\Omega).$$

In order to compare the assumptions of this result with those of Theorem 8.6.1, we first observe that (i) implies that there exist constants c and k with

$$|f(v)| \le c + k\,|v|^2\,.$$

Thus, in place of the lower bound in (iii) of Theorem 8.6.1, here we have an upper bound with the same asymptotic growth as $|v| \to \infty$. Thus, altogether, we are considering integrands with quadratic growth. In fact, it is also possible to consider variational integrands that asymptotically grow like $|v|^p$, with $1 < p < \infty$. The existence of a minimizer follows with similar techniques as described here, by working in the Banach space $H_0^{1,p}(\Omega)$ and exploiting a crucial geometric property of those particular Banach spaces, namely, that the unit ball is uniformly convex. The first steps of the regularity proof also do not change significantly, but higher regularity poses a problem for $p \ne 2$.

The lower bound in assumption (ii) above should be compared with the convexity assumption in Theorem 8.6.1. For $f \in C^2(\mathbb{R}^d)$, convexity means

$$\frac{\partial^2 f(v)}{\partial v^i \partial v^j} \xi_i \xi_j \ge 0 \quad \text{for all } \xi = (\xi_1, \ldots, \xi_d).$$

Thus, in contrast to the assumption in the regularity theorem, we are not summing here with respect i and j, and so this is a stronger assumption. On the other hand, we are not requiring a positive lower bound as in the regularity theorem, but only nonnegativity.

The existence of minimizers of variational problems is discussed in more detail in J. Jost–X. Li-Jost [14]. The minimizing scheme presented here is put in a broader context in J. Jost [11].

Summary

The Dirichlet principle consists in finding solutions of the Dirichlet problem

$$u = 0 \quad \text{in } \Omega,$$
$$u = g \quad \text{on } \partial\Omega,$$

by minimizing the Dirichlet integral

$$\int_\Omega |Du(x)|^2 dx$$

among all functions u with boundary values g in the function space $W^{1,2}(\Omega)$ (Sobolev space) (which turns out to be the appropriate space for this task). More generally, one may also treat the Poisson equation

$$\Delta u = f \quad \text{in } \Omega$$

this way, namely, minimizing

$$\int_\Omega |Du(x)|^2 \, dx + 2 \int_\Omega f(x) u(x) \, dx.$$

A minimizer then satisfies the equation

$$\int_\Omega Du(x) \, D\varphi(x) \, dx = 0$$

(respectively $\int_\Omega Du(x) D\varphi(x) \, dx + \int f(x)\varphi(x) \, dx = 0$ for the Poisson equation) for all $\varphi \in C_0^\infty(\Omega)$. If one manages to show that a minimizer u is regular (for example of class $C^2(\Omega)$), then this equation results from integrating the original differential equation (Laplace or Poisson equation, respectively) by parts. However, since the Sobolev space $W^{1,2}(\Omega)$ is considerably larger than the space $C^2(\Omega)$, we first need to show in the next chapter that a solution of this equation (called a "weak" differential equation) is indeed regular.

The Dirichlet principle also works for a more general class of elliptic equations, and it admits an abstract Hilbert space formulation.

Exercises

8.1 Show that the norm

$$\|u\| := \|u\|_{L^2(\Omega)} + \|Du\|_{L^2(\Omega)}$$

is equivalent to the norm $\|u\|_{W^{1,2}(\Omega)}$ (i.e., there are constants $0 < \alpha \le \beta < \infty$ satisfying

$$\alpha\|u\| \le \|u\|_{W^{1,2}(\Omega)} \le \beta\|u\| \quad \text{for all } u \in W^{1,2}(\Omega)).$$

Why does one prefer the norm $\|u\|_{W^{1,2}(\Omega)}$?

8.2 What would be a natural definition of k-times weak differentiablity? (The answer will be given in the next chapter, but you might wish to try yourself at this point to define Sobolev spaces $W^{k,2}(\Omega)$ of k-times weakly differentiably functions that are contained in $L^2(\Omega)$ together with all their weak derivatives and to prove results analogous to Theorem 8.2.1 and Corollary 8.2.1 for them.)

8.3 Consider a variational problem of the type

$$I(u) = \int_\Omega F(Du(x))dx$$

with a smooth function $F : \mathbb{R}^d \to \Omega$ satisfying an inequality of the form

$$|F(p)| \le c_1 |p|^2 + c_2 \quad \text{for all } p \in \mathbb{R}^d.$$

Derive the corresponding Euler–Lagrange equations for a minimizer (in the weak sense; cf. (8.4.4)). Try more generally to find conditions for integrands of the type $F(x, u(x), Du(x))$ that allow one to derive weak Euler–Lagrange equations for minimizers.

8.4 Following R. Courant, as a model problem for finite elements we consider the Poisson equation

$$\Delta u = f \quad \text{in } \Omega,$$
$$u = 0 \quad \text{on } \partial\Omega$$

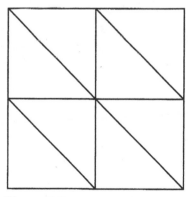

Figure 8.1.

in the unit square $\overline{\Omega} = [0, 1] \times [0, 1] \subset \mathbb{R}^2$. For $h = \frac{1}{2^n}$ ($n \in \mathbb{N}$), we subdivide $\overline{\Omega}$ into $\frac{1}{h^2}(= 2^{2n})$ subsquares of side length h, and each such square in turn is subdivided into two right-angled symmetric triangles by the diagonal from the upper left to the lower right vertex (see Figure 8.1). We thus obtain triangles Δ_i^h, $i = 1, \ldots, 2^{2n+1}$. What is the number of interior vertices p_j of this triangulation?

We consider the space of continuous triangular finite elements

$$S^h := \{\varphi \in C^0(\overline{\Omega}) : \varphi_{|\Delta_i^h} \quad \text{linear for all } i, \varphi = 0 \text{ on } \partial\overline{\Omega}\}.$$

The triangular elements φ_j with

$$\varphi_j(p_i) = \delta_{ij}$$

constitute a basis of S^h (proof?).
Compute

$$a_{ij} := \int_{\Omega} D\varphi_i \cdot D\varphi_j \quad \text{for all pairs } i, j$$

and establish the system of linear equations for the approximating solution of the Poisson equation in S^h, i.e., for the minimizer φ^h of

$$\int_{\Omega} |D\varphi|^2 + 2 \int_{\Omega} f\varphi$$

for $\varphi \in S^h$, with respect to the above basis φ_j of S^h (for that purpose, you have just computed the coefficients a_{ij}!).

9. Sobolev Spaces and L^2 Regularity Theory

9.1 General Sobolev Spaces. Embedding Theorems of Sobolev, Morrey, and John–Nirenberg

Definition 9.1.1: *Let $u : \Omega \to \mathbb{R}$ be integrable, $\boldsymbol{\alpha} := (\alpha_1, \ldots, \alpha_d)$,*

$$D_{\boldsymbol{\alpha}}\varphi := \left(\frac{\partial}{\partial x^1}\right)^{\alpha_1} \cdots \left(\frac{\partial}{\partial x^d}\right)^{\alpha_d} \varphi \quad for \ \varphi \in C^{|\boldsymbol{\alpha}|}(\Omega).$$

An integrable function $v : \Omega \to \mathbb{R}$ is called an $\boldsymbol{\alpha}$th weak derivative of u, in symbols $v = D_{\boldsymbol{\alpha}}u$, if

$$\int_\Omega \varphi v \, dx = (-1)^{|\boldsymbol{\alpha}|} \int_\Omega u D_{\boldsymbol{\alpha}}\varphi dx \quad for \ all \ \varphi \in C_0^{|\boldsymbol{\alpha}|}(\Omega). \qquad (9.1.1)$$

For $k \in \mathbb{N}$, $1 \le p < \infty$, we define the Sobolev space

$$W^{k,p}(\Omega) := \{u \in L^p(\Omega) : D_{\boldsymbol{\alpha}}u \ exists \ and \ is \ contained \ in \ L^p(\Omega) \ for \ all \ |\boldsymbol{\alpha}| \le k\},$$

$$\|u\|_{W^{k,p}(\Omega)} := \left(\sum_{|\boldsymbol{\alpha}|\le k} \int_\Omega |D_{\boldsymbol{\alpha}}u|^p\right)^{\frac{1}{p}}.$$

The spaces $H^{k,p}(\Omega)$ and $H_0^{k,p}(\Omega)$ are defined to be the closures of $C^\infty(\Omega)$ and $C_0^\infty(\Omega)$, respectively, with respect to $\|\cdot\|_{W^{k,p}(\Omega)}$. Occasionally, we shall employ the abbreviation $\|\cdot\|_p = \|\cdot\|_{L^p(\Omega)}$.

Concerning notation: The multi-index notation will be used in the present section only. Later on, for $u \in W^{1,p}(\Omega)$, first weak derivatives will be denoted by $D_i u$, $i = 1, \ldots, d$, as in Definition 8.2.1, and we shall denote the vector $(D_1 u, \ldots, D_d u)$ by Du. Likewise, for $u \in W^{2,p}(\Omega)$, second weak derivatives will be written $D_{ij}u$, $i, j = 1, \ldots, d$, and the matrix of second weak derivatives will be denoted by $D^2 u$.

As in Section 8.2, one proves the following lemma:

Lemma 9.1.1: $W^{k,p}(\Omega) = H^{k,p}(\Omega)$. *The space $W^{k,p}(\Omega)$ is complete with respect to $\|\cdot\|_{W^{k,p}(\Omega)}$, i.e., it is a Banach space.* $\qquad \square$

We now state the Sobolev embedding theorem:

Theorem 9.1.1:

$$H_0^{1,p}(\Omega) \subset \begin{cases} L^{\frac{dp}{d-p}}(\Omega) & \text{for } p < d, \\ C^0(\bar{\Omega}) & \text{for } p > d. \end{cases}$$

Moreover, for $u \in H_0^{1,p}(\Omega)$,

$$\|u\|_{\frac{dp}{d-p}} \leq c \|Du\|_p \qquad \text{for } p < d, \qquad (9.1.2)$$

$$\sup_\Omega |u| \leq c |\Omega|^{\frac{1}{d} - \frac{1}{p}} \cdot \|Du\|_p \quad \text{for } p > d, \qquad (9.1.3)$$

where the constant c depends on p and d only.

In order to better understand the content of the Sobolev embedding theorem, we first consider the scaling behavior of the expressions involved: Let $f \in H^{1,p}(\mathbb{R}^d) \cap L^q(\mathbb{R}^d)$. We look at the scaling $y = \lambda x$ (with $\lambda > 0$) and

$$f_\lambda(y) := f\left(\frac{y}{\lambda}\right) = f(x).$$

Then, with $y = \lambda x$,

$$\left(\int_{\mathbb{R}^d} |Df_\lambda(y)|^p \, dy\right)^{\frac{1}{p}} = \lambda^{\frac{d-p}{p}} \left(\int_{\mathbb{R}^d} |Df(x)|^p \, dx\right)^{\frac{1}{p}}$$

(note that on the left, the derivative is taken with respect to y, and on the right with respect to x; this explains the $-p$ in the exponent) and

$$\left(\int_{\mathbb{R}^d} |f_\lambda(y)|^q \, dy\right)^{\frac{1}{q}} = \lambda^{\frac{d}{q}} \left(\int_{\mathbb{R}^d} |f(x)|^q \, dx\right)^{\frac{1}{q}}.$$

Thus in the limit $\lambda \to 0$, $\|f_\lambda\|_{L^q}$ is controlled by $\|Df_\lambda\|_{L^p}$ if

$$\lambda^{\frac{d}{q}} \leq \lambda^{\frac{d-p}{p}} \quad \text{for } \lambda < 1$$

holds, i.e.,

$$\frac{d}{q} \geq \frac{d-p}{p},$$

i.e.,

$$q \leq \frac{dp}{d-p} \quad \text{if } p < d.$$

(We have implicitly assumed $\|Df\|_{L^p} > 0$ here, but you will easily convince yourself that this is the essential case of the embedding theorem.) We treat only the limit $\lambda \to 0$ here, since only for $\lambda \leq 1$ (for $f \in H_0^{1,p}(\mathbb{R}^d)$) do we have

$$\operatorname{supp} f_\lambda \subset \operatorname{supp} f,$$

and the Sobolev embedding theorem covers only the case where the functions have their support contained in a fixed bounded set Ω. Looking at the scaling properties for $\lambda \to \infty$, one observes that this assumption on the support is necessary for the theorem. The scaling properties for $p > d$ will be examined after Corollary 9.1.5.

Proof of Theorem 9.1.1: We shall first prove the inequalities (9.1.2) and (9.1.3) for $u \in C_0^1(\Omega)$. We put $u = 0$ on $\mathbb{R}^d \setminus \Omega$ again. As in the proof of Theorem 8.2.2,

$$|u(x)| \le \int_{-\infty}^{x^i} |D_i u(x^1, \ldots, x^{i-1}, \xi, x^{i+1}, \ldots, x^d)| \, d\xi \quad \text{with } x = (x^1, \ldots, x^d)$$

for $1 \le i \le d$, and hence

$$|u(x)|^d \le \prod_{i=1}^d \int_{-\infty}^{\infty} |D_i u| \, dx^i$$

and

$$|u(x)|^{\frac{d}{d-1}} \le \left(\prod_{i=1}^d \int_{-\infty}^{\infty} |D_i u| \, dx^i \right)^{\frac{1}{d-1}}.$$

It follows that

$$\int_{-\infty}^{\infty} |u(x)|^{\frac{d}{d-1}} \, dx^1 \le \left(\int_{-\infty}^{\infty} |D_1 u| \, dx^1 \right)^{\frac{1}{d-1}} \left(\prod_{i \ne 1} \int_{-\infty}^{\infty} \int_{-\infty}^{\infty} |D_i u| \, dx^i dx^1 \right)^{\frac{1}{d-1}},$$

where we have used (A.6) for $p_1 = \cdots = p_{d-1} = d - 1$. Iteratively, we obtain

$$\int_{\Omega} |u(x)|^{\frac{d}{d-1}} \, dx \le \left(\prod_{i=1}^d \int_{\Omega} |D_i u| \, dx \right)^{\frac{1}{d-1}},$$

and hence

$$\|u\|_{\frac{d}{d-1}} \le \left(\prod_{i=1}^d \int_{\Omega} |D_i u| \, dx \right)^{\frac{1}{d}} \le \frac{1}{d} \int_{\Omega} \sum_{i=1}^d |D_i u| \, dx,$$

since the geometric mean is not larger than the arithmetic one, and consequently

$$\|u\|_{\frac{d}{d-1}} \le \frac{1}{d} \|Du\|_1, \tag{9.1.4}$$

which is (9.1.2) for $p = 1$.

Applying (9.1.4) to $|u|^\gamma$ ($\gamma > 1$) ($|u|^\gamma$ is not necessarily contained in $C_0^1(\Omega)$, even if u is, but as will be explained at the end of the present proof, by an approximation argument, if shown for $C_0^1(\Omega)$, (9.1.4) continues to hold for $H_0^{1,1}$, and we shall choose γ such that for $u \in H_0^{1,p}(\Omega)$, we have $|u|^\gamma \in H_0^{1,1}(\Omega)$), we obtain

$$\||u|^\gamma\|_{\frac{d}{d-1}} \leq \frac{\gamma}{d} \int_\Omega |u|^{\gamma-1} |Du| \, dx \leq \frac{\gamma}{d} \left\| |u|^{\gamma-1} \right\|_q \cdot \|Du\|_p \quad \text{for } \frac{1}{p} + \frac{1}{q} = 1$$
(9.1.5)

applying Hölder's inequality (A.4). For $p < d$, $\gamma = \frac{(d-1)p}{d-p}$ satisfies

$$\frac{\gamma d}{d-1} = \frac{(\gamma-1)p}{p-1},$$

and (9.1.5) yields, taking $q = \frac{p}{p-1}$ into account,

$$\|u\|_{\frac{\gamma d}{d-1}}^\gamma \leq \frac{\gamma}{d} \|u\|_{\frac{\gamma d}{d-1}}^{\gamma-1} \cdot \|Du\|_p,$$

i.e.,

$$\|u\|_{\frac{\gamma d}{d-1}} \leq \frac{\gamma}{d} \|Du\|_p,$$

which is (9.1.2). In order to establish (9.1.3), we need the following generalization of Lemma 8.2.4:

Lemma 9.1.2: *For $\mu \in (0,1]$, $f \in L^1(\Omega)$ let*

$$(V_\mu f)(x) := \int_\Omega |x - y|^{d(\mu-1)} f(y) dy.$$

Let $1 \leq p \leq q \leq \infty$,

$$0 \leq \delta = \frac{1}{p} - \frac{1}{q} < \mu.$$

Then V_μ maps $L^p(\Omega)$ continuously to $L^q(\Omega)$, and for $f \in L^p(\Omega)$, we have

$$\|V_\mu f\|_q \leq \left(\frac{1-\delta}{\mu-\delta} \right)^{1-\delta} \omega_d^{1-\mu} |\Omega|^{\mu-\delta} \|f\|_p.$$
(9.1.6)

Proof: Let

$$\frac{1}{r} := 1 + \frac{1}{q} - \frac{1}{p} = 1 - \delta.$$

Then

$$\ell(x - y) := |x - y|^{d(\mu - 1)} \in L^r(\Omega),$$

and as in the proof of Lemma 8.2.4, we choose R such that $|\Omega| = |B(x, R)| = \omega_d R^d$, and we estimate as follows:

$$\|\ell\|_r = \left(\int_\Omega |x - y|^{\frac{d(\mu - 1)}{1 - \delta}} \, dy \right)^{1 - \delta}$$

$$\leq \left(\int_{B(x,R)} |x - y|^{\frac{d(\mu - 1)}{1 - \delta}} \, dy \right)^{1 - \delta}$$

$$= \left(\frac{1 - \delta}{\mu - \delta} \right)^{1 - \delta} \omega_d^{1 - \delta} R^{d(\mu - \delta)}$$

$$= \left(\frac{1 - \delta}{\mu - \delta} \right)^{1 - \delta} \omega_d^{1 - \mu} |\Omega|^{\mu - \delta}.$$

We write

$$\ell |f| = \ell^{r(1 - 1/p)} \left(\ell^r |f|^p \right)^{\frac{1}{q}} |f|^{p\delta},$$

and the generalized Hölder inequality (A.6) yields

$$|V_\mu f(x)|$$

$$\leq \left(\int_\Omega \ell^r(x - y) |f(y)|^p \, dy \right)^{\frac{1}{q}} \left(\int_\Omega \ell^r(x - y) dy \right)^{1 - \frac{1}{p}} \left(\int_\Omega |f(y)|^p \, dy \right)^\delta,$$

hence, integrating with respect to x and interchanging the integrations in the first integral, we obtain

$$\|V_\mu f\|_q \leq \sup_\Omega \left(\int \ell^r(x - y) dy \right)^{\frac{1}{r}} \|f\|_p \leq \left(\frac{1 - \delta}{\mu - \delta} \right)^{1 - \delta} \omega_d^{1 - \mu} |\Omega|^{\mu - \delta} \|f\|_p$$

by the above estimate for $\|\ell\|_r$.   □

In order to complete the proof of Theorem 9.1.1, we use (8.2.4), assuming first $u \in C_0^1(\Omega)$ as before, i.e.,

$$u(x) = \frac{1}{d\omega_d} \int_\Omega \sum_{i=1}^d \frac{(x^i - y^i)}{|x - y|^d} D_i u(y) dy \qquad (9.1.7)$$

for $x \in \Omega$. This implies

$$|u| \leq \frac{1}{d\omega_d} V_{\frac{1}{d}}(|D|). \qquad (9.1.8)$$

Inequality (9.1.6) for $q = \infty$, $\mu = 1/d$ then yields (9.1.3), again at this moment for $u \in C_0^1(\Omega)$ only.

If now $u \in H_0^{1,p}(\Omega)$, we approximate u in the $W^{1,p}$-norm by C_0^∞ functions u_n, and apply (9.1.2) and (9.1.3) to the difference $u_n - u_m$. It follows that (u_n) is a Cauchy sequence in $L^{dp/(d-p)}(\Omega)$ (for $p < d$) or $C^0(\bar\Omega)$ (for $p > d$), respectively. Thus u itself is contained in the same space and satisfies (9.1.2) or (9.1.3), respectively $\qquad\square$

Corollary 9.1.1:

$$H_0^{k,p}(\Omega) \subset \begin{cases} L^{\frac{dp}{d-kp}}(\Omega) & \text{for } kp < d, \\ C^m(\Omega) & \text{for } 0 \leq m < k - \frac{d}{p}. \end{cases}$$

Proof: The first embedding iteratively follows from Theorem 9.1.1, and the second one then from the first and the case $p > d$ in Theorem 9.1.1. $\qquad\square$

Corollary 9.1.2: *If $u \in H_0^{k,p}(\Omega)$ for some p and all $k \in \mathbb{N}$, then $u \in C^\infty(\Omega)$.* $\qquad\square$

The embedding theorems to follow will be used in Chapter 12 only. First we shall present another variant of the Sobolev embedding theorem. For a function $v \in L^1(\Omega)$, we define the mean of v on Ω as

$$\fint_\Omega v(x)dx := \frac{1}{|\Omega|} \int_\Omega v(x)dx,$$

$|\Omega|$ denoting the Lebesgue measure of Ω. We then have the following result:

Corollary 9.1.3: *Let $1 \leq p < d$ and $u \in H^{1,p}(B(x_0, R))$. Then*

$$\left(\fint_{B(x_0,R)} |u|^{\frac{dp}{d-p}} \right)^{\frac{d-p}{dp}} \leq c_0 \left(R^p \fint_{B(x_0,R)} |Du|^p + \fint_{B(x_0,R)} |u|^p \right)^{\frac{1}{p}}, \quad (9.1.9)$$

where c_0 depends on p and q only.

Proof: Without loss of generality, $x_0 = 0$. Likewise, we may assume $R = 1$, since we may consider the functions $\tilde{u}(x) = u(Rx)$ and check that the expressions in (9.1.9) scale in the right way. Thus, let $u \in H^{1,p}(B(0,1))$. We extend u to the ball $B(0,2)$, by putting

$$u(x) = u\left(\frac{x}{|x|^2} \right) \quad \text{for } |x| > 1.$$

This extension satisfies

$$\|u\|_{H^{1,p}(B(0,2))} \leq c_1 \|u\|_{H^{1,p}(B(0,1))}. \quad (9.1.10)$$

Now let $\eta \in C_0^\infty(B(0,2))$ with

$$\eta \geq 0, \quad \eta \equiv 1 \text{ on } B(0,1), \quad |D\eta| \leq 2.$$

Then $v = \eta u \in H_0^{1,p}(B(0,2))$, and by (9.1.2),

$$\left(\int_{B(0,2)} |v|^{\frac{dp}{d-p}} \right)^{\frac{d-p}{dp}} \leq c_2 \left(\int_{B(0,2)} |Dv|^p \right)^{\frac{1}{p}}. \tag{9.1.11}$$

Since

$$Dv = \eta Du + u D\eta,$$

from the properties of η, we deduce

$$|Dv|^p \leq c_3 \left(|Du|^p + |u|^p \right), \tag{9.1.12}$$

and hence with (9.1.10),

$$\int_{B(0,2)} |Dv|^p \leq c_4 \left(\int_{B(0,1)} |Du|^p + \int_{B(0,1)} |u|^p \right). \tag{9.1.13}$$

Since on the other hand

$$\int_{B(0,1)} |u|^{\frac{dp}{d-p}} \leq \int_{B(0,2)} |v|^{\frac{dp}{d-p}},$$

(9.1.9) follows from (9.1.11) and (9.1.13).   □

Later on (in Section 12.1), we shall need the following result of John and Nirenberg:

Theorem 9.1.2: *Let $B(y_0, R_0)$ be a ball in $\mathbb{R}^d$, $u \in W^{1,1}(B(y_0, R_0))$, and suppose that for all balls $B(y, R) \subset \mathbb{R}^d$,*

$$\int_{B(y,R) \cap B(y_0, R_0)} |Du| \leq R^{d-1}. \tag{9.1.14}$$

Then there exist $\alpha > 0$ and $\beta_0 < \infty$ satisfying

$$\int_{B(y_0, R_0)} e^{\alpha |u - u_0|} \leq \beta_0 R_0^d \tag{9.1.15}$$

with

$$u_0 = \frac{1}{\omega_d R_0^d} \int_{B(y_0, R_0)} u \quad \text{(mean of u on $B(y_0, R_0)$)}.$$

In particular,

$$\int_{B(y_0, R_0)} e^{\alpha u} \int_{B(y_0, R_0)} e^{-\alpha u} = \int_{B(y_0, R_0)} e^{\alpha (u - u_0)} \int_{B(y_0, R_0)} e^{-\alpha (u - u_0)} \leq \beta_0^2 R_0^{2d}.$$

$$\tag{9.1.16}$$

More generally, for a measurable set $B \subset \mathbb{R}^d$, and $u \in L^1(B)$, we denote the mean by

$$u_B := \frac{1}{|B|} \int_B u(y) dy, \qquad (9.1.17)$$

$|B|$ being the Lebesgue measure of B. In order to prepare the proof of Theorem 9.1.2, we start with a lemma:

Lemma 9.1.3: *Let $\Omega \subset \mathbb{R}^d$ be convex, $B \subset \Omega$ measurable with $|B| > 0$, $u \in W^{1,1}(\Omega)$. Then we have for almost all $x \in \Omega$,*

$$|u(x) - u_B| \leq \frac{(\operatorname{diam} \Omega)^d}{d |B|} \int_\Omega |x - z|^{1-d} |Du(z)| dz. \qquad (9.1.18)$$

Proof: As before, it suffices to prove the inequality for $u \in C^1(\Omega)$. Since Ω is convex, if x and y are contained in Ω, so is the straight line joining them, and we have

$$u(x) - u(y) = -\int_0^{|x-y|} \frac{\partial}{\partial r} u\left(x + r \frac{y-x}{|y-x|} \right) dr,$$

and thus

$$u(x) - u_B = \frac{1}{|B|} \int_B (u(x) - u(y)) dy$$

$$= -\frac{1}{|B|} \int_B \int_0^{|x-y|} \frac{\partial}{\partial r} u\left(x + r \frac{y-x}{|y-x|} \right) dr \, dy.$$

This implies

$$|u(x) - u_B| \leq \frac{1}{|B|} \frac{(\operatorname{diam} \Omega)^d}{d} \left| \int_{\substack{|\omega|=1 \\ x+r\omega\in\Omega}} \int_0^{|x-y|} \frac{\partial}{\partial r} u(x + r\omega) dr \, d\omega \right|, \qquad (9.1.19)$$

if instead of over B, we integrate over the ball $B(x, \operatorname{diam} \Omega)) \cap \Omega$, write $dy = \varrho^{d-1} d\omega \, d\varrho$ in polar coordinates, and integrate with respect to ϱ. Thus, as in the proofs of Theorems 1.2.1 and 8.2.2,

$$|u(x) - u_B| \leq \frac{1}{|B|} \frac{(\operatorname{diam} \Omega)^d}{d} \left| \int_0^{|x-y|} \int_{\partial B(x,r)\cap\Omega} \frac{1}{r^{d-1}} \frac{\partial u}{\partial \nu}(z) d\sigma(z) dr \right|$$

$$= \frac{1}{|B|} \frac{(\operatorname{diam} \Omega)^d}{d} \left| \int_\Omega \frac{1}{|x-z|^{d-1}} \sum_{i=1}^d \frac{\partial}{\partial z^i} u(z) \frac{x^i - z^i}{|x-z|} dz \right|$$

$$\leq \frac{(\operatorname{diam} \Omega)^d}{d |B|} \int_\Omega \frac{1}{|x-z|^{d-1}} |Du(z)| dz.$$

$\square$

We shall also need the following variant of Lemma 9.1.2:

Lemma 9.1.4: *Let $f \in L^1(\Omega)$, and suppose that for all balls $B(x_0, R) \subset \mathbb{R}^d$,*

$$\int_{\Omega \cap B(x_0, R)} |f| \leq K R^{d(1-\frac{1}{p})} \qquad (9.1.20)$$

with some fixed K. Moreover, let $p > 1$, $1/p < \mu$. Then

$$|(V_\mu f)(x)| \leq \frac{p-1}{\mu p - 1} (\operatorname{diam} \Omega)^{d(\mu - \frac{1}{p})} K \qquad (9.1.21)$$

$$\left((V_\mu f)(x) = \int_\Omega |x - y|^{d(\mu - 1)} f(y) dy \right).$$

Proof: We put $f = 0$ in the exterior of Ω. With $r = |x - y|$, then

$$|V_\mu f(x)| \leq \int_\Omega r^{d(\mu - 1)} |f(y)| \, dy$$

$$= \int_0^{\operatorname{diam} \Omega} r^{d(\mu - 1)} \int_{\partial B(x,r)} |f(z)| \, dz dr$$

$$= \int_0^{\operatorname{diam} \Omega} r^{d(\mu - 1)} \left(\frac{\partial}{\partial r} \int_{B(x,r)} |f(y)| \, dy \right) dr$$

$$= (\operatorname{diam} \Omega)^{d(\mu - 1)} \int_{B(x,\operatorname{diam} \Omega)} |f(y)| \, dy$$

$$+ d(1 - \mu) \int_0^{\operatorname{diam} \Omega} r^{d(\mu - 1) - 1} \int_{B(x,r)} |f(y)| \, dy dr$$

$$\leq K (\operatorname{diam} \Omega)^{d(\mu - 1) + d(1 - 1/p)}$$

$$+ K d(1 - \mu) \int_0^{\operatorname{diam} \Omega} r^{d(\mu - 1) - 1 + d(1 - 1/p)} dr \text{ by } (9.1.20)$$

$$= K \frac{1 - \frac{1}{p}}{\mu - \frac{1}{p}} (\operatorname{diam} \Omega)^{d(\mu - 1/p)}.$$

$\square$

Proof of Theorem 9.1.2: Because of (9.1.14), $f = |Du|$ satisfies the inequality (9.1.20) with $K = 1$ and $p = d$. Thus, by Lemma 9.1.4, for $\mu > 1/d$,

$$V_\mu(f)(x) = \int_{B(y_0, R_0)} |x - y|^{d(\mu - 1)} |f(y)| \, dy \leq \frac{d-1}{\mu d - 1} (2R_0)^{\mu d - 1}. \quad (9.1.22)$$

In particular, for $s \geq 1$ and $\mu = \frac{1}{d} + \frac{1}{ds}$,

$$V_{\frac{1}{d} + \frac{1}{ds}}(f) \leq (d - 1) s (2R_0)^{\frac{1}{s}}. \qquad (9.1.23)$$

By Lemma 9.1.2, we also have, for $s \geq 1$, $\mu = 1/ds$, $p = q = 1$,

$$\int_{B(y_0,R_0)} V_{\frac{1}{ds}}(f) \leq ds\omega_d^{1-1/ds} |B(y_0,R_0)|^{\frac{1}{ds}} \|f\|_{L^1(B(y_0,R_0))}$$

$$\leq ds\omega_d R_0^{\frac{1}{s}} R_0^{d-1} \qquad (9.1.24)$$

by (9.1.20), which, as noted, holds for $K = 1$ and $p = d$. Now

$$|x-y|^{1-d} = |x-y|^{d(\frac{1}{ds}-1)\frac{1}{s}} |x-y|^{d(\frac{1}{ds}+\frac{1}{d}-1)(1-\frac{1}{s})}, \qquad (9.1.25)$$

and from Hölder's inequality then

$$V_{\frac{1}{d}}(f) = \int \left(|x-y|^{d(\frac{1}{ds}-1)\frac{1}{s}} |f(y)|^{\frac{1}{s}}\right) \left(|x-y|^{d(\frac{1}{ds}+\frac{1}{d}-1)(1-\frac{1}{s})} |f(y)|^{1-\frac{1}{s}}\right) dy$$

$$\leq V_{\frac{1}{ds}}(f)^{\frac{1}{s}} V_{\frac{1}{d}+\frac{1}{ds}}(f)^{1-\frac{1}{s}}. \qquad (9.1.26)$$

With (9.1.23) and (9.1.24), this implies

$$\int_{B(y_0,R_0)} V_{\frac{1}{d}}(f)^s \leq ds\omega_d R_0^{d-1+\frac{1}{s}} (d-1)^{s-1} s^{s-1} (2R_0)^{\frac{s-1}{s}}$$

$$\leq 2d(d-1)^{s-1} s^s \omega_d R_0^d$$

$$= 2\frac{d}{d-1}\omega_d((d-1)s)^s R_0^d.$$

Thus

$$\int_{B(y_0,R_0)} \sum_{n=0}^{\infty} \frac{V_{\frac{1}{d}}(f)^n}{\gamma^n n!} \leq \frac{2d}{d-1}\omega_d R_0^d \sum_{n=0}^{\infty} \left(\frac{d-1}{\gamma}\right)^n \frac{n^n}{n!}$$

$$\leq cR_0^d, \text{ if } \frac{d-1}{\gamma} < \frac{1}{e},$$

i.e.,

$$\int_{B(y_0,R_0)} \exp\left(\frac{V_{1/d}(f)}{\gamma}\right) \leq cR_0^d. \qquad (9.1.27)$$

Now by Lemma 9.1.3

$$|u(x) - u_0| \leq \text{const } V_{\frac{1}{d}}(|Du|), \qquad (9.1.28)$$

and since we have proved (9.1.27) for $f = |Du|$, (9.1.15) follows. $\qquad \square$

Before concluding the present section, we would like to derive some further applications of the preceding lemmas, including the following version of the Poincaré inequality:

Corollary 9.1.4: *Let $\Omega \subset \mathbb{R}^d$ be convex, and $u \in W^{1,p}(\Omega)$. We then have for every measurable $B \subset \Omega$ with $|B| > 0$,*

$$\left(\int_\Omega |u - u_B|^p \right)^{\frac{1}{p}} \leq \frac{\omega_d^{1-\frac{1}{d}}}{|B|} |\Omega|^{\frac{1}{d}} (\operatorname{diam} \Omega)^d \left(\int_\Omega |Du|^p \right)^{\frac{1}{p}}. \qquad (9.1.29)$$

Proof: By Lemma 9.1.3,

$$|u(x) - u_B| \leq \frac{(\operatorname{diam} \Omega)^d}{d\,|B|} V_{\frac{1}{d}}(|Du|),$$

and by Lemma 9.1.2, then,

$$\left\| V_{\frac{1}{d}}(|Du|) \right\|_{L^p(\Omega)} \leq d\omega_d^{1-\frac{1}{d}} |\Omega|^{\frac{1}{d}} \|Du\|_{L^p(\Omega)},$$

and these two inequalities imply the claim. $\qquad\qquad\Box$

The next result is due to C.B. Morrey:

Theorem 9.1.3: *Assume $u \in W^{1,1}(\Omega)$, $\Omega \subset \mathbb{R}^d$, and that there exist constants $K < \infty$, $0 < \alpha < 1$, such that for all balls $B(x_0, R) \subset \mathbb{R}^d$,*

$$\int_{\Omega \cap B(x_0, R)} |Du| \leq KR^{d-1+\alpha}. \qquad (9.1.30)$$

Then we have for every ball $B(z, r) \subset \mathbb{R}^d$,

$$\operatorname*{osc}_{\Omega \cap B(z,r)} u := \sup_{x,y \in B(z,r) \cap \Omega} |u(x) - u(y)| \leq cKr^\alpha, \qquad (9.1.31)$$

with $c = c(d, \alpha)$.

Proof: We have

$$\operatorname*{osc}_{\Omega \cap B(z,r)} u \leq 2 \sup_{x \in B(z,r) \cap \Omega} |u(x) - u_{B(z,r)}|$$

$$\leq c_1 \int_{B(z,r)} |x - y|^{1-d} |Du(y)|\, dy$$

by Lemma 9.1.3, where c_1 depends on d only, and where we simply put $Du = 0$ on $\mathbb{R}^d \setminus \Omega$.

$$= c_1 V_{\frac{1}{d}}(|Du|)\,(x)$$

with the notation of Lemma 9.1.4. With

$$p = \frac{d}{1-\alpha}, \quad \text{i.e., } \alpha = 1 - \frac{d}{p},$$

and

$$\mu = \frac{1}{d} > \frac{1}{p},$$

$f = |Du|$ then satisfies the assumptions of Lemma 9.1.4, and the preceding estimate together with Lemma 9.1.4 (applied to $B(z,r)$ in place of Ω) then yields

$$\underset{\Omega \cap B(z,r)}{\operatorname{osc}} u \leq c_2 K (\operatorname{diam} B(z,r))^{1-\frac{d}{p}} = cKr^\alpha.$$

$\square$

Definition 9.1.2: *A function u defined on Ω is called α-Hölder continuous in Ω, for some $0 < \alpha < 1$, if for all $z \in \Omega$,*

$$\sup_{x \in \Omega} \frac{|u(x) - u(z)|}{|x - z|^\alpha} < \infty. \tag{9.1.32}$$

Notation: $u \in C^\alpha(\Omega)$. For $u \in C^\alpha(\Omega)$, we put

$$\|u\|_{C^\alpha(\Omega)} := \|u\|_{C^0(\Omega)} + \sup_{x,y \in \Omega} \frac{|u(x) - u(y)|}{|x - y|^\alpha}.$$

(For $\alpha = 1$, a function satisfying (9.1.32) is called Lipschitz continuous, and the corresponding space is denoted by $C^{0,1}(\Omega)$.)

If u satisfies the assumptions of Theorem 9.1.3, it thus turns out to be α-Hölder continuous on Ω; this follows by putting $r = \operatorname{dist}(z, \partial\Omega)$ in Theorem 9.1.3. The notion of Hölder continuity will play a crucial role in Chapters 11 and 12.

Theorem 9.1.3 now implies the following refinement, due to Morrey, of the Sobolev embedding theorem in the case $p > d$:

Corollary 9.1.5: *Let $u \in H_0^{1,p}(\Omega)$ with $p > d$. Then*

$$u \in C^{1-\frac{d}{p}}(\bar{\Omega}).$$

More precisely, for every ball $B(z,r) \subset \mathbb{R}^d$,

$$\underset{\Omega \cap B(z,r)}{\operatorname{osc}} u \leq cr^{1-\frac{d}{p}} \|Du\|_{L^p(\Omega)}, \tag{9.1.33}$$

where c depends on d and p only.

Once more, it helps in understanding the content of this embedding theorem if we take a look at the scaling properties of the norms involved: Let $f \in H^{1,p}(\mathbb{R}^d) \cap C^\alpha(\mathbb{R}^d)$ with $0 < \alpha < 1$. We again consider the scaling $y = \lambda x$ ($\lambda > 0$) and put

$$f_\lambda(y) = f(\lambda x).$$

Then

$$\frac{|f_\lambda(y_1) - f_\lambda(y_2)|}{|y_1 - y_2|^d} = \lambda^{-\alpha} \frac{|f(x_1) - f(x_2)|}{|x_1 - x_2|^\alpha} \qquad (y_i = \lambda x_i, \ i = 1, 2)$$

and thus

$$\|f_\lambda\|_{C^\alpha} = \lambda^{-\alpha} \|f\|_{C^\alpha},$$

and as has been computed above,

$$\|f_\lambda\|_{H^{1,p}} = \lambda^{\frac{d-p}{p}} \|f\|_{H^{1,p}}.$$

In the limit $\lambda \to 0$, thus $\|f_\lambda\|_{C^\alpha}$ is controlled by $\|Df_\lambda\|_{L^p}$, provided that

$$\lambda^{-\alpha} \leq \lambda^{\frac{d-p}{p}} \quad \text{for } \lambda < 1,$$

i.e.,

$$\alpha \leq 1 - \frac{d}{p} \quad \text{in the case } p > d.$$

Proof of Corollary 9.1.5: By Hölder's inequality

$$\int_{\Omega \cap B(x_0,R)} |Du| \leq |B(x_0,R)|^{1-\frac{1}{p}} \left(\int_{\Omega \cap B(x_0,R)} |Du|^p \right)^{\frac{1}{p}} \tag{9.1.34}$$

$$\leq c_3 \|Du\|_{L^p(\Omega)} R^{d\left(1-\frac{1}{p}\right)} \tag{9.1.35}$$

$$= c_3 \|Du\|_{L^p(\Omega)} R^{d-1+\left(1-\frac{d}{p}\right)}, \tag{9.1.36}$$

where c_3 depends on p and d only. Consequently, the assumptions of Theorem 9.1.3 hold. □

The following version of Theorem 9.1.3 is called "Morrey's Dirichlet growth theorem" and is frequently used for showing the regularity of minimizers of variational problems:

Corollary 9.1.6: *Let $u \in W^{1,2}(\Omega)$, and suppose there exist constants $K' < \infty$, $0 < \alpha < 1$ such that for all balls $B(x_0, R) \subset \mathbb{R}^d$,*

$$\int_{\Omega \cap B(x_0,R)} |Du|^2 \leq K' R^{d-2+2\alpha}. \tag{9.1.37}$$

Then $u \in C^\alpha(\bar{\Omega})$, and for all balls $B(z, r)$,

$$\operatorname*{osc}_{B(z,r) \cap \Omega} u \leq c(K')^{\frac{1}{2}} r^\alpha, \tag{9.1.38}$$

with c depending only on d and α.

Proof: By Hölder's inequality

$$\int_{\Omega \cap B(x_0,R)} |Du| \leq |B(x_0,R)|^{\frac{1}{2}} \left(\int_{\Omega \cap B(x_0,R)} |Du|^2 \right)^{\frac{1}{2}}$$

$$\leq c_4 (K')^{\frac{1}{2}} R^{d-1+\alpha}$$

by (9.1.37), with c_u depending on d only. Thus, the assumptions of Theorem 9.1.3 hold again.   □

Finally, later on (in Section 12.3), we shall use the following result of Campanato characterizing Hölder continuity in terms of L^p-approximability by means on balls:

Theorem 9.1.4: *Let $p \geq 1$, $d < \lambda \leq d+p$, and let $\Omega \subset \mathbb{R}^d$ be a bounded domain for which there exists some $\delta > 0$ with*

$$|B(x_0,r) \cap \Omega| \geq \delta r^d \quad \text{for all } x_0 \in \Omega, r > 0. \qquad (9.1.39)$$

Then a function $u \in L^p(\Omega)$ is contained in $C^\alpha(\Omega)$ for $\alpha = \frac{\lambda-d}{p}$ (or in $C^{0,1}(\Omega)$ in the case $\lambda = d+p$), precisely if there exists a constant $K < \infty$ with

$$\int_{B(x_0,r) \cap \Omega} |u(x) - u_{B(x_0,r)}|^p \, dx \leq K^p r^\lambda \quad \text{for all } x_0 \in \Omega, r > 0 \qquad (9.1.40)$$

(where for defining $u_{B(x_0,r)}$, we have extended u by 0 on $\mathbb{R}^d \setminus \Omega$).

Proof: Let $u \in C^\alpha(\Omega)$, $x \in \Omega \cap B(x_0,r)$. We then have

$$\left| u(x) - u_{B(x_0,R)} \right| \leq (2r)^\alpha \|u\|_{C^\alpha(\Omega)},$$

and hence

$$\int_{B(x_0,R) \cap \Omega} |u - u_{B(x_0,r)}|^p \leq c_5 \|u\|_{C^\alpha(\Omega)} r^{\alpha p + d},$$

whereby (9.1.40) is satisfied.

In order to prove the converse implication, we start with the following estimate for $0 < r < R$:

$$\left| u_{B(x_0,R)} - u_{B(x_0,r)} \right|^p \leq 2^{p-1} \left(\left| u(x) - u_{B(x_0,R)} \right|^p + \left| u(x) - u_{B(x_0,r)} \right|^p \right),$$

and thus, integrating with respect to x on $\Omega \cap B(x_0,r)$ and using (9.1.39),

$$\left| u_{B(x_0,R)} - u_{B(x_0,r)} \right|^p$$

$$\leq \frac{2^{p-1}}{\delta r^d} \left(\int_{B(x_0,r) \cap \Omega} |u - u_{B(x_0,R)}|^p + \int_{B(x_0,r) \cap \Omega} |u - u_{B(x_0,r)}|^p \right).$$

This implies

$$\left|u_{B(x_0,R)} - u_{B(x_0,r)}\right| \le c_6 K \frac{R^{\frac{\lambda}{p}}}{r^{\frac{d}{p}}}. \tag{9.1.41}$$

We put $R_i = \frac{R}{2^i}$ and obtain from (9.1.41)

$$\left|u_{B(x_0,R_i)} - u_{B(x_0,R_{i+1})}\right| \le c_7 K 2^{i\frac{d-\lambda}{p}} R^{\frac{\lambda-d}{p}}. \tag{9.1.42}$$

For $i < j$, this implies

$$\left|u_{B(x_0,R_i)} - u_{B(x_0,R_j)}\right| \le c_8 K R_i^{\frac{\lambda-d}{p}}. \tag{9.1.43}$$

Thus $\left(u_{B(x_0,R_i)}\right)_{i\in\mathbb{N}}$ constitutes a Cauchy sequence. Since (9.1.41) with $r_i = \frac{r}{2^i}$ also implies

$$\left|u_{B(x_0,R_i)} - u_{B(x_0,r_i)}\right| \le c_6 K \left(\frac{R}{r}\right)^{\frac{\lambda}{p}} r_i^{\frac{\lambda-d}{p}} \to 0 \quad \text{for } i \to \infty$$

because of $\lambda > d$, the limit of this Cauchy sequence does not depend on R. Since by Lemma A.4, $u_{B(x,r)}$ converges in L^1 for $r \to 0$ towards $u(x)$, in the limit $j \to \infty$, we obtain from (9.1.43)

$$\left|u_{B(x_0,R)} - u(x_0)\right| \le c_8 K R^{\frac{\lambda-d}{p}}. \tag{9.1.44}$$

Thus, $u_{B(x_0,R)}$ converges not only in L^1, but also uniformly towards u as $R \to 0$. Since for $R > 0$, $u_{B(x,R)}$ is continuous with respect x, then so is u.

It remains to show that u is α-Hölder continuous. For that purpose, let $x, y \in \Omega$, $R := |x - y|$. Then

$$\left|u(x) - u(y)\right| \le \left|u_{B(x,2R)} - u(x)\right| + \left|u_{B(x,2R)} - u_{B(y,2R)}\right|$$
$$+ \left|u(y) - u_{B(y,2R)}\right|. \tag{9.1.45}$$

Now

$$\left|u_{B(x,2R)} - u_{B(y,2R)}\right| \le \left|u_{B(x,2R)} - u(z)\right| + \left|u(z) - u_{B(y,2R)}\right|,$$

and integrating with respect to z on $B(x, 2R) \cap B(y, 2R) \cap \Omega$, we obtain

$$\left|u_{B(x,2R)} - u_{B(y,2R)}\right|$$
$$\le \frac{1}{|B(x,2R) \cap B(y,2R) \cap \Omega|} \left(\int_{B(x,2R)\cap\Omega} \left|u(z) - u_{B(x,2R)}\right| dz\right.$$
$$\left. + \int_{B(y,2R)\cap\Omega} \left|u(z) - u_{B(y,2R)}\right| dz\right)$$
$$\le \frac{c_9}{|B(x,2R) \cap B(y,2R) \cap \Omega|} K R^{\frac{\lambda-d}{p}+d}$$

by applying Hölder's inequality. Because of $R = |x - y|$,

$$B(x, R) \subset B(y, 2R),$$

and so by (9.1.39),

$$|B(x, 2R) \cap B(y, 2R) \cap \Omega| \geq |B(x, R) \cap \Omega| \geq \delta R^d.$$

We conclude that

$$\left| u_{B(x,2R)} - u_{B(y,2R)} \right| \leq c_{10} K R^{\frac{\lambda-d}{p}}. \tag{9.1.46}$$

Using (9.1.44) and (9.1.46), we obtain

$$|u(x) - u(y)| \leq c_{11} K |x - y|^{\frac{\lambda-d}{p}}, \tag{9.1.47}$$

which is Hölder continuity with exponent $\alpha = \frac{\lambda-d}{p}$. $\qquad\square$

Later on (in Section 12.3), we shall use the following local version of Campanato's theorem:

Corollary 9.1.7: *If for all $0 < r \leq R_0$ and all $x \in \Omega_0$, we have*

$$\int_{B(x_0,r)} \left| u - u_{B(x_0,r)} \right|^p \leq \gamma r^{d+p\alpha}$$

with constants γ and $0 < \alpha < 1$, then u is locally α-Hölder continuous in Ω_0 (this means that u is α-Hölder continuous in any $\Omega_1 \subset\subset \Omega_0$). $\qquad\square$

References for this section are Gilbarg–Trudinger [9] and Giaquinta [7].

9.2 L^2-Regularity Theory: Interior Regularity of Weak Solutions of the Poisson Equation

For $u : \Omega \to \mathbb{R}$, we define the difference quotient

$$\Delta_i^h u(x) := \frac{u(x + h e_i) - u(x)}{h} \quad (h \neq 0),$$

e_i being the ith unit vector of $\mathbb{R}^d$ ($i \in \{1, \ldots, d\}$).

Lemma 9.2.1: *Assume $u \in W^{1,2}(\Omega), \Omega' \subset\subset \Omega, |h| < \text{dist}(\Omega', \partial\Omega)$. Then $\Delta_i^h u \in L^2(\Omega')$ and*

$$\left\| \Delta_i^h u \right\|_{L^2(\Omega')} \leq \left\| D_i u \right\|_{L^2(\Omega)} \quad (i = 1, \ldots, d). \tag{9.2.1}$$

Proof: By an approximation argument, it again suffices to consider the case $u \in C^1(\Omega) \cap W^{1,2}(\Omega)$. Then

$$\Delta_i^h u(x) = \frac{u(x + he_i) - u(x)}{h}$$

$$= \frac{1}{h} \int_0^h D_i u(x^1, \ldots, x^{i-1}, x^i + \xi, x^{i+1}, \ldots, x^d) d\xi,$$

and with Hölder's inequality

$$\left|\Delta_i^h u(x)\right|^2 \leq \frac{1}{h} \int_0^h \left|D_i u(x_1, \ldots, x_i + \xi, \ldots, x_d)\right|^2 d\xi,$$

and thus

$$\int_{\Omega'} \left|\Delta_i^h u(x)\right|^2 dx \leq \frac{1}{h} \int_0^h \int_\Omega |D_i u|^2 \, dx d\xi = \int_\Omega |D_i u|^2 \, dx.$$

$\square$

Conversely, we have the following result:

Lemma 9.2.2: *Let $u \in L^2(\Omega)$, and suppose there exists $K < \infty$ with $\Delta_i^h u \in L^2(\Omega')$ and*

$$\left\|\Delta_i^h u\right\|_{L^2(\Omega')} \leq K \tag{9.2.2}$$

for all $h > 0$ and $\Omega' \subset\subset \Omega$ with $h < \text{dist}(\Omega', \partial\Omega)$. Then the weak derivative $D_i u$ exists and satisfies

$$\|D_i u\|_{L^2(\Omega)} \leq K. \tag{9.2.3}$$

Proof: For $\varphi \in C_0^1(\Omega)$ and $0 < h < \text{dist}(\text{supp}\,\varphi, \partial\Omega)$ ($\text{supp}\,\varphi$ is the closure of $\{x \in \Omega : \varphi(x) \neq 0\}$), we have

$$\int_\Omega \Delta_i^h u \, \varphi = - \int_\Omega u \Delta_i^{-h} \varphi \to - \int_\Omega u D_i \varphi,$$

as $h \to 0$. Thus, we also have

$$\left|\int_\Omega u D_i \varphi\right| \leq K \|\varphi\|_{L^2(\Omega)}.$$

Since $C_0^1(\Omega)$ is dense in $L^2(\Omega)$, we may thus extend

$$\varphi \mapsto - \int_\Omega u D_i \varphi$$

to a bounded linear functional on $L^2(\Omega)$. According to the Riesz representation theorem as quoted in Appendix 12.3, there then exists $v \in L^2(\Omega)$
with

$$\int_\Omega \varphi v = - \int_\Omega u D_i \varphi \quad \text{for all } \varphi \in C_0^1(\Omega).$$

Since this is precisely the equation defining $D_i u$, we must have $v = D_i u$. □

Theorem 9.2.1: *Let $u \in W^{1,2}(\Omega)$ be a weak solution of $\Delta u = f$ with $f \in$*
$L^2(\Omega)$. For any $\Omega' \subset\subset \Omega$, then $u \in W^{2,2}(\Omega')$, and

$$\|u\|_{W^{2,2}(\Omega')} \leq \text{const} \left(\|u\|_{L^2(\Omega)} + \|f\|_{L^2(\Omega)} \right), \tag{9.2.4}$$

where the constant depends only on $\delta := \text{dist}(\Omega', \partial\Omega)$. Furthermore, $\Delta u = f$
almost everywhere in Ω.

The content of Theorem 9.2.1 is twofold: First, there is a regularity result
saying that a weak solution of the Poisson equation is of class $W^{2,2}$ in the
interior, and second, we have an estimate for the $W^{2,2}$-norm. The proof will
yield both results at the same time. If the regularity result happens to be
known already, the estimate becomes much easier. That easier demonstration
of the estimate nevertheless contains the essential idea of the proof, and so
we present it first. To start with, we shall prove a lemma. The proof of that
lemma is typical for regularity arguments for weak solutions, and several of
the subsequent estimates will turn out to be variants of that proof. We thus
recommend that the reader study the following estimate very carefully.
Our starting point is the relation

$$\int_\Omega Du \cdot Dv = - \int_\Omega fv \quad \text{for all } v \in H_0^{1,2}(\Omega). \tag{9.2.5}$$

(Here, Du is the vector $(D_1 u, \ldots, D_d u)$.)
We need some technical preparation: We construct some $\eta \in C_0^1(\Omega)$ with
$0 \leq \eta \leq 1$, $\eta(x) = 1$ for $x \in \Omega'$ and $|D\eta| \leq \frac{2}{\delta}$. Such an η can be obtained
by mollification, i.e., by convolution with a smooth kernel as described in
Lemma A.2 in the Appendix, from the following function η_0:

$$\eta_0(x) := \begin{cases} 1 & \text{for } \text{dist}(x, \Omega') \leq \frac{\delta}{8}, \\ 0 & \text{for } \text{dist}(x, \Omega') \geq \frac{7\delta}{8}, \\ \frac{7}{6} - \frac{4}{3\delta} \text{dist}(x, \Omega') & \text{for } \frac{\delta}{8} \leq \text{dist}(x, \Omega') \leq \frac{7\delta}{8}. \end{cases}$$

Thus η_0 is a (piecewise) linear function of $\text{dist}(x, \Omega')$ interpolating between
Ω', where it takes the value 1, and the complement of Ω, where it is 0. This
is also the purpose of the cutoff function η. If one abandons the requirement
of continuous differentiability (which is not essential anyway), one may put
more simply

$$\eta(x) := \begin{cases} 1 & \text{for } x \in \Omega', \\ 0 & \text{for } \operatorname{dist}(x, \Omega') \geq \delta, \\ 1 - \frac{1}{\delta} \operatorname{dist}(x, \Omega') & \text{for } 0 \leq \operatorname{dist}(x, \Omega') \leq \delta \end{cases}$$

(note that $\operatorname{dist}(\Omega', \partial\Omega) \geq \delta$). It is not difficult to verify that $\eta \in H_0^{1,2}(\Omega)$, which suffices for the sequel. In (9.2.5), we now use the test function

$$v = \eta^2 u$$

with η of the type just presented. This yields

$$\int_\Omega \eta^2 |Du|^2 + 2 \int_\Omega \eta Du \cdot u D\eta = - \int_\Omega \eta^2 fu, \qquad (9.2.6)$$

and with the so-called Young inequality

$$\pm ab \leq \frac{\varepsilon}{2} a^2 + \frac{1}{2\varepsilon} b^2 \quad \text{for } a, b \in \mathbb{R}, \varepsilon > 0 \qquad (9.2.7)$$

used with $a = \eta|Du|$, $b = u|D\eta|$, $\varepsilon = \frac{1}{2}$ in the second integral, and with $a = \eta f$, $b = \eta u$, $\varepsilon = \delta^2$ in the integral on the right-hand side, we obtain

$$\int_\Omega \eta^2 |Du|^2 \leq \frac{1}{2} \int_\Omega \eta^2 |Du|^2 + 2 \int_\Omega |D\eta|^2 u^2 + \frac{1}{2\delta^2} \int_\Omega \eta^2 u^2 + \frac{\delta^2}{2} \int_\Omega \eta^2 f^2. \qquad (9.2.8)$$

We recall that $0 \leq \eta \leq 1$, $\eta = 1$ on Ω' to see that this yields

$$\int_{\Omega'} |Du|^2 \leq \int_\Omega \eta^2 |Du|^2 \leq \left(\frac{16}{\delta^2} + \frac{1}{\delta^2} \right) \int_\Omega u^2 + \delta^2 \int_\Omega f^2.$$

We record this inequality in the following lemma:

Lemma 9.2.3: Let u be a weak solution of $\Delta u = f$ with $f \in L^2(\Omega)$. We then have for any $\Omega' \subset\subset \Omega$,

$$\|Du\|^2_{L^2(\Omega')} \leq \frac{17}{\delta^2} \|u\|^2_{L^2(\Omega)} + \delta^2 \|f\|^2_{L^2(\Omega)}, \qquad (9.2.9)$$

where $\delta := \operatorname{dist}(\Omega', \partial\Omega)$. □

So far, we have not used that we are temporarily assuming $u \in W^{2,2}(\Omega')$ for any $\Omega' \subset\subset \Omega$. Now, however, we come to the estimate of the $W^{2,2}$-norm, so we shall need that assumption. Let $u \in W^{2,2}(\Omega') \cap W^{1,2}(\Omega)$ again satisfy

$$\int_\Omega Du \cdot Dv = - \int_\Omega fv \quad \text{for all } v \in H_0^{1,2}(\Omega). \qquad (9.2.10)$$

If $\operatorname{supp} v \subset\subset \Omega'$ (i.e., $v \in H_0^{1,2}(\Omega'')$ for some $\Omega'' \subset\subset \Omega'$), we may, assuming $u \in W^{2,2}(\Omega')$, integrate by parts in (9.2.10) to obtain

$$\int_{\Omega} (\sum_{i=1}^{d} D_i D_i u) v = \int_{\Omega} fv. \tag{9.2.11}$$

This in particular holds for all $v \in C_0^{\infty}(\Omega')$, and since $C_0^{\infty}(\Omega')$ is dense in $L^2(\Omega')$, (9.2.11) then also holds for $v \in L^2(\Omega')$, where we have put $v = 0$ in $\Omega \setminus \Omega'$.

We consider the matrix $D^2 u$ of the second weak derivatives of u and obtain

$$\int_{\Omega'} |D^2 u|^2 = \int_{\Omega'} \sum_{i,j=1}^{d} D_i D_j u \cdot D_i D_j u$$

$$= \int_{\Omega'} \sum_{i=2}^{d} D_i D_i u \cdot \sum_{i=1}^{d} D_j D_j u$$

+ boundary terms that we neglect for the moment (later on, they will be converted into interior terms with the help of cutoff functions),

by an integration by parts that will even require the assumption $u \in W^{3,2}(\Omega')$

$$= \int_{\Omega'} f \sum_{i=1}^{d} D_j D_j u$$

$$\leq \left(\int_{\Omega'} f^2 \right)^{\frac{1}{2}} \left(\int_{\Omega'} |D^2 u|^2 \right)^{\frac{1}{2}} \quad \text{by Hölder's inequality,} \tag{9.2.12}$$

and hence

$$\int_{\Omega'} |D^2 u|^2 \leq \int_{\Omega} f^2, \tag{9.2.13}$$

i.e.,

$$\|D^2 u\|_{L^2(\Omega')}^2 \leq \|f\|_{L^2(\Omega)}^2. \tag{9.2.14}$$

Taken together (9.2.9) and (9.2.14) yield

$$\|u\|_{W^{2,2}(\Omega')}^2 \leq (c_1(\delta) + 1) \|u\|_{L^2(\Omega)}^2 + 2 \|f\|_{L^2(\Omega)}^2. \tag{9.2.15}$$

We now come to the actual *Proof of Theorem 9.2.1*: Let

$$\Omega' \subset\subset \Omega'' \subset\subset \Omega, \quad \text{dist}(\Omega'', \partial\Omega) \geq \frac{\delta}{4}, \quad \text{dist}(\Omega', \partial\Omega'') \geq \frac{\delta}{4}.$$

We again use

$$\int_{\Omega} Du \cdot Dv = -\int_{\Omega} f \cdot v \quad \text{for all } v \in H_0^{1,2}(\Omega). \tag{9.2.16}$$

In the sequel, we consider v with

$$\operatorname{supp} v \subset\subset \Omega''$$

and choose $h > 0$ with

$$2h < \operatorname{dist}(\operatorname{supp} v, \partial\Omega'').$$

In (9.2.16), we may then also insert $\Delta_i^h v$ ($i \in \{1, \ldots, d\}$) in place of v. We obtain

$$\int_{\Omega''} D\Delta_i^h u \cdot Dv = \int_{\Omega''} \Delta_i^h(Du) \cdot Dv = -\int_{\Omega''} Du \cdot \Delta_i^h Dv$$

$$= -\int_{\Omega''} Du \cdot D\left(\Delta_i^h v\right) \qquad (9.2.17)$$

$$= \int_{\Omega''} f\Delta_i^h v \quad \leq \|f\|_{L^2(\Omega)} \cdot \|Dv\|_{L^2(\Omega'')}$$

by Lemma 9.2.1 and the choice of h. As described above, let $\eta \in C_0^1(\Omega'')$, $0 \leq \eta \leq 1$, $\eta(x) = 1$ for $x \in \Omega'$, $|D\eta| \leq 8/\delta$. We put

$$v := \eta^2 \Delta_i^h u.$$

From (9.2.17), we obtain

$$\int_{\Omega''} \left|\eta D\Delta_i^h u\right|^2 = \int_{\Omega''} D\Delta_i^h u \cdot Dv - 2\int_{\Omega''} \eta D\Delta_i^h u \cdot \Delta_i^h u D\eta$$

$$\leq \|f\|_{L^2(\Omega)} \left\|D\left(\eta^2 \Delta_i^h u\right)\right\|_{L^2(\Omega'')}$$

$$+ 2\left\|\eta D\Delta_i^h u\right\|_{L^2(\Omega'')} \left\|\Delta_i^h u D\eta\right\|_{L^2(\Omega'')}.$$

With Young's inequality (9.2.7) and employing Lemma 9.2.1 (recall the choice of h), we hence obtain

$$\left\|\eta D\Delta_i^h u\right\|_{L^2(\Omega'')}^2 \leq 2\|f\|_{L^2(\Omega)}^2 + \frac{1}{4}\left\|\eta D\Delta_i^h u\right\|_{L^2(\Omega'')}^2$$

$$+ \frac{1}{4}\left\|\eta D\Delta_i^h u\right\|_{L^2(\Omega'')}^2 + 8\sup|D\eta|^2 \|D_i u\|_{L^2(\Omega'')}^2.$$

The essential point in employing Young's inequality here is that the expression $\left\|\eta D\Delta_i^h u\right\|_{L^2(\Omega'')}^2$ occurs on the right-hand side with a smaller coefficient than on the left-hand side, and so the contribution on the right-hand side can be absorbed in the left-hand side. Because of $\eta \equiv 1$ on Ω' and $(a^2 + b^2)^{\frac{1}{2}} \leq a + b$ with Lemma 9.2.2, as $h \to \infty$, we obtain

$$\left\|D^2 u\right\|_{L^2(\Omega')} \leq \operatorname{const} \left(\|f\|_{L^2(\Omega)} + \frac{1}{\delta}\|Du\|_{L^2(\Omega'')}\right). \qquad (9.2.18)$$

Lemma 9.2.3 (with Ω'' in place of Ω') now implies

$$\|Du\|_{L^2(\Omega'')} \leq c_1 \left(\frac{1}{\delta} \|u\|_{L^2(\Omega)} + \delta \|f\|_{L^2(\Omega)} \right) \qquad (9.2.19)$$

with some constant c_1. Inequality (9.2.4) then follows from (9.2.18) and (9.2.19). □

If f happens to be even of class $W^{1,2}(\Omega)$, in (9.2.5) we may insert $D_i v$ in place of v to obtain

$$\int_\Omega D(D_i u) \cdot Dv = - \int_\Omega D_i f \cdot v.$$

Theorem 9.2.1 then implies $D_i u \in W^{2,2}(\Omega')$, i.e., $u \in W^{3,2}(\Omega')$. In this manner, we iteratively obtain the following theorem:

Theorem 9.2.2: Let $u \in W^{1,2}(\Omega)$ be a weak solution of $\Delta u = f$, $f \in W^{k,2}(\Omega)$. For any $\Omega' \subset\subset \Omega$ then $u \in W^{k+2,2}(\Omega')$, and

$$\|u\|_{W^{k+2,2}(\Omega')} \leq \mathrm{const} \left(\|u\|_{L^2(\Omega)} + \|f\|_{W^{k,2}(\Omega)} \right),$$

where the constant depends on d, h, and $\mathrm{dist}(\Omega', \partial\Omega)$.

Corollary 9.2.1: If $u \in W^{1,2}(\Omega)$ is a weak solution of $\Delta u = f$ with $f \in C^\infty(\Omega)$, then also $u \in C^\infty(\Omega)$.

Proof: From Theorem 9.2.2 and Corollary 9.1.2. □

At the end of this section, we wish to record once more a fundamental observation concerning elliptic regularity theory as encountered in the present section for the first time and to be encountered many more times in the subsequent sections. For any u contained in the Sobolev space $W^{2,2}(\Omega)$, we have the trivial estimate

$$\|u\|_{L^2(\Omega)} + \|\Delta u\|_{L^2(\Omega)} \leq \mathrm{const} \|u\|_{W^{2,2}(\Omega)}$$

(where Δu is to be understood as the sum of the weak pure second derivatives of u). Elliptic regularity theory yields an estimate in the opposite direction; according to Theorem 9.2.1, we have

$$\|u\|_{W^{2,2}(\Omega')} \leq \mathrm{const}(\|u\|_{L^2(\Omega)} + \|\Delta u\|_{L^2(\Omega)}) \quad \text{for } \Omega' \subset\subset \Omega.$$

Thus Δu and some lower order term already control all second derivatives of u. Lemma 9.2.3 shall be interpreted in this sense as well.

The Poincaré inequality states that for every $u \in H_0^{1,2}(\Omega)$,

$$\|u\|_{L^2(\Omega)} \leq \mathrm{const} \|Du\|_{L^2(\Omega)},$$

while for a harmonic $u \in W^{1,2}(\Omega)$, we have the estimate in the opposite direction,

$$\|Du\|_{L^2(\Omega')} \leq \|u\|_{L^2(\Omega)}$$

(for $\Omega' \subset\subset \Omega$).

In this sense, in elliptic regularity theory one has estimates in both directions, one direction resulting from general embedding theorems, and the other one from the elliptic equation. Combining both directions often allows iteration arguments for proving even higher regularity, as we have seen in the present section and as we shall have ample occasion to witness in subsequent sections.

9.3 Boundary Regularity and Regularity Results for Solutions of General Linear Elliptic Equations

With the help of Dirichlet's principle, we have found weak solutions of

$$\Delta u = f \quad \text{in } \Omega$$

with

$$u - g \in H_0^{1,2}(\Omega)$$

for given $f \in L^2(\Omega)$, $g \in H^{1,2}(\Omega)$. In the previous section, we have seen that in the interior of Ω, u is as regular as f allows. It is then natural to ask whether u is regular at $\partial\Omega$ as well, provided that g and $\partial\Omega$ satisfy suitable regularity conditions. A preliminary observation is that a solution of the above Dirichlet problem possesses a global bound that depends only on f and g:

Lemma 9.3.1: *Let u be a weak solution of $\Delta u = f$, $u - g \in H_0^{1,2}(\Omega)$ in the bounded region Ω. Then*

$$\|u\|_{W^{1,2}(\Omega)} \leq c \left(\|g\|_{W^{1,2}(\Omega)} + \|f\|_{L^2(\Omega)} \right), \tag{9.3.1}$$

where the constant c depends only on the Lebesgue measure $|\Omega|$ of Ω and on d.

Proof: We insert the test function $v = u - g$ into the weak differential equation

$$\int_\Omega Du \cdot Dv = -\int_\Omega fv \quad \text{for all } v \in H_0^{1,2}(\Omega)$$

to obtain

$$\int_\Omega |Du|^2 = \int Du \cdot Dg - \int fu + \int fg$$
$$\leq \frac{1}{2}\int |Du|^2 + \frac{1}{2}\int |Dg|^2 + \frac{1}{\varepsilon}\int f^2 + \frac{\varepsilon}{2}\int u^2 + \frac{\varepsilon}{2}\int g^2$$

for any $\varepsilon > 0$, by Young's inequality, and hence

$$\|Du\|_{L^2}^2 \leq \varepsilon\|u\|_{L^2}^2 + \|Dg\|_{L^2}^2 + \frac{2}{\varepsilon}\|f\|_{L^2}^2 + \varepsilon\|g\|_{L^2}^2,$$

i.e.,

$$\|Du\|_{L^2} \leq \sqrt{\varepsilon}\|u\|_{L^2} + \|Dg\|_{L^2} + \sqrt{\frac{2}{\varepsilon}}\|f\|_{L^2} + \sqrt{\varepsilon}\|g\|_{L^2}. \tag{9.3.2}$$

Obviously,

$$\|u\|_{L^2} \leq \|u - g\|_{L^2} + \|g\|_{L^2}, \tag{9.3.3}$$

and by the Poincaré inequality

$$\|u - g\|_{L^2} \leq \left(\frac{|\Omega|}{\omega_d}\right)^{\frac{1}{d}}(\|Du\|_{L^2} + \|Dg\|_{L^2}). \tag{9.3.4}$$

Altogether, it follows that

$$\|Du\|_{L^2} \leq \sqrt{\varepsilon}\left(\frac{|\Omega|}{\omega_d}\right)^{\frac{1}{d}}\|Du\|_{L^2} + \left(1 + \sqrt{\varepsilon}\left(\frac{|\Omega|}{\omega_d}\right)^{\frac{1}{d}}\right)\|Dg\|_{L^2}$$
$$+ 2\sqrt{\varepsilon}\|g\|_{L^2} + \sqrt{\frac{2}{\varepsilon}}\|f\|_{L^2}.$$

We now choose

$$\varepsilon = \frac{1}{4}\left(\frac{\omega_d}{|\Omega|}\right)^{\frac{2}{d}},$$

i.e.,

$$\sqrt{\varepsilon}\left(\frac{|\Omega|}{\omega_d}\right)^{\frac{1}{d}} = \frac{1}{2},$$

and obtain

$$\|Du\|_{L^2} \leq 3\|Dg\|_{L^2} + 2\left(\frac{\omega_d}{|\Omega|}\right)^{\frac{1}{d}}\|g\|_{L^2} + \sqrt{2}\cdot 4\left(\frac{|\Omega|}{\omega_d}\right)^{\frac{1}{d}}\|f\|_{L^2}. \tag{9.3.5}$$

Inequalities (9.3.3)–(9.3.5) then also yield an estimate for $\|u\|_{L^2}$, and (9.3.1) follows. $\square$

We also wish to convince ourselves that we can reduce our considerations to the case $u \in H_0^{1,2}(\Omega)$. Namely, we simply consider $\bar{u} := u - g \in H_0^{1,2}(\Omega)$, which satisfies

$$\Delta\bar{u} = \Delta u - \Delta g = f - \Delta g = \bar{f} \tag{9.3.6}$$

in the weak sense. Here, we are assuming $g \in W^{2,2}(\Omega)$, and thus, for $\bar{u} \in H_0^{1,2}(\Omega)$, we obtain the equation

$$\Delta\bar{u} = \bar{f} \tag{9.3.7}$$

with $\bar{f} \in L^2(\Omega)$, again in the weak sense. Since the $W^{2,2}$-norm of u can be estimated by those of $\bar{u}$ and g, it thus suffices to consider vanishing boundary values. We consequently assume that $u \in H_0^{1,2}(\Omega)$ is a weak solution of $\Delta u = f$ in Ω.

We now consider a special situation; namely, we assume that in the vicinity of a given point $x_0 \in \partial\Omega$, $\partial\Omega$ contains a piece of a hyperplane; for example, without loss of generality, $x_0 = 0$ and

$$\partial\Omega \cap \mathring{B}(0, R) = \left\{ (x^1, \ldots, x^{d-1}, 0) \right\} \cap \mathring{B}(0, R)$$

(here, $\mathring{B}(0, R) = \{ x \in \mathbb{R}^d : |x| < R \}$ is the interior of the ball $B(0, R)$) for some $R > 0$. Let

$$B^+(0, R) := \left\{ (x^1, \ldots, x^d) \in \mathring{B}(0, R) : x^d > 0 \right\} \subset \Omega.$$

If now $\eta \in C_0^1(\mathring{B}(0, R))$, we have

$$\eta^2 u \in H_0^{1,2}(B^+(0, R)),$$

because we are assuming that u vanishes on $\partial\Omega \cap \mathring{B}(0, R)$ in the Sobolev space sense. If now $1 \le i \le d - 1$ and $|h| < \text{dist}(\text{supp}\,\eta, \partial\mathring{B}(0, R))$, we also have

$$\eta^2 \Delta_i^h u \in H_0^{1,2}(B^+(0, R)).$$

Thus, we may proceed as in the proof of Theorem 9.2.1, in order to show that

$$D_{ij}u \in L^2\left(\mathring{B}\left(0, \frac{R}{2}\right)\right) \tag{9.3.8}$$

with a corresponding estimate, provided that i and j are not both equal to d. However, since, from our differential equation we have

$$D_{dd}u = f - \sum_{j=1}^{d-1} D_{jj}u, \tag{9.3.9}$$

we then also obtain

$$D_{dd}u \in L^2\left(\mathring{B}\left(0, \frac{R}{2}\right)\right),$$

and thus the desired regularity result

$$u \in W^{2,2}\left(\mathring{B}\left(0, \frac{R}{2}\right)\right),$$

as well as the corresponding estimate.

In order to treat the general case, we have to require suitable assumptions for $\partial\Omega$.

Definition 9.3.1: *An open and bounded set $\Omega \subset \mathbb{R}^d$ is of class C^k ($k = 0, 1, 2, \ldots, \infty$) if for any $x_0 \in \partial\Omega$ there exist $r > 0$ and a bijective map $\phi : \mathring{B}(x_0, r) \to \phi(\mathring{B}(x_0, r)) \subset \mathbb{R}^d$ ($\mathring{B}(x_0, r) = \{y \in \mathbb{R}^d : |x_0 - y| < r\}$) with the following properties:*

(i) $\phi(\Omega \cap \mathring{B}(x_0, r)) \subset \{(x^1, \ldots, x^d) : x^d > 0\}$.
(ii) $\phi(\partial\Omega \cap \mathring{B}(x_0, r)) \subset \{(x^1, \ldots, x^d) : x^d = 0\}$.
(iii) ϕ and ϕ^{-1} are of class C^k.

Remark: This means that $\partial\Omega$ is a $(d-1)$-dimensional submanifold of $\mathbb{R}^d$ of differentiability class C^k.

Definition 9.3.2: *Let $\Omega \subset \mathbb{R}^d$ be of class C^k, as defined in Definition 9.3.1. We say that $g : \bar{\Omega} \to \mathbb{R}$ is of class $C^l(\bar{\Omega})$ for $l \leq k$ if $g \in C^l(\Omega)$ and if for any $x_0 \in \partial\Omega$ and ϕ as in Definition 9.3.1,*

$$g \circ \phi^{-1} : \{(x^1, \ldots, x^d) : x^d \geq 0\} \to \mathbb{R}$$

is of class C^l.

The crucial idea for boundary regularity is to consider, instead of u, local functions $u \circ \phi^{-1}$ with ϕ as in Definition 9.3.1. As we have argued at the beginning of this section, we may assume that the prescribed boundary values are $g = 0$. Then $u \circ \phi^{-1}$ is defined on some half-ball, and we may therefore carry over the interior regularity theory as just described. However, in general $u \circ \phi^{-1}$ no longer satisfies the Laplace equation. It turns out, however, that $u\circ\phi^{-1}$ satisfies a more general differential equation that is structurally similar to the Laplace equation and for which one may derive interior regularity in a similar manner.

We have derived a corresponding transformation formula already in Section 8.4. Thus $w = u \circ \phi^{-1}$ satisfies a differential equation (8.4.11), i.e.,

$$\frac{1}{\sqrt{g}} \sum_{J=1}^{d} \left(\frac{\partial}{\partial\xi^j}\left(\sqrt{g}\sum_{i=1}^{d} g^{ij}\frac{\partial w}{\partial\xi^i}\right)\right) = 0, \tag{9.3.10}$$

where the positive definite matrix g^{ij} is computed from ϕ and its derivatives (cf. (8.4.7)).

We shall consider an even more general class of elliptic differential equations:

$$Lu := \sum_{i,j=1}^{d} \frac{\partial}{\partial x^j} \left(a^{ij}(x) \frac{\partial}{\partial x^i} u(x) \right) + \sum_{j=1}^{d} \frac{\partial}{\partial x^j} \left(b^j(x)u(x) \right)$$

$$+ \sum_{i=1}^{d} c^i(x) \frac{\partial}{\partial x^i} u(x) + d(x)u(x)$$

$$= f(x). \tag{9.3.11}$$

We shall need two essential assumptions:

(A1) (Ellipticity) There exists some $\lambda > 0$ with

$$\sum_{i,j=1}^{d} a^{ij}(x)\xi_i\xi_j \geq \lambda \left| \xi \right|^2 \quad \text{for all } x \in \Omega, \xi \in \mathbb{R}^d.$$

(A2) (Boundedness) There exists some $M < \infty$ with

$$\sup_{x \in \Omega, i, j} \left(\left| a^{ij}(x) \right|, \left| b^i(x) \right|, \left| c(x) \right|, \left| d(x) \right| \right) \leq M.$$

A function u is called a weak solution of the Dirichlet problem

$$Lu = f \quad \text{in } \Omega \quad (f \in L^2(\Omega) \text{ given}),$$

$$u - g \in H_0^{1,2}(\Omega),$$

if for all $v \in H_0^{1,2}(\Omega)$,

$$\int_{\Omega} \left\{ \sum_{i,j} a^{ij}(x)D_iu(x)D_jv(x) + \sum_j b^j(x)u(x)D_jv(x) \right.$$

$$\left. - \left(\sum_i c^i(x)D_iu(x) + d(x)u(x) \right) v(x) \right\} dx = - \int_{\Omega} f(x)v(x)dx. \tag{9.3.12}$$

In order to become a little more familiar with (9.3.12), we shall first try to find out what happens if we insert our test functions that proved successful for the weak Poisson equation, namely, $v = \eta^2 u$ and $v = u - g$. Here η is a cutoff function as described in Section 9.2 with respect to $\Omega' \subset\subset \Omega$. With $v = \eta^2 u$, (9.3.12) then becomes

$$\int_{\Omega} \left\{ \sum \eta^2 a^{ij} D_iu D_ju + 2 \sum \eta a^{ij} u D_iu D_j\eta + \sum \eta^2 b^j u D_ju \right. \tag{9.3.13}$$

$$\left. + 2 \sum u^2 b^j \eta D_j\eta - \sum \eta^2 c^i u D_iu - d\eta^2 u^2 \right\} = - \int f\eta^2 u.$$

Analogously to (9.2.8), using Young's inequality, this time of the form

$$\sum a^{ij} a_i b_j \leq \frac{\varepsilon}{2} \sum a^{ij} a_i a_j + \frac{1}{2\varepsilon} \sum a^{ij} b_i b_j \qquad (9.3.14)$$

for $\varepsilon > 0$, $(a_1, \ldots, a_d), (b_1, \ldots, b_d) \in \mathbb{R}^d$, and a positive definite matrix $(a^{ij})_{i,j=1,\ldots,d}$, we thence obtain the following inequality:

$$\int \eta^2 |Du|^2 \leq \frac{1}{\lambda} \int \eta^2 \sum a^{ij} D_i u D_j u$$
$$\leq \frac{\varepsilon M}{\lambda} \int |Du|^2 \eta^2 + c_1(\varepsilon, \lambda, M, d) \int \eta^2 u^2 \qquad (9.3.15)$$
$$+ c_2(\delta, \lambda, M, d) \int u^2 |D\eta|^2 + \frac{\delta^2}{2} \int \eta^2 f^2,$$

where $\varepsilon > 0$ remains to be chosen appropriately, and $\delta = \text{dist}(\Omega', \partial\Omega)$, with constants c_1, c_2 that depend only on the indicated quantities. Of course, we have used (A1) and (A2) here. With $\varepsilon = \frac{\lambda}{2M}$, this yields

$$\int_{\Omega'} |Du|^2 \leq c_3(\delta, \lambda, M, d) \int_\Omega u^2 + \delta^2 \int_\Omega f^2, \qquad (9.3.16)$$

where we have also used the properties of η. This is the analogue of Lemma 9.2.3. The global bound of Lemma 9.3.1, however, does not admit a direct generalization. If we insert the test function $u - g$ in (9.3.12), we obtain only (as usual, employing Young's inequality in order to absorb all the terms containing derivatives into the positive definite leading term)

$$\int_\Omega |Du|^2 \leq \frac{1}{\lambda} \int \sum a^{ij} D_i u D_j u$$
$$\leq c_4(\lambda, M, d, |\Omega|) \left(\|g\|_{W^{1,2}}^2 + \|f\|_{L^2(\Omega)}^2 + \|u\|_{L^2(\Omega)}^2 \right). \qquad (9.3.17)$$

Thus, the additional term $\|u\|_{L^2(\Omega)}^2$ appears in the right-hand side. That this is really necessary can already be seen from the differential equation

$$u''(t) + \kappa^2 u(t) = 0 \quad \text{for } 0 < t < \pi,$$
$$u(0) = u(\pi) = 0, \qquad (9.3.18)$$

with $\kappa > 0$. Namely, for $\kappa \in \mathbb{N}$, we have the solutions

$$u(t) = b \sin(\kappa t)$$

with $b \in \mathbb{R}$ arbitrary, and these solutions obviously cannot be controlled solely by the right-hand side of the differential equation and the boundary values, because those are all zero. The local interior regularity theory of Section 9.2, however, remains fully valid. Namely, we have the following theorem:

Theorem 9.3.1: *Let $u \in W^{1,2}(\Omega)$ be a weak solution of $Lu = f$; i.e., let (9.3.12) hold. Let the ellipticity assumption (A1) hold. Moreover, let all coefficients $a^{ij}(x), \ldots, d(x)$ as well as $f(x)$ be of class C^∞. Then also $u \in C^\infty(\Omega)$.*

Remark: Regularity is a local result. Since we assume that all coefficients are C^∞, in particular, on every $\Omega' \subset\subset \Omega$, we have a bound of type (A2), with the constant M depending on Ω' here, however.

Let us discuss the *Proof of Theorem 9.3.1:* We first reduce the proof to the case $b^j, c^i, d \equiv 0$, i.e., to the regularity of weak solutions of

$$Mu := \sum_{i,j} \frac{\partial}{\partial x^j}\left(a^{ij}(x)\frac{\partial}{\partial x^i}u(x)\right) = f(x). \qquad (9.3.19)$$

For that purpose, we simply rewrite

$$Lu = f$$

as

$$Mu = -\sum \frac{\partial}{\partial x^j}(b^j(x)u(x)) - \sum c^i(x)\frac{\partial}{\partial x^i}u(x) - d(x)u(x) + f(x). \qquad (9.3.20)$$

We then prove the following theorem:

Theorem 9.3.2: *Let $u \in W^{1,2}(\Omega)$ be a weak solution of $Mu = f$ with $f \in W^{k,2}(\Omega)$. Assume (A1), and that the coefficients $a^{ij}(x)$ of M are of class $C^{k+1}(\Omega)$. Then for every $\Omega' \subset\subset \Omega$,*

$$u \in W^{k+2,k}(\Omega').$$

If

$$\left\|a^{ij}\right\|_{C^{k+1}(\Omega')} \leq M_k \quad \text{for all } i, j, \qquad (9.3.21)$$

then

$$\|u\|_{W^{k+2,k}(\Omega')} \leq c\left(\|u\|_{L^2(\Omega)} + \|f\|_{W^{k,2}(\Omega)}\right) \qquad (9.3.22)$$

with $c = c(d, \lambda, k, M_k, \text{dist}(\Omega', \partial\Omega))$.

The Sobolev embedding theorem then implies that in case $a^{ij}, f \in C^\infty$, any solution of $Mu = f$ is of class C^∞ as well. The corresponding regularity for solutions of $Lu = f$, as claimed in Theorem 9.3.1 can then be obtained through the following important iteration argument: Since we assume $u \in W^{1,2}(\Omega)$, the right-hand side of (9.3.20) is in $L^2(\Omega)$. According to Theorem 9.3.2, for $k = 0$, then $u \in W^{2,2}(\Omega)$. This in turn implies that the right-hand side of (9.3.20) is in $W^{1,2}(\Omega)$. Thus, we may apply Theorem 9.3.2

for $k = 1$ to obtain $u \in W^{3,2}(\Omega)$. But then, the right-hand side is in $W^{2,2}(\Omega)$; hence $u \in W^{4,2}(\Omega)$, and so on.

In that manner we deduce $u \in W^{m,2}(\Omega)$ for all $m \in \mathbb{N}$, and by the Sobolev embedding theorem, hence that u is in $C^\infty(\Omega)$.

We shall not display all details of the *Proof of Theorem 9.3.2* here, since this represents a generalization of the reasoning given in Section 9.2 that only needs a more cumbersome notation, but no new ideas. We have already seen how such a generalization works when we inserted the test function $\eta^2 u$ in (9.3.12). The only additional ingredient is certain rules for manipulating difference quotients, like the product rule

$$\Delta_l^k(ab)(x) = \frac{1}{h}\left(a(x + he_l)b(x + he_l) - a(x)b(x)\right)$$
$$= a(x + he_l)\Delta_l^h b(x) + \left(\Delta_l^h a(x)\right)b(x). \tag{9.3.23}$$

For example,

$$\Delta_l^h\left(\sum_{i=1}^d a^{ij}(x)D_i u(x)\right) = \sum_i \left(a^{ij}(x + he_l)\Delta_l^h D_i u(x) + \Delta_l^h a^{ij}(x)D_i u(x)\right). \tag{9.3.24}$$

As before, we use $\Delta_l^{-h}v$ as a test function in place of v, and in the case $\operatorname{supp} v \subset\subset \Omega''$, $2h < \operatorname{dist}(\operatorname{supp} v, \partial\Omega'')$, we obtain

$$\int_{\Omega''}\sum_{i,j}\Delta_l^h\left(a^{ij}(x)D_i u(x)\right)D_j v(x)dx = \int f(x)\Delta_l^{-h}v(x)dx. \tag{9.3.25}$$

With (9.3.23) and Lemma 9.2.1, this yields

$$\int_{\Omega''}\sum_{i,j}a^{ij}(x + he_l)D_i\Delta_l^h u(x)D_j v(x)dx$$
$$\leq c_5(d, M_1)\left(\|u\|_{W^{1,2}(\Omega'')} + \|f\|_{L^2(\Omega)}\right)\|Dv\|_{L^2(\Omega'')}, \tag{9.3.26}$$

i.e., an analogue of (9.2.17). Since because of the ellipticity condition (A1), we have the estimate

$$\lambda\int_\Omega \left|\eta D\Delta_l^h u(x)\right|^2 dx \leq \int_\Omega \eta^2\sum_{i,j}a^{ij}(x + he_l)\Delta_l^h D_i u(x)\Delta_l^h D_j u(x)dx,$$

we can then proceed as in the proofs of Theorems 9.2.1 and 9.2.2. Readers so inclined should face no difficulties in supplying the details.   □

We now return to the question of boundary regularity and state a theorem:

Theorem 9.3.3: *Let u be a weak solution of $Mu = f$ in Ω with $u - g \in H_0^{1,2}(\Omega)$. As always, suppose (A1). Let $f \in W^{k,2}(\Omega)$, $g \in W^{k+2,2}(\Omega)$. Let Ω be of class C^{k+2}, and let the coefficients of M be of class $C^{k+1}(\bar{\Omega})$ (in the sense of Definition 9.3.1). Then*

$$u \in W^{k+2,2}(\Omega),$$

and we have the estimate

$$\|u\|_{W^{k+2,2}(\Omega)} \leq c \left(\|f\|_{W^{k,2}(\Omega)} + \|g\|_{W^{k+2,2}(\Omega)} \right),$$

with c depending on λ, d, and Ω, and on C^{k+1}-bounds for the a^{ij}.

Proof: As explained at the beginning of this section, we may assume that $\partial\Omega$ is locally a hyperplane, by considering the composition $u \circ \phi^{-1}$ in place of u, where ϕ is a diffeomorphism of the type described in Definition 9.3.1. Namely, by (8.4.12) our equation $Mu = f$ gets transformed into an equation

$$\tilde{M}\tilde{u} = \tilde{f}$$

of the same type, with estimates for the coefficients of $\tilde{M}$ following from those for the a^{ij} as well as estimates for the derivatives of ϕ. We have already explained above how to obtain estimates for u in that particular geometric situation. We let this suffice here, instead of offering tedious details without new ideas.   □

Remark: As a reference for the regularity theory of weak solutions, we recommend Gilbarg–Trudinger [9].

9.4 Extensions of Sobolev Functions and Natural Boundary Conditions

Most of our preceding results have been formulated for the spaces $H_0^{k,p}(\Omega)$ only, but not for the general Sobolev spaces $W^{k,p}(\Omega) = H^{k,p}(\Omega)$. A technical reason for this is that the mollifications that we have frequently employed use the values of the given function in some full ball about the point under consideration, and this cannot be done at a boundary point if the function is defined only in the domain Ω, perhaps up to its boundary, but not in the exterior of Ω. Thus, it seems natural to extend a given Sobolev function on a domain Ω in $\mathbb{R}^d$ to all of $\mathbb{R}^d$, or at least to some larger domain that contains the closure of Ω in its interior. The problem then is to guarantee that the extended function maintains all the weak differentiability properties of the original function. It turns out that for this to be successfully resolved, we need to impose certain regularity conditions on $\partial\Omega$ as in Definition 9.3.1. In the spirit of that definition, we thus start with the model situation of the domain

$$\mathbb{R}^d_+ := \left\{ (x^1, \dots, x^d) \in \mathbb{R}^d, x^d > 0 \right\}.$$

If now $u \in C^k(\overline{\mathbb{R}^d_+})$, we define an extension via

$$E_0 u(x) := \begin{cases} u(x) & \text{for } x^d \geq 0, \\ \sum_{j=1}^{k} a_j u(x^1, \dots, x^{d-1}, -\frac{1}{j} x^d) & \text{for } x^d < 0, \end{cases} \tag{9.4.1}$$

where the a_j are chosen such that

$$\sum_{j=1}^{k} a_j \left(-\frac{1}{j} \right)^{\nu} = 1 \quad \text{for } \nu = 0, \dots, k-1. \tag{9.4.2}$$

One readily verifies that the system (9.4.2) is uniquely solvable for the a_j (the determinant of this system is a Vandermonde determinant that is nonzero). One moreover verifies, and this of course is the reason for the choice of the a_j, that the derivatives of $E_0 u$ up to order $k-1$ coincide with the corresponding ones of u on the hyperplane $\{x^d = 0\}$, and that the derivatives of order k are bounded whenever those of u are. Thus

$$E_0 u \in C^{k-1,1}(\mathbb{R}^d), \tag{9.4.3}$$

where $C^{l,1}(\Omega)$ is defined as the space of l-times continuously differentiable functions on Ω whose lth derivatives are Lipschitz continuous, i.e.,

$$\sup_{x \in \Omega} \frac{|v(x) - v(x_0)|}{|x - x_0|} < \infty$$

for any such derivative v and $x_0 \in \Omega$ (see also Definition 11.1.1 below).

If now Ω is a domain of class C^k in the sense of Definition 9.3.1, and if $u \in C^k(\bar{\Omega})$ (see Definition 9.3.2), we may locally straighten out the boundary with a C^k-diffeomorphism ϕ^{-1}, extend the functions $u \circ \phi^{-1}$ with the above operator E_0, and then take $E_0(u \circ \phi^{-1}) \circ \phi$. This function then defines a local extension of class $C^{k-1,1}$ of u across $\partial\Omega$. In order to obtain a global extension, we simply patch these local extensions together with the help of a partition of unity. This is easy, and the reader may know this construction already, but for completeness, we present the details. We assume that Ω is a bounded domain of class C^k. Thus, $\partial\Omega$ is compact, and so it may be covered by finitely many sets of the type $\Omega \cap \mathring{B}(x_0, r)$ on which a local diffeomorphism with the properties specified in Definition 9.3.1 exists.

We call these sets Ω_ν, $\nu = 1, \dots, n$, and the corresponding diffeomorphisms ϕ_ν. In addition, we may find an open set $\Omega_0 \subset \Omega$, with $\partial\Omega \cap \bar{\Omega}_0 = \emptyset$, so that

$$\Omega \subset \bigcup_{\nu=0}^{m} \Omega_\nu.$$

We then let φ_ν, $\nu = 0, \ldots, m$, be a partition of unity subordinate to this covering of Ω and put

$$Eu := \varphi_0 u + \sum_{\nu=1}^{m} E_0 \left((\varphi_\nu u) \circ \phi_\nu^{-1} \right) \circ \phi_\nu.$$

This then extends u as a $C^{k-1,1}$ function to some open neighborhood Ω' of $\bar{\Omega}$. By taking a $C_0^\infty(\mathbb{R}^d)$ function η with $\eta \equiv 1$ on Ω, $\eta \equiv 0$ in $\mathbb{R}^d \setminus \Omega'$, one may then also extend u to the $C^{k-1,1}(\mathbb{R}^d)$ function ηEu. In fact, this extension lies in $C_0^{k-1,1}(\Omega')$.

This was for C^k-functions, but it may be extended to Sobolev functions by approximation. Again considering the model situation of $\mathbb{R}_+^d$, we observe that $u \in W^{k,p}(\mathbb{R}_+^d)$ can be approximated by the translated mollifications

$$u_h(x + 2he_d) = \frac{1}{h^d} \int_{y^d > 0} u(y) \varrho \left(\frac{x + 2he_d - y}{h} \right) dy$$

for $h \to 0$ ($h > 0$) (here, e_d is the dth unit vector in $\mathbb{R}^d$). The limit for $h \to 0$ of the extensions $Eu(x + 2he_d)$ then yields the extension $Eu(x)$. One readily verifies that $Eu \in W^{k,p}(\Omega')$ for some domain Ω' containing $\bar{\Omega}$ (for the detailed argument, one needs the extension lemma (Lemma 8.2.2), which obviously holds for all p, not just for $p = 2$) in order to handle the possible discontinuity of the highest-order derivatives along $\partial \Omega$ in the above construction), and that

$$\|Eu\|_{W^{k,p}(\Omega')} \leq C \|u\|_{W^{k,p}(\Omega)} \tag{9.4.4}$$

for some constant C depending on Ω (via bounds on the maps ϕ, ϕ^{-1} from Definition 9.3.1) and k. As above, by multiplying by a C_0^∞ function η with $\eta \equiv 1$ on Ω, $\eta \equiv 0$ outside Ω', we may even assume

$$Eu \in H_0^{k,p}(\Omega'). \tag{9.4.5}$$

Equipped with our extension operator E, we may now extend the embedding theorems from the Sobolev spaces $H_0^{k,p}(\Omega)$ to the spaces $W^{k,p}(\Omega)$, if Ω is a C^k-domain. Namely, if $u \in W^{k,p}(\Omega)$, we consider $Eu \in H_0^{k,p}(\Omega')$, which then is contained in $L^{\frac{dp}{d-kp}}(\Omega')$ for $kp < d$, and in $C^m(\Omega')$, respectively, for $0 \leq m < k - \frac{d}{p}$, according to Corollary 9.1.1, and thus in $L^{\frac{dp}{d-kp}}(\Omega)$ or $C^m(\Omega)$, by restriction from Ω' to Ω. Since $Eu = u$ on Ω, we have thus proved the following version of the Sobolev embedding theorem:

Theorem 9.4.1: *Let $\Omega \subset \mathbb{R}^d$ be a bounded domain of class C^k. Then*

$$W^{k,p}(\Omega) \subset \begin{cases} L^{\frac{dp}{d-kp}}(\Omega) & \text{for } kp < d, \\ C^m(\bar{\Omega}) & \text{for } 0 \leq m < k - \frac{d}{p}. \end{cases} \tag{9.4.6}$$

$\square$

In the same manner, we may extend the compactness theorem of Rellich:

Theorem 9.4.2: *Let $\Omega \subset \mathbb{R}^d$ be a bounded domain of class C^1. Then any sequence $(u_n)_{n \in \mathbb{N}}$ that is bounded in $W^{1,2}(\Omega)$ contains a subsequence that converges in $L^2(\Omega)$.* $\square$

The preceding version of the Sobolev embedding theorem allows us to put our previous existence and regularity results together to obtain a very satisfactory treatment of the Poisson equation in the smooth setting:

Theorem 9.4.3: *Let $\Omega \subset \mathbb{R}^d$ be a bounded domain of class C^∞, and let $g \in C^\infty(\partial \Omega)$, $f \in C^\infty(\bar{\Omega})$. Then the Dirichlet problem*

$$\Delta u = f \quad in \ \Omega,$$
$$u = g \quad on \ \partial \Omega,$$

possesses a (unique) solution u of class $C^\infty(\bar{\Omega})$.

Proof: As explained in the beginning of Section 9.3, we may restrict ourselves to the case where $g = 0$, by considering $\bar{u} = u - g$ in place of u, where we have extended g as a C^∞-function to all of $\bar{\Omega}$. (Since Ω is bounded, C^∞-functions on $\bar{\Omega}$ are contained in all Sobolev spaces $W^{k,p}(\bar{\Omega})$.)

In Section 8.3, we have seen how Dirichlet's principle produces a weak solution $u \in H_0^{1,2}(\Omega)$ of $\Delta u = f$. We have already observed in Corollary 8.3.1 that such a u is smooth in Ω, but of course this follows also from the more general approach of Section 9.2, as stated in Corollary 9.2.1. Regularity up to the boundary, i.e., the result that $u \in C^\infty(\bar{\Omega})$, finally follows from the Sobolev estimates of Theorem 9.3.3 together with the embedding theorem (Theorem 9.4.1). $\square$

Of course, analogous statements can be stated and proved with the concepts and methods developed here in the C^k-case, for any $k \in \mathbb{N}$. In this setting, however, a somewhat more refined result will be obtained below in Theorem 11.3.1.

Likewise, the results extend to more general elliptic operators. Combining Corollary 8.5.2 with Theorem 9.3.3 and Theorem 9.4.1, we obtain the following theorem:

Theorem 9.4.4: *Let $\Omega \subset \mathbb{R}^d$ be a bounded domain of class C^∞. Let the functions a^{ij} $(i,j = 1,\ldots,d)$ and c be of class C^∞ in Ω and satisfy the assumptions (A)–(D) of Section 8.5, and let $f \in C^\infty(\Omega)$, $g \in C^\infty(\partial \Omega)$ be given. Then the Dirichlet problem*

$$\sum_{i,j=1}^{d} \frac{\partial}{\partial x^i} \left(a^{ij}(x) \frac{\partial}{\partial x^j} u(x) \right) - c(x)u(x) = f(x) \quad in \ \Omega,$$

$$u(x) = g(x) \quad on \ \partial\Omega,$$

admits a (unique) solution of class $C^{\infty}(\bar{\Omega})$. □

It is instructive to compare this result with Theorem 11.3.2 below.

We now address a question that the curious reader may already have wondered about. Namely, what happens if we consider the weak differential equation

$$\int_{\Omega} Du \cdot Dv + \int_{\Omega} fv = 0 \quad (f \in L^2(\Omega)) \tag{9.4.7}$$

for all $v \in W^{1,2}(\Omega)$, and not only for those in $H_0^{1,2}(\Omega)$? A solution u again has to be as regular as f and Ω allow, and in fact, the regularity proofs become simpler, since we do not need to restrict our test functions to have vanishing boundary values. In particular we have the following result:

Theorem 9.4.5: *Let (9.4.7) be satisfied for all* $v \in W^{1,2}(\Omega)$, *on some* C^{∞}-*domain* Ω, *for some function* $f \in C^{\infty}(\bar{\Omega})$. *Then also*

$$u \in C^{\infty}(\bar{\Omega}).$$

The *Proof* follows the scheme presented in Section 9.3. We obtain differentiability results on the boundary $\partial\Omega$ (note that here we conclude that u is smooth even on the boundary and not only in Ω as in Theorem 9.3.1) by applying the version stated in Theorem 9.4.1 of the Sobolev embedding theorem. □

In Section 9.5 we shall need regularity results for solutions of

$$\int_{\Omega} Du \cdot Dv + \mu \int_{\Omega} u \cdot v = 0 \quad (\mu \in \mathbb{R}), \quad for \ all \ v \in W^{1,2}(\Omega). \tag{9.4.8}$$

We can apply the iteration scheme described in Section 9.3 to establish the following corollary:

Corollary 9.4.1: *Let* u *be a solution of (9.4.8), for all* $v \in W^{1,2}(\Omega)$. *If the domain* Ω *is of class* C^{∞}, *then* $u \in C^{\infty}(\bar{\Omega})$. □

We return to the equation

$$\int_{\Omega} Du \cdot Dv + \int_{\Omega} fv = 0$$

on a C^{∞}-domain Ω, for $f \in C^{\infty}(\bar{\Omega})$. Since u is smooth up to the boundary by Theorem 9.4.5, we may integrate by parts to obtain

$$-\int_\Omega \Delta u \cdot v + \int_{\partial\Omega} \frac{\partial u}{\partial n} \cdot v + \int_\Omega fv = 0 \quad \text{for all } v \in W^{1,2}(\Omega). \qquad (9.4.9)$$

We know from our discussion of the weak Poisson equation that already if (9.4.7) holds for all $v \in H_0^{1,2}(\Omega)$, then, since u is smooth, necessarily

$$\Delta u = f \quad \text{in } \Omega. \qquad (9.4.10)$$

Equation (9.4.9) then implies

$$\int_{\partial\Omega} \frac{\partial u}{\partial n} \cdot v = 0 \quad \text{for all } v \in W^{1,2}(\Omega).$$

This then implies

$$\frac{\partial u}{\partial n} = 0 \quad \text{on } \partial\Omega. \qquad (9.4.11)$$

Thus, u satisfies a homogeneous Neumann boundary condition. Since this boundary condition arises from (9.4.7) when we do not impose any restrictions on v, it then is also called a natural boundary condition.

We add some further easy observations (which have already been made in Section 1.1): If u is a solution, so is $u + c$, for any $c \in \mathbb{R}$. Thus, in contrast to the Dirichlet problem, a solution of the Neumann problem is not unique. On the other hand, a solution does not always exist. Namely, we have

$$-\int_\Omega \Delta u + \int_\Omega \frac{\partial u}{\partial n} = 0,$$

and therefore, using $v \equiv 1$ in (9.4.9), we obtain the condition

$$\int_\Omega f = 0 \qquad (9.4.12)$$

on f as a necessary condition for the solvability of (9.4.9), hence of (9.4.7). It is not hard to show that this condition is also sufficient, but we do not pursue that point here.

Again, the preceding considerations about the regularity of solutions of the Neumann problem extend to more general elliptic operators, in the same manner as in Section 9.3. This is straightforward.

Finally, one may also consider inhomogeneous Neumann boundary conditions; for simplicity, we consider only the Laplace equation, i.e., assume $f = 0$ in the above.

A solution of

$$\Delta u = 0 \quad \text{in } \Omega,$$

$$\frac{\partial u}{\partial n} = h \quad \text{on } \partial\Omega, \text{ for some given smooth function } h \text{ on } \partial\Omega, \qquad (9.4.13)$$

can then be obtained by minimizing

$$\frac{1}{2} \int_\Omega |Du|^2 - \int_{\partial\Omega} hu \quad \text{in } W^{1,2}(\Omega). \qquad (9.4.14)$$

Here, a necessary (and sufficient) condition for solvability is

$$\int_{\partial\Omega} h = 0. \qquad (9.4.15)$$

In contrast to the inhomogeneous Dirichlet boundary condition, here the boundary values do not constrain the space in which we seek a minimizer, but rather enter into the functional to be minimized. Again, a weak solution u, i.e., satisfying

$$\int_\Omega Du \cdot Dv - \int_{\partial\Omega} hv = 0 \quad \text{for all } v \in W^{1,2}(\Omega), \qquad (9.4.16)$$

is determined up to a constant and is smooth up to the boundary, assuming, of course, that $\partial\Omega$ is smooth as before.

9.5 Eigenvalues of Elliptic Operators

In this textbook, at several places (see Sections 4.1, 5.2, 5.3, 6.1), we have already encountered expansions in terms of eigenfunctions of the Laplace operator. These expansions, however, served as heuristic motivations only, since we did not show the convergence of these expansions. It is the purpose of the present section to carry this out and to study the eigenvalues of the Laplace operator systematically. In fact, our reasoning will also apply to elliptic operators in divergence form,

$$Lu = \sum_{i,j=1}^d \frac{\partial}{\partial x^j} \left(a^{ij}(x) \frac{\partial}{\partial x^i} u(x) \right), \qquad (9.5.1)$$

for which the coefficients $a^{ij}(x)$ satisfy the assumptions stated in Section 9.3 and are smooth in Ω. Nevertheless, since we have already learned in this chapter how to extend the theory of the Laplace operator to such operators, here we shall carry out the analysis only for the Laplace operator. The indicated generalization we shall leave as an easy exercise. We hope that this strategy has the pedagogical advantage of concentrating on the really essential features.

Let Ω be an open and bounded domain in $\mathbb{R}^d$. The eigenvalue problem for the Laplace operator consists in finding nontrivial solutions of

$$\Delta u(x) + \lambda u(x) = 0 \quad \text{in } \Omega, \qquad (9.5.2)$$

for some constant λ, the eigenvalue in question. Here one also imposes some boundary conditions on u. In the light of the preceding, it seems natural to require the Dirichlet boundary condition

$$u = 0 \quad \text{on } \partial\Omega. \tag{9.5.3}$$

For many applications, however, it is more natural to have the Neumann boundary condition

$$\frac{\partial u}{\partial n} = 0 \quad \text{on } \partial\Omega \tag{9.5.4}$$

instead, where $\frac{\partial}{\partial n}$ denotes the derivative in the direction of the exterior normal. Here, in order to make this meaningful, one needs to impose certain restrictions, for example, as in Section 1.1, that the divergence theorem is valid for Ω. For simplicity, as in the preceding section, we shall assume that Ω is a C^∞-domain in treating Neumann boundary conditions. In any case, we shall treat the eigenvalue problem for either type of boundary condition.

As with many questions in the theory of PDEs, the situation becomes much clearer when a more abstract approach is developed. Thus, we shall work in some Hilbert space H; for the Dirichlet case, we choose

$$H = H_0^{1,2}(\Omega), \tag{9.5.5}$$

while for the Neumann case, we take

$$H = W^{1,2}(\Omega). \tag{9.5.6}$$

In either case, we shall employ the L^2-product

$$\langle f, g \rangle := \int_\Omega f(x)g(x)dx$$

for $f, g \in L^2(\Omega)$, and we shall also put

$$\|f\| := \|f\|_{L^2(\Omega)} = \langle f, f \rangle^{\frac{1}{2}}.$$

It is important to realize that we are not working here with the scalar product of our Hilbert space H, but rather with the scalar product of another Hilbert space, namely $L^2(\Omega)$, into which H is compactly embedded by Rellich's theorem (Theorems 8.2.2 and 9.4.2).

Another useful point in the sequel is the symmetry of the Laplace operator,

$$\langle \Delta\varphi, \psi \rangle = -\langle D\varphi, D\psi \rangle = \langle \varphi, \Delta\psi \rangle \tag{9.5.7}$$

for all $\varphi, \psi \in C_0^\infty(\Omega)$, as well as for $\varphi, \psi \in C^\infty(\Omega)$ with $\frac{\partial\varphi}{\partial n} = 0 = \frac{\partial\psi}{\partial n}$ on $\partial\Omega$. This symmetry will imply that all eigenvalues are real.

We now start our eigenvalue search with

$$\lambda := \inf_{u \in H \setminus \{0\}} \frac{\langle Du, Du \rangle}{\langle u, u \rangle} \quad \left(= \inf_{u \in H \setminus \{0\}} \frac{\|Du\|^2_{L^2(\Omega)}}{\|u\|^2_{L^2(\Omega)}} \right). \qquad (9.5.8)$$

We wish to show that (because the expression in (9.5.8) is scaling invariant, in the sense that it is not affected by replacing u by cu for some nonzero constant c) this infimum is realized by some $u \in H$ with

$$\Delta u + \lambda u = 0.$$

We first observe that (because the expression in (9.5.8) is scaling invariant, in the sense that it is not affected by replacing u by cu for some constant c) we may restrict our attention to those u that satisfy

$$\|u\|_{L^2(\Omega)} (= \langle u, u \rangle) = 1. \qquad (9.5.9)$$

We then let $(u_n)_{n \in \mathbb{N}} \subset H$ be a minimizing sequence with $\langle u_n, u_n \rangle = 1$, and thus

$$\lambda = \lim_{n \to \infty} \langle Du_n, Du_n \rangle. \qquad (9.5.10)$$

Thus, $(u_n)_{n \in \mathbb{N}}$ is bounded in H, and by the compactness theorem of Rellich (Theorems 8.2.2 and 9.4.2), a subsequence, again denoted by u_n, converges to some limit u in $L^2(\Omega)$ that then also satisfies $\|u\|_{L^2(\Omega)} = 1$. In fact, since

$$\|D(u_n - u_m)\|^2_{L^2(\Omega)} + \|D(u_n + u_m)\|^2_{L^2(\Omega)}$$
$$= 2 \|Du_n\|^2_{L^2(\Omega)} + 2 \|Du_m\|^2_{L^2(\Omega)} \quad \text{for all } n, m \in \mathbb{N},$$

and

$$\|D(u_n + u_m)\|^2_{L^2(\Omega)} \geq \lambda \|u_n + u_m\|^2_{L^2(\Omega)} \quad \text{by definition of } \lambda,$$

we obtain

$$\|Du_n - Du_m\|^2_{L^2(\Omega)} \leq 2 \|Du_n\|^2_{L^2(\Omega)} + 2 \|Du_m\|^2_{L^2(\Omega)}$$
$$- \lambda \|u_n + u_m\|^2_{L^2(\Omega)}. \qquad (9.5.11)$$

Since by choice of the sequence $(u_n)_{n \in \mathbb{N}}$, $\|Du_n\|^2_{L^2(\Omega)}$ and $\|Du_m\|^2_{L^2(\Omega)}$ converge to λ, and $\|u_n + u_m\|^2_{L^2(\Omega)}$ converges to 4, since the u_n converge in $L^2(\Omega)$ to an element u of norm 1, the right-hand side of (9.5.11) converges to 0, and so then does the left-hand side. This, together with the L^2-convergence, implies that $(u_n)_{n \in \mathbb{N}}$ is a Cauchy sequence even in H, and so it also converges to u in H. Thus

$$\frac{\langle Du, Du \rangle}{\langle u, u \rangle} = \lambda. \tag{9.5.12}$$

In the Dirichlet case, the Poincaré inequality (Theorem 8.2.2) implies

$$\lambda > 0.$$

At this point, the assumption enters that Ω as a domain is connected. In the Neumann case, we simply take any nonzero constant c, which now is an element of $H \setminus \{0\}$, to see that

$$0 \leq \lambda \leq \frac{\langle Dc, Dc \rangle}{\langle c, c \rangle} = 0,$$

i.e.,

$$\lambda = 0.$$

Following standard conventions for the enumeration of eigenvalues, we put

$$\lambda =: \lambda_1 \qquad \text{in the Dirichlet case,}$$
$$\lambda =: \lambda_0 (= 0) \quad \text{in the Neumann case,}$$

and likewise $u =: u_1$ and $u =: u_0$, respectively.

Let us now assume that we have iteratively determined $((\lambda_0, u_0)), (\lambda_1, u_1),$ $\ldots, (\lambda_{m-1}, u_{m-1})$, with

$$(\lambda_0 \leq) \lambda_1 \leq \cdots \leq \lambda_{m-1},$$
$$u_i \in L^2(\Omega) \cap C^\infty(\Omega),$$

$$u_i = 0 \quad \text{on } \partial\Omega \quad \text{in the Dirichlet case, and}$$
$$\frac{\partial u_i}{\partial n} = 0 \quad \text{on } \partial\Omega \quad \text{in the Neumann case,}$$

$$\langle u_i, u_j \rangle = \delta_{ij} \quad \text{for all } i, j \leq m-1$$

$$\Delta u_i + \lambda_i u_i = 0 \quad \text{in } \Omega \quad \text{for } i \leq m-1. \tag{9.5.13}$$

We define

$$H_m := \{v \in H : \langle v, u_i \rangle = 0 \quad \text{for } i \leq m-1\}$$

and

$$\lambda_m := \inf_{u \in H_m \setminus \{0\}} \frac{\langle Du, Du \rangle}{\langle u, u \rangle}. \tag{9.5.14}$$

Since $H_m \subset H_{m-1}$, the infimum over the former space cannot be smaller than the one over the latter, i.e.,

$$\lambda_m \geq \lambda_{m-1}. \tag{9.5.15}$$

Note that H_m is a Hilbert space itself, being the orthogonal complement of a finite-dimensional subspace of the Hilbert space H. Therefore, with the previous reasoning, we may find $u_m \in H_m$ with $\|u_m\|_{L^2(\Omega)} = 1$ and

$$\lambda_m = \frac{\langle Du_m, Du_m \rangle}{\langle u_m, u_m \rangle}. \tag{9.5.16}$$

We now want to verify the smoothness of u_m and equation (9.5.13) for $i = m$.
From (9.5.14), (9.5.16), for all $\varphi \in H_m$, $t \in \mathbb{R}$,

$$\frac{\langle D(u_m + t\varphi), D(u_m + t\varphi) \rangle}{\langle u_m + t\varphi, u_m + t\varphi \rangle} \geq \lambda_m,$$

where we choose $|t|$ so small that the denominator is bounded away from 0. This expression then is differentiable w.r.t. t near $t = 0$ and has a minimum at 0. Hence the derivative vanishes at $t = 0$, and we get

$$\begin{aligned}
0 &= \frac{\langle Du_m, D\varphi \rangle}{\langle u_m, u_m \rangle} - \frac{\langle Du_m, Du_m \rangle}{\langle u_m, u_m \rangle} \frac{\langle u_m, \varphi \rangle}{\langle u_m, u_m \rangle} \\
&= \langle Du_m, D\varphi \rangle - \lambda_m \langle u_m, \varphi \rangle \quad \text{for all } \varphi \in H_m.
\end{aligned}$$

In fact, this relation even holds for all $\varphi \in H$, because for $i \leq m - 1$,

$$\langle u_m, u_i \rangle = 0$$

and

$$\langle Du_m, Du_i \rangle = \langle Du_i, Du_m \rangle = \lambda_i \langle u_i, u_m \rangle = 0,$$

since $u_m \in H_i$. Thus, u_m satisfies

$$\int_\Omega Du_m \cdot D\varphi - \lambda_m \int_\Omega u_m \varphi = 0 \quad \text{for all } \varphi \in H. \tag{9.5.17}$$

By Theorem 9.3.1 and Corollary 9.4.1, respectively, u_m is smooth, and so we obtain from (9.5.17)

$$\Delta u_m + \lambda_m u_m = 0 \quad \text{in } \Omega.$$

As explained in the preceding section, we also have

$$\frac{\partial u_m}{\partial n} = 0 \quad \text{on } \partial\Omega$$

in the Neumann case. In the Dirichlet case, we have of course

$$u_m = 0 \quad \text{on } \partial\Omega$$

(this holds pointwise if $\partial\Omega$ is smooth, as explained in Section 9.4; for a general, not necessarily smooth, $\partial\Omega$, this relation is valid in the sense of Sobolev).

Theorem 9.5.1: *Let $\Omega \subset \mathbb{R}^d$ be connected, open and bounded. Then the eigenvalue problem*

$$\Delta u + \lambda u = 0, \quad u \in H_0^{1,2}(\Omega)$$

has countably many eigenvalues

$$0 < \lambda_1 < \lambda_2 \leq \cdots \leq \lambda_m \leq \cdots$$

with

$$\lim_{m \to \infty} \lambda_m = \infty$$

and pairwise L^2-orthonormal eigenfunctions u_i and $\langle Du_i, Du_i \rangle = \lambda_i$. Any $v \in L^2(\Omega)$ can be expanded in terms of these eigenfunctions,

$$v = \sum_{i=1}^{\infty} \langle v, u_i \rangle u_i \quad \text{(and thus } \langle v, v \rangle = \sum_{i=1}^{\infty} \langle v, u_i \rangle^2 \text{)}, \tag{9.5.18}$$

and if $v \in H_0^{1,2}(\Omega)$, we also have

$$\langle Dv, Dv \rangle = \sum_{i=1}^{\infty} \lambda_i \langle v, u_i \rangle^2. \tag{9.5.19}$$

Theorem 9.5.2: *Let $\Omega \subset \mathbb{R}^d$ be bounded, open, and of class C^∞. Then the eigenvalue problem*

$$\Delta u + \lambda u = 0, \quad u \in W^{1,2}(\Omega)$$

has countably many eigenvalues

$$0 = \lambda_0 \leq \lambda_1 \leq \cdots \leq \lambda_m \leq \cdots$$

with

$$\lim_{n \to \infty} \lambda_m = \infty$$

and pairwise L^2-orthonormal eigenfunctions u_i that satisfy

$$\frac{\partial u_i}{\partial n} = 0 \quad on \ \partial\Omega.$$

Any $v \in L^2(\Omega)$ can be expanded in terms of these eigenfunctions

$$v = \sum_{i=0}^{\infty} \langle v, u_i \rangle u_i \quad (and \ thus \ \langle v, v \rangle = \sum_{i=0}^{\infty} \langle v, u_i \rangle^2), \qquad (9.5.20)$$

and if $v \in W^{1,2}(\Omega)$, also

$$\langle Dv, Dv \rangle = \sum_{i=1}^{\infty} \lambda_i \langle v, u_i \rangle^2. \qquad (9.5.21)$$

Remark: Those $v \in L^2(\Omega)$ that are not contained in H can be characterized by the fact that the expression on the right-hand side of (9.5.19) or (9.5.21) diverges.

The *Proofs* of Theorems 9.5.1 and 9.5.2 are now easy: We first check

$$\lim_{m \to \infty} \lambda_m = \infty.$$

Indeed, otherwise,

$$\|Du_m\| \leq c \quad \text{for all } m \text{ and some constant c.}$$

By Rellich's theorem again, a subsequence of (u_m) would then be a Cauchy sequence in $L^2(\Omega)$. This, however, is not possible, since the u_m are pairwise L^2-orthonormal.

It remains to prove the expansion. For $v \in H$ we put

$$\beta_i := \langle v, u_i \rangle$$

and

$$v_m := \sum_{i \leq m} \beta_i u_i, \quad w_m := v - v_m.$$

Thus, v_m is the orthogonal projection of v onto H_m, and w_m then is orthogonal to H_m; hence

$$\langle w_m, u_i \rangle = 0 \quad \text{for } i \leq m.$$

Thus also

$$\langle Dw_m, Dw_m \rangle \geq \lambda_{m+1} \langle w_m, w_m \rangle$$

and

$$\langle Dw_m, Du_i \rangle = \lambda_i \langle u_i, w_m \rangle = 0.$$

These orthogonality relations imply

$$\langle w_m, w_m \rangle = \langle v, v \rangle - \langle v_m, v_m \rangle,$$
$$\langle Dw_m, Dw_m \rangle = \langle Dv, Dv \rangle - \langle Dv_m, Dv_m \rangle, \qquad (9.5.22)$$

and then

$$\langle w_m, w_m \rangle \leq \frac{1}{\lambda_{m+1}} \langle Dv, Dv \rangle,$$

which converges to 0 as the λ_m tend to ∞. Thus, the remainder w_m converges to 0 in L^2, and so

$$v = \lim_{m \to \infty} v_m = \sum_i \langle v, u_i \rangle u_i \quad \text{in } L^2(\Omega).$$

Also,

$$Dv_m = \sum_{i \leq m} \beta_i Du_i,$$

and hence

$$\langle Dv_m, Dv_m \rangle = \sum_{i \leq m} \beta_i^2 \langle Du_i, Du_i \rangle \quad (\text{since } \langle Du_i, Du_j \rangle = 0 \quad \text{for } i \neq j)$$
$$= \sum_{i \leq m} \lambda_i \beta_i^2.$$

Since $\langle Dv_m, Dv_m \rangle \leq \langle Dv, Dv \rangle$ by (9.5.22) and the λ_i are nonnegative, this series then converges, and then for $m < n$,

$$\langle Dw_m - Dw_n, Dw_m - Dw_n \rangle = \langle Dv_n - Dv_m, Dv_n - Dv_m \rangle$$
$$= \sum_{i=m+1}^{n} \lambda_i \beta_i^2 \to 0 \quad \text{for } m, n \to \infty,$$

and so $(Dw_m)_{m \in \mathbb{N}}$ is a Cauchy sequence in L^2, and so w_m converges in H, and the limit is the same as the L^2-limit, namely 0. Therefore, we get (9.5.19) and (9.5.21), namely

$$\langle Dv, Dv \rangle = \lim_{m \to \infty} \langle Dv_m, Dv_m \rangle = \sum_i \lambda_i \beta_i^2.$$

The eigenfunctions $(u_m)n \in \mathbb{N}$ thus are an L^2-orthonormal sequence. The closure of the span of the u_m then is a Hilbert space contained in $L^2(\Omega)$ and containing H. Since H (in fact, even $C_0^\infty(\Omega) \cap H$, see the Appendix) is dense in $L^2(\Omega)$, this Hilbert space then has to be all of $L^2(\Omega)$. So, the expansions

(9.5.18), (9.5.20) are valid for all $v \in L^2(\Omega)$.
The strict inequality $\lambda_1 < \lambda_2$ in the Dirichlet case will be proved in Theorem
9.5.4 below. $\square$

A moment's reflection also shows that the above procedure produces all
the eigenvalues of Δ on H, and that any eigenfunction is a linear combination
of the u_i.

An easy consequence of the theorems is the following sharp version of the
Poincaré inequality (cf. Theorem 8.2.2).

Corollary 9.5.1: *For $v \in H_0^{1,2}(\Omega)$,*

$$\lambda_1 \langle v, v \rangle \leq \langle Dv, Dv \rangle \tag{9.5.23}$$

*where λ_1 is the first Dirichlet eigenvalue according to Theorem 9.5.1.
For $v \in H^{1,2}(\Omega)$ with $\frac{\partial v}{\partial \nu}$ on $\partial\Omega$*

$$\lambda_1 \langle v - \bar{v}, v - \bar{v} \rangle \leq \langle Dv, Dv \rangle \tag{9.5.24}$$

*where λ_1 now is the first Neumann eigenvalue according to Theorem 9.5.2,
and $\bar{v} := \frac{1}{\|\Omega\|} \int_\Omega v(x) dx$ is the average of v on Ω ($\|\Omega\|$ is the Lebesgue measure
of Ω). Moreover, if such a v with vanishing Neumann boundary values is of
class $H^{2,2}(\Omega)$, then also*

$$\lambda_1 \langle Dv, Dv \rangle \leq \langle \Delta v, \Delta v \rangle, \tag{9.5.25}$$

λ_1 again being the first Neumann eigenvalue.

Proof: The inequalities (9.5.23), (9.5.24) readily follow from (9.5.14), noting
that in the second case, $v - \bar{v}$ is orthogonal to the constants, the eigenfunctions
for $\lambda_0 = 0$, since

$$\int_\Omega (v(x) - \bar{v}) dx = 0. \tag{9.5.26}$$

As an alternative, and in order to obtain also (9.5.25), we note that
$Dv = D(v - \bar{v})$, $\Delta v = \Delta(v - \bar{v})$, and

$$\langle v - \bar{v}, v - \bar{v} \rangle = \sum_{i=1}^{\infty} \langle v, u_i \rangle^2, \tag{9.5.27}$$

that is, the term for $i = 0$ disappears from the expansion because $v - \bar{v}$ is
orthogonal to the constant eigenfunction u_0. Using

$$\langle Dv, Dv \rangle = \sum_{i=1}^{\infty} \lambda_i \langle v, u_i \rangle^2$$

$$\langle \Delta v, \Delta v \rangle = \sum_{i=1}^{\infty} \lambda_i^2 \langle v, u_i \rangle^2$$

and $\lambda_1 \leq \lambda_i$ then yields (9.5.24), (9.5.25). $\square$

More generally, we can derive Courant's minimax principle for the eigenvalues of Δ:

Theorem 9.5.3: *Under the above assumptions, let P^k be the collection of all k-dimensional linear subspaces of the Hilbert space H. Then the kth eigenvalue of Δ (i.e., λ_k in the Dirichlet case, λ_{k-1} in the Neumann case) is characterized as*

$$\max_{L \in P^{k-1}} \min \left\{ \frac{\langle Du, Du \rangle}{\langle u, u \rangle} : \begin{array}{l} u \neq 0, u \text{ orthogonal to } L, \\ \text{i.e., } \langle u, v \rangle = 0 \quad \text{for all } v \in L \end{array} \right\}, \qquad (9.5.28)$$

or dually as

$$\min_{L \in P^k} \max \left\{ \frac{\langle Du, Du \rangle}{\langle u, u \rangle} : u \in L \setminus \{0\} \right\}. \qquad (9.5.29)$$

Proof: We have seen that

$$\lambda_m = \min \left\{ \frac{\langle Du, Du \rangle}{\langle u, u \rangle} : u \neq 0, u \text{ orthogonal to the } u_i \text{ with } i \leq m - 1 \right\}. \qquad (9.5.30)$$

It is also clear that

$$\lambda_m = \max \left\{ \frac{\langle Du, Du \rangle}{\langle u, u \rangle} : u \neq 0 \text{ linear combination of } u_i \text{ with } i \leq m \right\}, \qquad (9.5.31)$$

and in fact, this minimum is realized if u is a multiple of the mth eigenfunction u_m, because $\lambda_i = \frac{\langle Du_i, Du_i \rangle}{\langle u_i, u_i \rangle} \leq \lambda_m$ for $i \leq m$ and the u_i are pairwise orthogonal.

Now let L be another linear subspace of H of the same dimension as the span of the u_i, $i \leq m$. Let L be spanned by vectors v_i, $i \leq m$. We may then find some $v = \sum \alpha_j v_j \in L$ with

$$\langle v, u_i \rangle = \sum_j \alpha_j \langle v_j, u_i \rangle = 0 \quad \text{for } i \leq m - 1. \qquad (9.5.32)$$

(This is a system of homogeneous linearly independent equations for the α_j, with one fewer equation than unknowns, and so it can be solved.) Inserting (9.5.32) into the expansion (9.5.19) or (9.5.21), we obtain

$$\frac{\langle Dv, Dv \rangle}{\langle v, v \rangle} = \frac{\sum_{j=m}^{\infty} \lambda_j \langle v, u_j \rangle^2}{\sum_{j=m}^{\infty} \langle v, u_j \rangle^2} \geq \lambda_m.$$

Therefore,

$$\max_{v \in L \setminus \{0\}} \frac{\langle Dv, Dv \rangle}{\langle v, v \rangle} \geq \lambda_m,$$

and (9.5.29) follows. Suitably dualizing the preceding argument, which we leave to the reader, yields (9.5.28). $\square$

While for certain geometrically simple domains, like balls and cubes, one may determine the eigenvalues explicitly, for a general domain, it is a hopeless endeavor to attempt an exact computation of its eigenvalues. One therefore needs approximation schemes, and the minimax principle of Courant suggests one such method, the Rayleigh–Ritz scheme. For that scheme, one selects linearly independent functions $w_1, \ldots, w_k \in H$, which then span a linear subspace L, and seeks the critical values, and in particular the maximum of

$$\frac{\langle Dw, Dw \rangle}{\langle w, w \rangle} \quad \text{for } w \in L.$$

With

$$a_{ij} := \langle Dw_i, Dw_j \rangle, \quad A := (a_{ij})_{i,j=1,\ldots,k},$$
$$b_{ij} := \langle w_i, w_j \rangle, \qquad B := (b_{ij})_{i,j=1,\ldots,k},$$

for

$$w = \sum_{j=1} c_j w_j,$$

then

$$\frac{\langle Dw, Dw \rangle}{\langle w, w \rangle} = \frac{\sum_{i,j=1}^{k} a_{ij} c_i c_j}{\sum_{i,j=1}^{k} b_{ij} c_i c_j},$$

and the critical values are given by the solutions $\mu_1, \ldots, \mu_k$ of

$$\det(A - \mu B) = 0.$$

These values $\mu_1, \ldots, \mu_k$ then are taken as approximations of the first k eigenvalues; in particular, if they are ordered such that μ_k is the largest among them, that value is supposed to approximate the kth eigenvalue. One then tries to optimize with respect to the choice of the functions $w_1, \ldots, w_k$; i.e., one tries to make μ_k as small as possible, according to (9.5.29), by suitably choosing $w_1, \ldots, w_k$.

The characerizations (9.5.28) and (9.5.29) of the eigenvalues have many further useful applications. The basis of those applications is the following simple remark: In (9.5.29), we take the maximum over all $u \in H$ that are contained in some subspace L. If we then enlarge H to some Hilbert space H', then H' contains more such subspaces than H, and so the minimum over all of them cannot increase.

Formally, if we put $P^k(H) := \{k - \text{dimensional linear subspaces of } H\}$, then, if $H \subset H'$, it follows that $P^k(H) \subset P^k(H')$, and so

$$\min_{L \in P^k(H)} \max_{u \in L \setminus \{0\}} \frac{\langle Du, Du \rangle}{\langle u, u \rangle} \geq \min_{L' \in P^k(H')} \max_{u \in L' \setminus \{0\}} \frac{\langle Du, Du \rangle}{\langle u, u \rangle}. \tag{9.5.33}$$

Corollary 9.5.2: *Under the above assumptions, we let $0 < \lambda_1^D \leq \lambda_2^D \leq \cdots$ be the Dirichlet eigenvalues, and $0 = \lambda_0^N < \lambda_1^N \leq \lambda_2^N \leq \cdots$ be the Neumann eigenvalues. Then*

$$\lambda_{j-1}^N \leq \lambda_j^D \quad \text{for all } j.$$

Proof: The Hilbert space for the Dirichlet case, namely $H_0^{1,2}(\Omega)$, is a subspace of that for the Neumann case, namely $W^{1,2}(\Omega)$, and so (9.5.33) applies.   □

The next result states that the eigenvalues decrease if the domain is enlarged:

Corollary 9.5.3: *Let $\Omega_1 \subset \Omega_2$ be bounded open subsets of $\mathbb{R}^d$. We denote the eigenvalues for the Dirichlet case of the domain Ω by $\lambda_k(\Omega)$. Then*

$$\lambda_k(\Omega_2) \leq \lambda_k(\Omega_1) \quad \text{for all } k. \tag{9.5.34}$$

Proof: Any $v \in H_0^{1,2}(\Omega_1)$ can be extended to a function $\tilde{v} \in H_0^{1,2}(\Omega_2)$, simply by putting

$$\tilde{v}(x) = \begin{cases} v(x) & \text{for } x \in \Omega_1, \\ 0 & \text{for } x \in \Omega_2 \setminus \Omega_1. \end{cases}$$

Lemma 8.2.2 tells us that indeed $\tilde{v} \in H_0^{1,2}(\Omega_2)$. Thus, the Hilbert space employed for Ω_1 is contained in that for Ω_2, and the principle (9.5.33) again implies the result for the Dirichlet case.   □

Remark: Corollary 9.5.3 is not in general valid for the Neumann case. A first idea to show a result in that case is to extend functions $v \in W^{1,2}(\Omega_1)$ to Ω_2 by the extension operator E constructed in Section 9.4. However, this operator does not preserve the norm: In general, $\|Ev\|_{W^{1,2}(\Omega_2)} > \|v\|_{W^{1,2}(\Omega_1)}$, and so this does not represent $W^{1,2}(\Omega_1)$ as a Hilbert subspace of $W^{1,2}(\Omega_2)$. This difficulty makes the Neumann case more involved, and we omit it here.

The next result concerns the first eigenvalue λ_1 of Δ with Dirichlet boundary conditions:

Theorem 9.5.4: *Let λ_1 be the first eigenvalue of Δ on the open and bounded domain $\Omega \subset \mathbb{R}^d$ with Dirichlet boundary conditions. Then λ_1 is a simple eigenvalue, meaning that the corresponding eigenspace is one-dimensional. Moreover, an eigenfunction u_1 for λ_1 has no zeros in Ω, and so it is either everywhere positive or negative in Ω.*

Proof: Let

$$\Delta u_1 + \lambda_1 u_1 = 0 \quad \text{in } \Omega.$$

By Corollary 8.2.2, we know that $|u_1| \in W^{1,2}(\Omega)$, and

$$\frac{\langle D|u_1|, D|u_1|\rangle}{\langle |u_1|, |u_1|\rangle} = \frac{\langle Du_1, Du_1\rangle}{\langle u_1, u_1\rangle} = \lambda_1.$$

Therefore, $|u_1|$ also minimizes

$$\frac{\langle Du, Du\rangle}{\langle u, u\rangle},$$

and by the reasoning leading to Theorem 9.5.1, it must also be an eigenfunction with eigenvalue λ_1. Therefore, it is a nonnegative solution of

$$\Delta u + \lambda u = 0 \quad \text{in } \Omega,$$

and by the strong maximum principle (Theorem 9.1.2), it cannot assume a nonpositive interior minimum. Thus, it cannot become 0 in Ω, and so it is positive in Ω. This, however, implies that the original function u_1 cannot become 0 either. Thus, u_1 is of a fixed sign.

This argument applies to all eigenfunctions with eigenvalue λ_1. Since two functions v_1, v_2 neither of which changes sign in Ω cannot satisfy

$$\int_\Omega v_1(x)v_2(x)dx = 0,$$

i.e., cannot be L^2-orthogonal, the space of eigenfunctions for λ_1 is one-dimensional. $\qquad\square$

The classical text on eigenvalue problems is Courant–Hilbert [4].

Remark: More generally, Courant's nodal set theorem holds: Let $\Omega \subset \mathbb{R}^d$ be open and bounded, with Dirichlet eigenvalues $0 < \lambda_1 < \lambda_2 \leq \ldots$ and corresponding eigenfunctions $u_1, u_2, \ldots$. We call

$$\Gamma^k := \{x \in \Omega : u_k(x) = 0\}$$

the nodal set of u_k. The complement $\Omega \setminus \Gamma^k$ then has at most k components.

Summary

In this chapter we have introduced Sobolev spaces as spaces of integrable functions that are not necessarily differentiable in the classical sense, but do possess so-called generalized or weak derivatives that obey the rules for integration by parts. Embedding theorems relate Sobolev spaces to spaces of L^p-functions or of continuous, Hölder continuous, or differentiable functions.

The weak solutions of the Laplace and Poisson equations, obtained in Chapter 8 by Dirichlet's principle, naturally lie in such Sobolev spaces. In this chapter, embedding theorems allow us to show that weak solutions are regular, i.e., differentiable of any order, and hence also solutions in the classical sense.

Based on Rellich's theorem, we have treated the eigenvalue problem for the Laplace operator and shown that any L^2-function admits an expansion in terms of eigenfunctions of the Laplace operator.

Exercises

9.1 Let $u : \Omega \to \mathbb{R}$ be integrable, and let α, β be multi-indices. Show that if two of the weak derivatives $D_{\alpha+\beta}u, D_\alpha D_\beta u, D_\beta D_\alpha u$ exist, then the third one also exists, and all three of them coincide.

9.2 Let $u, v \in W^{1,1}(\Omega)$ with $uv, uDv + vDu \in L^1(\Omega)$. Then $uv \in W^{1,1}(\Omega)$ as well, and the weak derivative satisfies the product rule

$$D(uv) = uDv + vDu.$$

(For the proof, it is helpful to first consider the case where one of the two functions is of class $C^1(\Omega)$.)

9.3 For $m \geq 2, 1 \leq q \leq m/2, u \in H_0^{2,\frac{m}{q+1}}(\Omega) \cap L^{\frac{m}{q-1}}(\Omega)$ we have $u \in H^{1,\frac{m}{q}}(\Omega)$ and

$$\|Du\|_{L^{\frac{m}{q}}(\Omega)}^2 \leq \text{const } \|u\|_{L^{\frac{m}{q-1}}(\Omega)} \, \|D^2 u\|_{L^{\frac{m}{q+1}}(\Omega)}.$$

(Hint: For $p = \frac{m}{q}$,

$$|D_i u|^p = D_i(uD_i u|D_i u|^{p-2}) - uD_i(D_i u|D_i u|^{p-2}).$$

The first term on the right-hand side disappears upon integration over Ω for $u \in C_0^\infty(\Omega)$ (approximation argument!), and for the second one, we utilize the formula

$$D_i(v|v|^{p-2}) = (p-1)(D_i v)|v|^{p-2}.$$

Finally, you need the following version of Hölder's inequality

$$\|u_1 u_2 u_3\|_{L^1(\Omega)} \leq \|u_1\|_{L^{p_1}(\Omega)} \|u_2\|_{L^{p_2}(\Omega)} \|u_3\|_{L^{p_3}(\Omega)}$$

for $u_i \in L^{p_i}(\Omega), \frac{1}{p_1} + \frac{1}{p_2} + \frac{1}{p_3} = 1$ (proof!).)

9.4 Let

$$\Omega_1 := \mathring{B}(0,1) \subset \mathbb{R}^d,$$

$$\Omega_2 := \mathbb{R}^d \setminus \mathring{B}(0,1),$$

i.e., the d-dimensional unit ball and its complement. For which values of k, p, d, α is

$$f(x) := |x|^\alpha$$

in $W^{k,p}(\Omega_1)$ or $W^{k,p}(\Omega_2)$?

9.5 Prove the following version of the Sobolev embedding theorem: Let $u \in W^{k,p}(\Omega), \Omega' \subset\subset \Omega \subset \mathbb{R}^d$. Then

$$u \in \begin{cases} L^{\frac{dp}{d-kp}}(\Omega') & \text{for } kp < d, \\ C^m(\overline{\Omega'}) & \text{for } 0 \leq m < k - d/p. \end{cases}$$

9.6 State and prove a generalization of Corollary 9.1.5 for $u \in W^{k,p}(\Omega)$ that is analogous to Exercise 8.5.

9.7 Supply the details of the proof of Theorem 9.3.2 (This may sound like a dull exercise after what has been said in the text, but in order to understand the techniques for estimating solutions of PDEs, a certain drill in handling additional lower-order terms and variable coefficients may be needed.)

9.8 Carry out the eigenvalue analysis for the Laplace operator under periodic boundary conditions as defined in §1.1. In particular, state and prove an analogue of Theorems 9.5.1 and 9.5.2.

10. Strong Solutions

10.1 The Regularity Theory for Strong Solutions

We start with an elementary observation: Let $v \in C_0^3(\Omega)$. Then

$$\|D^2 v\|_{L^2(\Omega)}^2 = \int_\Omega \sum_{i,j=1}^d v_{x^i x^j} v_{x^i x^j} = -\int_\Omega \sum_{i,j=1}^d v_{x^i x^j x^i} v_{x^j}$$

$$= \int_\Omega \sum_{i=1}^d v_{x^i x^i} \sum_{j=1}^d v_{x^j x^j} = \|\Delta v\|_{L^2(\Omega)}^2 . \tag{10.1.1}$$

Thus, the L^2-norm of Δv controls the L^2-norms of all second derivatives of v. Therefore, if v is a solution of the differential equation

$$\Delta v = f,$$

the L^2-norm of f controls the L^2-norm of the second derivatives of v. This is a result in the spirit of elliptic regularity theory as encountered in Section 9.2 (cf. Theorem 9.2.1). In the preceding computation, however, we have assumed that, firstly, v is thrice continuously differentiable, and secondly, that it has compact support. The aim of elliptic regularity theory, however, is to deduce such regularity results, and also, one typically encounters non-vanishing boundary terms on $\partial\Omega$. Thus, our assumptions are inappropriate, and we need to get rid of them. This is the content of this section.

We shall first discuss an elementary special case of the Calderon-Zygmund inequality. Let $f \in L^2(\Omega)$, Ω open and bounded in $\mathbb{R}^d$. We define the Newton potential of f as

$$w(x) := \int_\Omega \Gamma(x,y) f(y) dy \tag{10.1.2}$$

using the fundamental solution constructed in Section 1.1,

$$\Gamma(x,y) = \begin{cases} \frac{1}{2\pi} \log |x-y| & \text{for } d = 2, \\ \frac{1}{d(2-d)\omega_d} |x-y|^{2-d} & \text{for } d > 2. \end{cases}$$

Theorem 10.1.1: *Let $f \in L^2(\Omega)$ and let w be the Newton potential of f. Then $w \in W^{2,2}(\Omega)$, $\Delta w = f$ almost everywhere in Ω, and*

$$\left\| D^2 w \right\|_{L^2(\mathbb{R}^d)} = \|f\|_{L^2(\Omega)} \tag{10.1.3}$$

(w is called a strong solution of $\Delta w = f$, because this equation holds almost everywhere).

Proof: We first assume $f \in C_0^\infty(\Omega)$. Then $w \in C^\infty(\mathbb{R}^d)$. Let $\Omega \subset\subset \Omega_0$, Ω_0 bounded with a smooth boundary. We first wish to show that for $x \in \Omega$,

$$\frac{\partial^2}{\partial x^i \partial x^j} w(x) = \int_{\Omega_0} \frac{\partial^2}{\partial x^i \partial x^j} \Gamma(x,y)(f(y) - f(x)) dy$$
$$+ f(x) \int_{\partial \Omega_0} \frac{\partial}{\partial x^i} \Gamma(x,y) \nu^j \, do(y), \tag{10.1.4}$$

where $\nu = (\nu^1, \dots, \nu^d)$ is the exterior normal and $do(y)$ yields the induced measure on $\partial \Omega_0$. This is an easy consequence of the fact that

$$\left| \frac{\partial^2}{\partial x^i \partial x^j} \Gamma(x,y)(f(y) - f(x)) \right| \leq \text{const} \frac{1}{|x-y|^d} |f(y) - f(x)|$$
$$\leq \text{const} \frac{1}{|x-y|^{d-1}} \|f\|_{C^1} .$$

In other words, the singularity under the integral sign is integrable. (Namely, one simply considers

$$v_\varepsilon(x) = \int \frac{\partial}{\partial x^i} \Gamma(x,y) \eta_\varepsilon(y) f(y) dy,$$

with $\eta_\varepsilon(y) = 0$ for $|y| \leq \varepsilon$, $\eta_\varepsilon(y) = 1$ for $|y| \geq 2\varepsilon$ and $|D\eta_\varepsilon| \leq \frac{2}{\varepsilon}$, and shows that as $\varepsilon \to 0$, $D_j v_\varepsilon$ converges to the right-hand side of (10.1.4).)

Remark: Equation (10.1.4) continues to hold for a Hölder continuous f, cf. Section 11.1 below, since in that case, one can estimate the integrand by

$$\text{const} \frac{1}{|x-y|^{d-\alpha}} \|f\|_{C^\alpha}$$

$(0 < \alpha < 1)$.

Since

$$\Delta \Gamma(x,y) = 0 \quad \text{for all } x \neq y,$$

for $\Omega_0 = B(x,R)$, R sufficiently large, from (10.1.4) we obtain

$$\Delta w(x) = \frac{1}{d\omega_d R^{d-1}} f(x) \int_{|x-y|=R} \sum_{i=1}^{d} \nu^i(y)\nu^i(y)\, do(y) = f(x). \quad (10.1.5)$$

Thus, if f has compact support, so does Δw; let the latter be contained in the interior of $B(0, R)$. Then

$$\int_{B(0,R)} \sum_{i,j=1}^{d} \left(\frac{\partial^2}{\partial x^i \partial x^j} w \right)^2 = - \int_{B(0,R)} \sum_i \frac{\partial}{\partial x^i} w \frac{\partial}{\partial x^i} f$$

$$+ \int_{\partial B(0,R)} Dw \cdot \frac{\partial}{\partial \nu} Dw\, do(y)$$

$$= \int_{B(0,R)} (\Delta w)^2 \qquad (10.1.6)$$

$$+ \int_{\partial B(0,R)} Dw \cdot \frac{\partial}{\partial \nu} Dw\, do(y).$$

As $R \to \infty$, Dw behaves like R^{1-d}, $D^2 w$ like R^{-d}, and therefore, the integral on $\partial B(0, R)$ converges to zero for $R \to \infty$. Because of (10.1.5), (10.1.6) then yields (10.1.3).

In order to treat the general case $f \in L^2(\Omega)$, we argue that by Theorem 8.2.2, for $f \in C_0^\infty(\Omega)$ the $W^{1,2}$-norm of w can be controlled by the L^2-norm of f.[1] We then approximate $f \in L^2(\Omega)$ by $(f_n) \in C_0^\infty(\Omega)$. Applying (10.1.3) to the differences $(w_n - w_m)$ of the Newton potentials w_n of f_n, we see that the latter constitute a Cauchy sequence in $W^{2,2}(\Omega)$. The limit w again satisfies (10.1.3), and since L^2-functions are defined almost everywhere, $\Delta w = f$ holds almost everywhere, too. □

The above considerations can also be used to provide a proof of Theorem 9.2.1. We recall that result:

Theorem 10.1.2: Let $u \in W^{1,2}(\Omega)$ be a weak solution of $\Delta u = f$, with $f \in L^2(\Omega)$. Then $u \in W^{2,2}(\Omega')$, for every $\Omega' \subset\subset \Omega$, and

$$\|u\|_{W^{2,2}(\Omega')} \leq \text{const} \left(\|u\|_{L^2(\Omega)} + \|f\|_{L^2(\Omega)} \right), \quad (10.1.7)$$

with a constant depending only on d, Ω, and Ω'. Moreover,

$$\Delta u = f \quad \text{almost everywhere in } \Omega.$$

Proof: As before, we first consider the case $u \in C^3(\Omega)$. Let $B(x, R) \subset \Omega$, $\sigma \in (0, 1)$, and let $\eta \in C_0^3(B(x, R))$ be a cutoff function with

---

[1] See the proof of Lemma 8.3.1.

$$0 \le \eta(y) \le 1,$$
$$\eta(y) = 1 \quad \text{for } y \in B(x, \sigma R),$$
$$\eta(y) = 0 \quad \text{for } y \in \mathbb{R}^d \setminus B\left(x, \frac{1+\sigma}{2} \cdot R\right),$$
$$|D\eta| \le \frac{4}{(1-\sigma)R},$$
$$|D^2\eta| \le \frac{16}{(1-\sigma)^2 R^2}.$$

We put

$$v := \eta u.$$

Then $v \in C_0^3(B(x, R))$, and (10.1.1) implies

$$\left\|D^2 v\right\|_{L^2(B(x,R))} = \left\|\Delta v\right\|_{L^2(B(x,R))}. \tag{10.1.8}$$

Now,

$$\Delta v = \eta \Delta u + 2Du \cdot D\eta + u \Delta \eta,$$

and thus

$$\left\|D^2 u\right\|_{L^2(B(x,\sigma R))} \le \left\|D^2 v\right\|_{L^2(B(x,R))}$$
$$\le \text{const}\left(\|f\|_{L^2(B(x,R))} + \frac{1}{(1-\sigma)R}\|Du\|_{L^2(B(x,\frac{1+\sigma}{2}\cdot R))}\right.$$
$$\left. + \frac{1}{(1-\sigma)^2 R^2}\|u\|_{L^2(B(x,R))}\right). \tag{10.1.9}$$

Now let $\xi \in C_0^1(B(x, R))$ be a cutoff function with

$$0 \le \xi(y) \le 1,$$
$$\xi(y) = 1 \quad \text{for } y \in B\left(x, \frac{1+\sigma}{2}R\right),$$
$$|D\xi| \le \frac{4}{(1-\sigma)R}.$$

Putting $w = \xi^2 u$ and using that u is a weak solution of $\Delta u = f$, we obtain

$$\int_{B(x,R)} Du \cdot D(\xi^2 u) = -\int_{B(x,R)} f\,\xi^2 u,$$

hence

$$\int_{B(x,R)} \xi^2 |Du|^2 = -2 \int_{B(x,R)} \xi u Du \cdot D\xi - \int_{B(x,R)} f \xi^2 u$$

$$\leq \frac{1}{2} \int_{B(x,R)} \xi^2 |Du|^2 + 2 \int_{B(x,R)} u^2 |D\xi|^2$$

$$+ (1-\sigma)^2 R^2 \int_{B(x,R)} f^2 + \frac{1}{(1-\sigma)^2 R^2} \int_{B(x,R)} u^2.$$

Thus, we have an estimate for $\|\xi Du\|_{L^2(B(x,R))}$, and also

$$\|Du\|_{L^2\left(B\left(x,\frac{1+\sigma}{2}R\right)\right)} \leq \|\xi Du\|_{L^2(B(x,R))}$$

$$\leq \text{const} \left(\frac{1}{(1-\sigma)R} \|u\|_{L^2(B(x,R))} \right. \tag{10.1.10}$$

$$\left. + (1-\sigma)R \|f\|_{L^2(B(x,R))} \right).$$

Inequalities (10.1.9) and (10.1.10) yield

$$\|D^2 u\|_{L^2(B(x,\sigma R))} \leq \text{const} \left(\|f\|_{L^2(B(x,R))} + \frac{1}{(1-\sigma)^2 R^2} \|u\|_{L^2(B(x,R))} \right). \tag{10.1.11}$$

In (10.1.11) we put $\sigma = \frac{1}{2}$, and we cover Ω' by a finite number of balls $B(x, R/2)$ with $R \leq \text{dist}(\Omega', \partial\Omega)$ and obtain (10.1.7) for $u \in C^3(\Omega)$.

For the general case $u \in W^{1,2}(\Omega)$, we consider the mollifications u_h defined in Appendix 12.3. Thus, let $0 < h < \text{dist}(\Omega', \partial\Omega)$. Then

$$\int_\Omega Du_h \cdot Dv = - \int f_h v, \quad \text{for all } v \in H_0^{1,2}(\Omega),$$

and since $u_h \in C^\infty(\Omega)$, also

$$\Delta u_h = f_h.$$

By Lemma A.3,

$$\|u_h - u\|, \quad \|f_h - f\|_{L^2(\Omega)} \to 0.$$

In particular, the u_h and the f_h satisfy the Cauchy property in $L^2(\Omega)$. We apply (10.1.7) for $u_{h_1} - u_{h_2}$ to obtain

$$\|u_{h_1} - u_{h_2}\|_{W^{2,2}(\Omega')} \leq \text{const} \left(\|u_{h_1} - u_{h_2}\|_{L^2(\Omega)} + \|f_{h_1} - f_{h_2}\|_{L^2(\Omega)} \right).$$

Thus, the u_h satisfy the Cauchy property in $W^{2,2}(\Omega')$. Consequently, the limit u is in $W^{2,2}(\Omega')$ and satisfies (10.1.7). $\qquad \square$

If now $f \in W^{1,2}(\Omega)$, then, because $u \in W^{2,2}(\Omega')$ for all $\Omega' \subset\subset \Omega$, $D_i u$ is a weak solution of $\Delta D_i u = D_i f$ in Ω'. We then obtain $D_i u \in W^{2,2}(\Omega'')$ for all $\Omega'' \subset\subset \Omega'$, i.e., $u \in W^{3,2}(\Omega'')$. Iteratively, we thus obtain a new proof of Theorem 9.2.2, which we now recall:

Theorem 10.1.3: *Let $u \in W^{1,2}(\Omega)$ be a weak solution of $\Delta u = f$. Then $u \in W^{k+2,2}(\Omega_0)$ for all $\Omega_0 \subset\subset \Omega$, and*

$$\|u\|_{W^{k+2,2}(\Omega_0)} \leq \text{const} \left(\|u\|_{L^2(\Omega)} + \|f\|_{W^{k,2}(\Omega)} \right),$$

with a constant depending on k, d, Ω, and Ω_0. $\square$

In the same manner, we also obtain a new proof of Corollary 9.2.1:

Corollary 10.1.1: *Let $u \in W^{1,2}(\Omega)$ be a weak solution of $\Delta u = f$, for $f \in C^\infty(\Omega)$. Then $u \in C^\infty(\Omega)$.*

Proof: Theorems 10.1.3 and 9.1.2. $\square$

10.2 A Survey of the L^p-Regularity Theory and Applications to Solutions of Semilinear Elliptic Equations

The results of the preceding section are valid not only for the exponent $p = 2$, but in fact for any $1 < p < \infty$. We wish to explain this result in the present section. The basis of this L^p-regularity theory is the Calderon–Zygmund inequality, which we shall only quote here without proof:

Theorem 10.2.1: *Let $1 < p < \infty$, $f \in L^p(\Omega)$ ($\Omega \subset \mathbb{R}^d$ open and bounded), and let w be the Newton potential (10.1.1) of f. Then $w \in W^{2,p}(\Omega)$, $\Delta w = f$ almost everywhere in Ω, and*

$$\left\| D^2 w \right\|_{L^p(\Omega)} \leq c(d,p) \|f\|_{L^p(\Omega)}, \tag{10.2.1}$$

with the constant $c(d,p)$ depending only on the space dimension d and the exponent p.

In contrast to the case $p = 2$, i.e., Theorem 10.1.1 above, where $c(d,2) = 1$ for all d and the proof is elementary, the proof of the general case is relatively involved; we refer the reader to Bers–Schechter [1] or Gilbarg–Trudinger [9].

The Calderon–Zygmund inequality yields a generalization of Theorem 10.1.2:

Theorem 10.2.2: *Let $u \in W^{1,1}(\Omega)$ be a weak solution of $\Delta u = f$, $f \in L^p(\Omega)$, $1 < p < \infty$, i.e.,*

$$\int Du \cdot D\varphi = - \int f\varphi \quad \text{for all } \varphi \in C_0^\infty(\Omega). \qquad (10.2.2)$$

Then $u \in W^{2,p}(\Omega')$ for any $\Omega' \subset\subset \Omega$, and

$$\|u\|_{W^{2,p}(\Omega')} \leq \text{const} \left(\|u\|_{L^p(\Omega)} + \|f\|_{L^p(\Omega)} \right), \qquad (10.2.3)$$

with a constant depending on p, d, Ω', and Ω. Also,

$$\Delta u = f \quad \text{almost everywhere in } \Omega. \qquad (10.2.4)$$

We do not provide a complete proof of this result either. This time, however, we shall present at least a sketch of the *proof:*
Apart from the fact that (10.1.8) needs to be replaced by the inequality

$$\left\| D^2 v \right\|_{L^p(B(x,R))} \leq \text{const.} \|\Delta v\|_{L^p(B(x,r))} \qquad (10.2.5)$$

coming from the Calderon–Zygmund inequality (Theorem 10.2.1), we may first proceed as in the proof of Theorem 10.1.2 and obtain the estimate

$$\left\| D^2 v \right\|_{L^p(B(x,R))} \leq \text{const} \Bigg(\|f\|_{L^p(B(x,R))} + \frac{1}{(1-\sigma)R} \|Du\|_{L^p(B(x,\frac{1+\sigma}{2}R))}$$
$$+ \frac{1}{(1-\sigma)^2 R^2} \|u\|_{L^p(B(x,r))} \Bigg) \qquad (10.2.6)$$

for $0 < \sigma < 1$, $B(x,R) \subset \Omega$. The second part of the proof, namely the estimate of $\|Du\|_{L^p}$, however, is much more difficult for $p \neq 2$ than for $p = 2$. One needs an interpolation argument. For details, we refer to Gilbarg–Trudinger [9] or Giaquinta [8]. This ends our sketch of the proof.

The reader may now get the impression that the L^p-theory is a technically subtle, but perhaps essentially useless, generalization of the L^2-theory. The L^p-theory becomes necessary, however, for treating many nonlinear PDEs. We shall now discuss an example of this. We consider the equation

$$\Delta u + \Gamma(u)|Du|^2 = 0 \qquad (10.2.7)$$

with a smooth Γ. We also require that $\Gamma(u)$ be bounded. This holds if we assume that Γ itself is bounded, or if we know already that our (weak) solution u is bounded.

Equation (10.2.7) occurs as the Euler–Lagrange equation of the variational problem

$$I(u) := \int_\Omega g(u(x))|Du(x)|^2 \, dx \to \min, \qquad (10.2.8)$$

with a smooth g that satisfies the inequalities

$$0 < \lambda \le g(v) \le \Lambda < \infty, \ |g'(v)| \le k < \infty \tag{10.2.9}$$

(g' is the derivative of g), with constants λ, Λ, k, for all v.

In order to derive the Euler–Lagrange equation for (10.2.8), as in Section 8.4, for $\varphi \in H_0^{1,2}(\Omega)$, $t \in \mathbb{R}$, we consider

$$I(u + t\varphi) = \int_\Omega g(u + t\varphi) |D(u + t\varphi)|^2 \, dx.$$

In that case,

$$\frac{d}{dt} I(u + t\varphi)_{|t=0} = \int \left\{ 2g(u) \sum_i D_i u D_i \varphi + g'(u) |Du|^2 \varphi \right\} dx$$

$$= \int \left(-2g(u) \Delta u - 2 \sum_i D_i g(u) D_i u + g'(u) |Du|^2 \right) \varphi \, dx$$

$$= \int \left(-2g(u) \Delta u - g'(u) |Du|^2 \right) \varphi \, dx$$

after integrating by parts and assuming for the moment $u \in C^2$.

The Euler–Lagrange equation stems from requiring that this expression vanish for all $\varphi \in H_0^{1,2}(\Omega)$, which is the case, for example, if u minimizes $I(u)$ with respect to fixed boundary values. Thus, that equation is

$$\Delta u + \frac{g'(u)}{2g(u)} |Du|^2 = 0. \tag{10.2.10}$$

With $\Gamma(u) := \frac{g'(u)}{2g(u)}$, we have (10.2.7).

In order to apply the L^p-theory, we assume that u is a weak solution of (10.2.7) with

$$u \in W^{1,p_1}(\Omega) \quad \text{for some } p_1 > d \tag{10.2.11}$$

(as always, $\Omega \subset \mathbb{R}^d$, and so d is the space dimension).

The assumption (10.2.11) might appear rather arbitrary. It is typical for nonlinear differential equations, however, that some such hypothesis is needed. Although one may show in the present case[2] that any weak solution u of class $W^{1,2}(\Omega)$ is also contained in $W^{1,p}(\Omega)$ for all p, in structurally similar cases, for example if u is vector-valued instead of scalar-valued (so that in place of a single equation, we have a system of—typically coupled—equations of the type (10.2.7)), there exist examples of solutions of class $W^{1,2}(\Omega)$ that are not contained in any of the spaces $W^{1,p}(\Omega)$ for $p > 2$. In other words, for nonlinear equations, one typically needs a certain initial regularity of the solution before the linear theory can be applied.

---

[2] See Ladyzhenskya and Ural'tseva [17] or the remarks in Section 12.3 below.

In order to apply the L^p-theory to our solution u of (10.2.7), we put

$$f(x) := -\Gamma(u(x))|Du(x)|^2. \tag{10.2.12}$$

Because of (10.2.11) and the boundedness of $\Gamma(u)$, then

$$f \in L^{p_1/2}(\Omega), \tag{10.2.13}$$

and u satisfies

$$\Delta u = f \quad \text{in } \Omega. \tag{10.2.14}$$

By Theorem 10.2.2,

$$u \in W^{2,p_1/2}(\Omega') \quad \text{for any } \Omega' \subset\subset \Omega. \tag{10.2.15}$$

By the Sobolev embedding theorem (Corollary 9.1.1, Corollary 9.1.3, and Exercise 2 of Chapter 9),

$$u \in W^{1,p_2}(\Omega') \quad \text{for any } \Omega' \subset\subset \Omega, \tag{10.2.16}$$

with

$$p_2 = \frac{d\frac{p_1}{2}}{d - \frac{p_1}{2}} > p_1 \quad \text{because of } p_1 > d. \tag{10.2.17}$$

Thus,

$$f \in L^{\frac{p_2}{2}}(\Omega') \quad \text{for all } \Omega' \subset\subset \Omega, \tag{10.2.18}$$

and we can apply Theorem 10.2.2 and the Sobolev embedding theorem once more, to obtain

$$u \in W^{2,\frac{p_2}{2}} \cap W^{1,p_3}(\Omega') \quad \text{with } p_3 = \frac{d\frac{p_2}{2}}{d - \frac{p_2}{2}} > p_2 \tag{10.2.19}$$

for all $\Omega' \subset\subset \Omega''$. Iterating this procedure, we finally obtain

$$u \in W^{2,q}(\Omega') \quad \text{for all } q. \tag{10.2.20}$$

We now differentiate (10.2.7), in order to obtain an equation for $D_i u$, $i = 1, \ldots, d$:

$$\Delta D_i u + \Gamma'(u) D_i u |Du|^2 + 2\Gamma(u) \sum_i D_j u D_{ij} u = 0. \tag{10.2.21}$$

This time, we put

$$f := -\Gamma'(u) D_i u |Du|^2 - 2\Gamma(u) \sum_i D_j u D_{ij} u. \tag{10.2.22}$$

Then

$$|f| \leq \text{const}\,(|Du|^3 + |Du||D^2u|),$$

and because of (10.2.20) thus

$$f \in L^p(\Omega') \quad \text{for all } p.$$

This means that $v := D_i u$ satisfies

$$\Delta v = f \quad \text{with } f \in L^p(\Omega') \quad \text{for all } p. \tag{10.2.23}$$

By Theorem 10.2.2, we infer

$$v \in W^{2,p}(\Omega') \quad \text{for all } p,$$

i.e.,

$$u \in W^{3,p}(\Omega') \quad \text{for all } p. \tag{10.2.24}$$

We differentiate the equation again, to obtain equations for $D_{ij}u$ $(i,j = 1,\ldots,d)$, apply Theorem 10.2.2, conclude that $u \in W^{4,p}(\Omega')$, etc. Iterating the procedure again (this time with higher-order derivatives instead of higher exponents) and applying the Sobolev embedding theorem (Corollary 9.1.2), we obtain the following result:

Theorem 10.2.3: *Let* $u \in W^{1,p_1}(\Omega)$, *for* $p_1 > d$ $(\Omega \subset \mathbb{R}^d)$, *be a weak solution of*

$$\Delta u + \Gamma(u)|Du|^2 = 0$$

where Γ *is smooth and* $\Gamma(u)$ *is bounded. Then*

$$u \in C^\infty(\Omega).$$

$\square$

The principle of the preceding iteration process is to use the information about the solution u derived in one step as structural information about the equation satisfied by u in the next step, in order to obtain improved information about u. In the example discussed here, we use this information in the right-hand side of the equation, but in Chapter 12 we shall see other instances. Such iteration processes are typical and essential tools in the study of nonlinear PDEs. Usually, to get the iteration started, one needs to know some initial regularity of the solution, however.

Summary

A function u from the Sobolev space $W^{2,2}(\Omega)$ is called a strong solution of

$$\Delta u = f$$

if that equation holds for almost all x in Ω.

In this chapter we show that weak solutions of the Poisson equation are strong solutions as well. This makes an alternative approach to regularity theory possible.

More generally, for a weak solution $u \in W^{1,1}(\Omega)$ of

$$\Delta u = f,$$

where $f \in L^p(\Omega)$, one may utilize the Calderon–Zygmund inequality to get the L^p-estimate for all $\Omega \subset\subset \Omega$,

$$\|u\|_{W^{2,p}(\Omega')} \leq \text{const} \left(\|u\|_{L^p(\Omega)} + \|f\|_{L^p(\Omega)}\right).$$

This is valid for all $1 < p < \infty$ (but not for $p = 1$ or $p = \infty$).

This estimate is useful for iteration methods for the regularity of solutions of nonlinear elliptic equations. For example, any solution u of

$$\Delta u + \Gamma(u)|Du|^2 = 0$$

with regular Γ is of class $C^\infty(\Omega)$, provided that it satisfies the initial regularity

$$u \in W^{1,p}(\Omega) \quad \text{for some } p > d \ (= \text{space dimension}).$$

Exercises

10.1 Using the theorems discussed in Section 10.2, derive the following result: Let $u \in W^{1,2}(\Omega)$ be a weak solution of

$$\Delta u = f,$$

with $f \in W^{k,p}(\Omega)$ for some $k \geq 2$ and some $1 < p < \infty$. Then $u \in W^{k+2,p}(\Omega_0)$ for all $\Omega_0 \subset\subset \Omega$, and

$$\|u\|_{W^{k+2,p}(\Omega_0)} \leq \text{const} \left(\|u\|_{L^1(\Omega)} + \|u\|_{W^{k,p}(\Omega)}\right).$$

20.2 Consider the map

$$u : B(0,1)(\subset \mathbb{R}^d) \to \mathbb{R}^d,$$

$$x \mapsto \frac{x}{|x|}.$$

Show that for $d \geq 3$, $u \in W^{1,2}(B(0,1), \mathbb{R}^d)$ (this means that all components of u are of class $W^{1,2}$). Show, moreover, that u is a weak solution of the following system of PDEs:

$$\Delta u^\alpha + u^\alpha \sum_{i,\beta=1}^{d} |D_i u^\beta|^2 = 0 \quad \text{for } \alpha = 1, \ldots, d.$$

Since u is not continuous, we see that solutions of systems of semilinear elliptic equations need not be regular.

11. The Regularity Theory of Schauder and the Continuity Method (Existence Techniques IV)

11.1 C^α-Regularity Theory for the Poisson Equation

In this chapter we shall need the fundamental concept of Hölder continuity, which we now recall from Section 9.1:

Definition 11.1.1: *Let $f : \Omega \to \mathbb{R}$, $x_0 \in \Omega$, $0 < \alpha < 1$. The function f is called Hölder continuous at x_0 with exponent α if*

$$\sup_{x \in \Omega} \frac{|f(x) - f(x_0)|}{|x - x_0|^\alpha} < \infty. \tag{11.1.1}$$

Moreover, f is called Hölder continuous in Ω if it is Hölder continuous at each $x_0 \in \Omega$ (with exponent α); we write $f \in C^\alpha(\Omega)$. If (11.1.1) holds for $\alpha = 1$, then f is called Lipschitz continuous at x_0. Similarly, $C^{k,\alpha}(\Omega)$ is the space of those $f \in C^k(\Omega)$ whose kth derivative is Hölder continuous with exponent α.

We define a seminorm by

$$|f|_{C^\alpha(\Omega)} := \sup_{x,y \in \Omega} \frac{|f(x) - f(y)|}{|x - y|^\alpha}. \tag{11.1.2}$$

We define

$$\|f\|_{C^\alpha(\Omega)} = \|f\|_{C^0(\Omega)} + |f|_{C^\alpha(\Omega)}$$

and

$$\|f\|_{C^{k,\alpha}(\Omega)}$$

as the sum of $\|f\|_{C^k(\Omega)}$ and the Hölder seminorms of all kth partial derivatives of f. As in Definition 11.1.1, in place of $C^{0,\alpha}$, we usually write C^α. The following result is elementary:

Lemma 11.1.1: *If $f_1, f_2 \in C^\alpha(G)$ on $G \subset \mathbb{R}^d$, then $f_1 f_2 \in C^\alpha(G)$, and*

$$|f_1 f_2|_{C^\alpha(G)} \le \left(\sup_G |f_1|\right) |f_2|_{C^\alpha(G)} + \left(\sup_G |f_2|\right) |f_1|_{C^\alpha(G)}.$$

Proof:

$$\frac{|f_1(x)f_2(x) - f_1(y)f_2(y)|}{|x-y|^\alpha} \le \frac{|f_1(x) - f_1(y)|}{|x-y|^\alpha}|f_2(x)| + \frac{|f_2(x) - f_2(y)|}{|x-y|^\alpha}|f_1(x)|,$$

which directly implies the claim.   □

Theorem 11.1.1: *As always, let $\Omega \subset \mathbb{R}^d$ be open and bounded,*

$$u(x) := \int_\Omega \Gamma(x,y)f(y)dy, \tag{11.1.3}$$

where Γ is the fundamental solution defined in Section 1.1.

(a) If $f \in L^\infty(\Omega)$ (i.e., $\sup_{x \in \Omega}|f(x)| < \infty$),[1] then $u \in C^{1,\alpha}(\Omega)$, and

$$\|u\|_{C^{1,\alpha}(\Omega)} \le c_1 \sup |f| \quad \text{for } \alpha \in (0,1). \tag{11.1.4}$$

(b) If $f \in C_0^\alpha(\Omega)$, then $u \in C^{2,\alpha}(\Omega)$, and

$$\|u\|_{C^{2,\alpha}(\Omega)} \le c_2 \|f\|_{C^\alpha(\Omega)} \quad \text{for } 0 < \alpha < 1. \tag{11.1.5}$$

The constants in (11.1.4) and (11.1.5) depend on α, d, and $|\Omega|$.

Proof: (a) Up to a constant factor, the first derivatives of u are given by

$$v^i(x) := \int_\Omega \frac{x^i - y^i}{|x-y|^d} f(y)dy \quad (i = 1, \dots, d).$$

From this formula,

$$|v^i(x_1) - v^i(x_2)| \le \sup_\Omega |f| \cdot \int_\Omega \left| \frac{x_1^i - y^i}{|x_1 - y|^d} - \frac{x_2^i - y^i}{|x_2 - y|^d} \right| dy. \tag{11.1.6}$$

By the intermediate value theorem, on the line from x_1 to x_2 there exists some x_3 with

$$\left| \frac{x_1^i - y^i}{|x_1 - y|^d} - \frac{x_2^i - y^i}{|x_2 - y|^d} \right| \le \frac{c_3 |x_1 - x_2|}{|x_3 - y|^d}. \tag{11.1.7}$$

We put $\delta := 2|x_1 - x_2|$. Since Ω is bounded, we can find $R > 0$ with $\Omega \subset B(x_3, R)$, and we replace the integral on Ω in (11.1.6) by the integral on $B(x_3, R)$, and we decompose the latter as

$$\int_{B(x_3,R)} = \int_{B(x_3,\delta)} + \int_{B(x_3,R)\backslash B(x_3,\delta)} = I_1 + I_2, \tag{11.1.8}$$

---

[1] "sup" here is the essential supremum, as explained in Appendix 12.3.

where without loss of generality, we may take $\delta < R$. We have

$$I_1 \leq 2 \int_{B(x_3,\delta)} \frac{1}{|x_3 - y|^{d-1}} dy = 2\omega_d \delta \qquad (11.1.9)$$

and by (11.1.7)

$$I_2 \leq c_4 \delta(\log R - \log \delta), \qquad (11.1.10)$$

and hence

$$I_1 + I_2 \leq c_5 |x_1 - x_2|^\alpha \quad \text{for any } \alpha \in (0,1).$$

This proves (a), because obviously, we also have

$$|v^i(x)| \leq c_6 \sup_\Omega |f|. \qquad (11.1.11)$$

(b) Up to a constant factor, the second derivatives of u are given by

$$w^{ij}(x) = \int \left(|x - y|^2 \delta_{ij} - d \left(x^i - y^i \right) \left(x^j - y^j \right) \right) \frac{1}{|x - y|^{d+2}} f(y) \, dy;$$

however, we still need to show that this integral is finite if our assumption $f \in C_0^\alpha(\Omega)$ holds. This will also follow from our subsequent considerations.

We first put $f(x) = 0$ for $x \in \mathbb{R}^d \setminus \Omega$; this does not affect the Hölder continuity of f. We write

$$K(x - y) := \left(|x - y|^2 \delta_{ij} - d \left(x^i - y^i \right) \left(x^j - y^j \right) \right) \frac{1}{|x - y|^{d+2}}$$

$$= \frac{\partial}{\partial x^j} \left(\frac{x^i - y^i}{|x - y|^d} \right).$$

We have

$$\int_{R_1 < |y| < R_2} K(y) dy = \int_{|y|=R_2} \frac{y^j}{R_2} \cdot \frac{y^i}{|y|^d} - \int_{|y|=R_1} \frac{y^j}{R_1} \cdot \frac{y^i}{|y|^d} \qquad (11.1.12)$$

$$= 0,$$

since $\frac{y^i}{|y|^d}$ is homogeneous of degree $1 - d$. Thus also

$$\int_{\mathbb{R}^d} K(y) dy = 0. \qquad (11.1.13)$$

We now write

$$w^{ij}(x) = \int_{\mathbb{R}^d} K(x - y) f(y) dy \qquad (11.1.14)$$

$$= \int_{\mathbb{R}^d} (f(y) - f(x)) K(x - y) dy$$

by (11.1.13). As before, on the line from x_1 to x_2 there is some x_3 with

$$|K(x_1 - y) - K(x_2 - y)| \leq \frac{c_7 |x_1 - x_2|}{|x_3 - y|^{d+1}}. \tag{11.1.15}$$

We again put

$$\delta := 2 |x_1 - x_2|$$

and write (cf. (11.1.14))

$$w^{ij}(x_1) - w^{ij}(x_2)$$
$$= \int_{\mathbb{R}^d} \{(f(y) - f(x_1)) K(x_1 - y) - (f(y) - f(x_2)) K(x_2 - y)\} \, dy$$
$$= I_1 + I_2, \tag{11.1.16}$$

where I_1 denotes the integral on $B(x_1, \delta)$, and I_2 that on $\mathbb{R}^d \setminus B(x_1, \delta)$. Since $|f(y) - f(x)| \leq \|f\|_{C^\alpha} \cdot |x - y|^\alpha$, it follows that

$$|I_1| \leq \|f\|_{C^\alpha} \int_{B(x_1,\delta)} \{K(x_1 - y) |x_1 - y|^\alpha - K(x_2 - y) |x_2 - y|^\alpha\} \, dy$$
$$\leq c_8 \|f\|_{C^\alpha} \cdot \delta^\alpha. \tag{11.1.17}$$

Moreover,

$$I_2 = \int_{\mathbb{R}^d \setminus B(x_1,\delta)} (f(x_2) - f(x_1)) K(x_1 - y) \, dy$$
$$+ \int_{\mathbb{R}^d \setminus B(x_1,\delta)} (f(y) - f(x_2)) (K(x_1 - y) - K(x_2 - y)) \, dy, \tag{11.1.18}$$

and the first integral vanishes because of (11.1.12). Employing (11.1.15), and since for $y \in \mathbb{R}^d \setminus B(x_1, \delta)$,

$$\frac{1}{|x_3 - y|^{d+1}} \leq \frac{c_9}{|x_1 - y|^{d+1}},$$

it follows that

$$|I_2| \leq c_{10} \delta \|f\|_{C^\alpha} \int_{\mathbb{R}^d \setminus B(x_1,\delta)} |x_1 - y|^{\alpha-d-1} \leq c_{11} \delta^\alpha \|f\|_{C^\alpha}. \tag{11.1.19}$$

Inequality (11.1.5) then follows from (11.1.16), (11.1.17), (11.1.19).   □

Theorem 11.1.2: *As always, let $\Omega \subset \mathbb{R}^d$ be open and bounded, and $\Omega_0 \subset\subset \Omega$. Let u be a weak solution of $\Delta u = f$ in Ω.*

(a) If $f \in C^0(\Omega)$, then $u \in C^{1,\alpha}(\Omega)$, and

$$\|u\|_{C^{1,\alpha}(\Omega_0)} \leq c_{12} \left(\|f\|_{C^0(\Omega)} + \|u\|_{L^2(\Omega)} \right). \qquad (11.1.20)$$

(b) If $f \in C^\alpha(\Omega)$, then $u \in C^{2,\alpha}(\Omega)$, and

$$\|u\|_{C^{2,\alpha}(\Omega)} \leq c_{13} \left(\|f\|_{C^\alpha(\Omega)} + \|u\|_{L^2(\Omega)} \right). \qquad (11.1.21)$$

Remark: The restriction $0 < \alpha < 1$ is essential for Theorem 11.1.2, as well as for the subsequent results. For example, in some neighborhood of 0, the function

$$u\left(x^1, x^2\right) = |x^1| \, |x^2| \log \left(|x^1| + |x^2|\right)$$

satisfies the inequality

$$|u| + |\Delta u| \leq \text{const},$$

while the mixed second derivative $\frac{\partial^2 u}{\partial x^1 \partial x^2}$ behaves like

$$\log \left(|x^1| + |x^2|\right).$$

Consequently, the $C^{1,1}$-norm of u cannot be controlled by pointwise bounds for $f := \Delta u$ and u.

Proof: We demonstrate the estimates (11.1.20) and (11.1.21) first under the assumption $u \in C^{2,\alpha}(\Omega)$. We may cover Ω_0 by finitely many balls that are contained in Ω. Therefore, it suffices to verify the estimates for the case

$$\Omega_0 = B(0, r),$$
$$\Omega = B(0, R), \quad 0 < r < R < \infty.$$

Let $0 < R_1 < R_2 < R$. We choose some $\eta \in C_0^\infty(B(0, R_2))$ with $0 \leq \eta \leq 1$, $\eta(x) = 1$ for $|x| \leq R_1$, and

$$\|\eta\|_{C^{k,\alpha}(B(0,R_2))} \leq c_{14}(R_2 - R_1)^{-k-\alpha}. \qquad (11.1.22)$$

We put

$$\phi := \eta u. \qquad (11.1.23)$$

Then ϕ vanishes outside of $B(0, R_2)$, and by (1.1.7),

$$\phi(x) = \int_\Omega \Gamma(x, y) \Delta \phi(y) dy. \qquad (11.1.24)$$

Here,

$$\Delta\phi = \eta\Delta u + 2Du \cdot D\eta + u\Delta\eta, \tag{11.1.25}$$

and so

$$\|\Delta\phi\|_{C^0} \le \|\Delta u\|_{C^0} + c_{15}\|\eta\|_{C^2} \cdot \|u\|_{C^1}, \tag{11.1.26}$$

and by Lemma 11.1.1

$$\|\Delta\phi\|_{C^\alpha} \le c_{16}\|\eta\|_{C^{2,\alpha}}(\|\Delta u\|_{C^\alpha} + \|u\|_{C^{1,\alpha}}), \tag{11.1.27}$$

where all norms are computed on $B(0, R_2)$. From Theorem 11.1.1 and (11.1.26) and (11.1.27), we obtain

$$\|\phi\|_{C^{1,\alpha}} \le c_{17}(\|\Delta u\|_{C^0} + \|\eta\|_{C^2}\|u\|_{C^1}) \tag{11.1.28}$$

and

$$\|\phi\|_{C^{2,\alpha}} \le c_{18}\|\eta\|_{C^{2,\alpha}}(\|\Delta u\|_{C^\alpha} + \|u\|_{C^{1,\alpha}}), \tag{11.1.29}$$

respectively. Since $u(x) = \phi(x)$ for $|x| \le R_1$, and recalling (11.1.22), we obtain

$$\|u\|_{C^{1,\alpha}(B(0,R_1))} \le c_{19}\left(\|\Delta u\|_{C^0(B(0,R_2))} + \frac{1}{(R_2 - R_1)^2}\|u\|_{C^1(B(0,R_2))}\right) \tag{11.1.30}$$

and

$$\|u\|_{C^{2,\alpha}(B(0,R_1))} \le c_{20}\frac{1}{(R_2 - R_1)^{2+\alpha}}\left(\|\Delta u\|_{C^\alpha(B(0,R_2))} + \|u\|_{C^{1,\alpha}(B(0,R_2))}\right) \tag{11.1.31}$$

respectively.

We now interrupt the proof for some auxiliary results:

Lemma 11.1.2:
a) There exists a constant c_a such that for every $\rho > 0$ and any function $v \in C^1(B(0, \rho))$

$$\|v\|_{C^0(B(0,\rho))} \le \|Dv\|_{C^0(B(0,\rho))} + c_a\|v\|_{L^2(B(0,\rho))}. \tag{11.1.32}$$

b) There exists a constant c_b such that for every $\rho > 0$ and any function $v \in C^{1,\alpha}(B(0, \rho))$

$$\|v\|_{C^1(B(0,\rho))} \le |Dv|_{C^\alpha(B(0,\rho))} + c_b\|v\|_{L^2(B(0,\rho))} \tag{11.1.33}$$

(here, $|Dv|_{C^\alpha}$ is the Hölder seminorm defined in (11.1.2)).

Proof: If a) did not hold, for every $n \in \mathbb{N}$, we could find a radius ρ_n and a function $v_n \in C^1(B(0, \rho_n))$ with

$$1 = \|v_n\|_{C^0(B(0,\rho_n))} \geq \|Dv_n\|_{C^0(B(0,\rho_n))} + n\|v_n\|_{L^2(B(0,\rho_n))}. \quad (11.1.34)$$

We first consider the case where the radii ρ_n stay bounded for $n \to \infty$ in which case we may assume that they converge towards some radius ρ_0 and we can consider everything on the fixed ball $B(0, \rho_0)$.

Thus, in that situation, we have a sequence $v_n \in C^1(B(0, \rho_0))$ for which $\|v_n\|_{C^1(B(0,\rho_0))}$ is bounded. This implies that the v_n are equicontinuous. By the theorem of Arzela-Ascoli, after passing to a subsequence, we can assume that the v_n converge uniformly towards some $v_0 \in C^0(B(0, \rho))$ with $\|v_0\|_{C^0(B(0,\rho_0))} = 1$. But (11.1.34) would imply $\|v_0\|_{L^2(B(0,\rho_0))} = 0$, hence $v \equiv 0$, a contradiction.

It remains to consider the case where the ρ_n tend to ∞. In that case, we use (11.1.34) to choose points $x_n \in B(0, \rho_n)$ with

$$|v_n(x_n)| \geq \frac{1}{2}\|v_n\|_{C^0(B(0,\rho_n))} = \frac{1}{2}. \quad (11.1.35)$$

We then consider $w_n(x) := v_n(x + x_n)$ so that $w_n(0) \geq \frac{1}{2}$ while (11.1.34) holds for w_n on some fixed neighborhood of 0. We then apply the Arzela-Ascoli argument to the w_n to get a contradiction as before.

b) is proved in the same manner. The crucial point now is that for a sequence v_n for which the norms $\|v_n\|_{C^{1,\alpha}}$ are uniformly bounded, both the v_n and their first derivatives are equicontinuous. $\qquad \square$

Lemma 11.1.3:

a) For $\varepsilon > 0$, there exists $M(\varepsilon) (< \infty)$ such that for all $u \in C^1(B(0,1))$

$$\|u\|_{C^0(B(0,1))} \leq \varepsilon\|u\|_{C^1(B(0,1))} + M(\varepsilon)\|u\|_{L^2(B(0,1))} \quad (11.1.36)$$

for all $u \in C^{1,\alpha}$. For $\varepsilon \to 0$,

$$M(\varepsilon) \leq \text{const.}\,\varepsilon^{-d}. \quad (11.1.37)$$

b) For every $\alpha \in (0,1)$ and $\varepsilon > 0$, there exists $N(\varepsilon) (< \infty)$ such that for all $u \in C^{1,\alpha}(B(0,1))$

$$\|u\|_{C^1(B(0,1))} \leq \varepsilon\|u\|_{C^{1,\alpha}(B(0,1))} + N(\varepsilon)\|u\|_{L^2(B(0,1))} \quad (11.1.38)$$

for all $u \in C^{1,\alpha}$. For $\varepsilon \to 0$,

$$N(\varepsilon) \leq \text{const.}\,\varepsilon^{-\frac{d+1}{\alpha}}. \quad (11.1.39)$$

c) For every $\alpha \in (0,1)$ and $\varepsilon > 0$, there exists $Q(\varepsilon) (< \infty)$ such that for all $u \in C^{2,\alpha}(B(0,1))$

$$\|u\|_{C^{1,\alpha}(B(0,1))} \leq \varepsilon \|u\|_{C^{2,\alpha}(B(0,1))} + Q(\varepsilon) \|u\|_{L^2(B(0,1))} \tag{11.1.40}$$

for all $u \in C^{1,\alpha}$. For $\varepsilon \to 0$,

$$Q(\varepsilon) \leq \text{const.} \, \varepsilon^{-d-1-\alpha}. \tag{11.1.41}$$

Proof: We rescale:

$$u_\rho(x) := u(\frac{x}{\rho}), \ u_\rho : B(0,\rho) \to \mathbb{R}. \tag{11.1.42}$$

(11.1.36) then is equivalent to

$$\|u_\rho\|_{C^0(B(0,\rho))} \leq \varepsilon\rho \|u_\rho\|_{C^1(B(0,\rho))} + M(\varepsilon)\rho^{-d} \|u\|_{L^2(B(0,\rho))}. \tag{11.1.43}$$

We choose ρ such that $\varepsilon\rho = 1$, that is, $\rho = \varepsilon^{-1}$ and apply a) of Lemma 11.1.2. This shows (11.1.43), and a) follows.
For b), we shall show

$$\|Du\|_{C^0(B(0,1))} \leq \varepsilon |Du|_{C^\alpha(B(0,1))} + N(\varepsilon) \|u\|_{L^2(B(0,1))}. \tag{11.1.44}$$

Combining this with a) then shows the claim. We again rescale by (11.1.42). This transforms (11.1.44) into

$$\|Du_\rho\|_{C^0(B(0,\rho))} \leq \varepsilon\rho^\alpha |Du|_{C^\alpha(B(0,\rho))} + N(\varepsilon)\rho^{-d-1} \|u\|_{L^2(B(0,\rho))}. \tag{11.1.45}$$

We choose ρ such that $\varepsilon\rho^\alpha = 1$, that is, $\rho = \varepsilon^{-\frac{1}{\alpha}}$ and apply a) of Lemma 11.1.2. This shows (11.1.45) and completes the proof of b).
c) is proved in the same manner. $\qquad\qquad\qquad\qquad\qquad\qquad\square$

We now continue the *proof* of Theorem 11.1.2:
For homogeneous polynomials $p(t), q(t)$, we define

$$A_1 := \sup_{0 \leq r \leq R} p(R-r) \|u\|_{C^{1,\alpha}(B(0,r))},$$

$$A_2 := \sup_{0 \leq r \leq R} q(R-r) \|u\|_{C^{2,\alpha}(B(0,r))}.$$

For the proof of (a), we choose R_1 such that

$$A_1 \leq 2p(R-R_1) \|u\|_{C^{1,\alpha}(B(0,R_1))}, \tag{11.1.46}$$

and for (b), such that

$$A_2 \leq 2q(R-R_1) \|u\|_{C^{2,\alpha}(B(0,R_1))}. \tag{11.1.47}$$

(In general, the R_1 of (11.1.46) will not be the same as that of (11.1.47).)
Then (11.1.30) and (11.1.38) imply

$$A_1 \leq c_{21}\, p(R - R_1)\Big(\|\Delta u\|_{C^0(B(0,R_2))} + \frac{\varepsilon}{(R_2 - R_1)^2}\, \|u\|_{C^{1,\alpha}(B(0,R_2))}$$

$$+ \frac{1}{(R_2 - R_1)^2}\, N(\varepsilon)\, \|u\|_{L^2(B(0,R_2))}\Big)$$

$$\leq c_{22}\, \frac{p(R - R_1)}{p(R - R_2)} \cdot \frac{\varepsilon}{(R_2 - R_1)^2} \cdot A_1$$

$$+ c_{23}\, p(R - R_1)\, \|\Delta u\|_{C^0(B(0,R_2))} + c_{24}\, N(\varepsilon)\, \frac{p(R - R_1)}{(R_2 - R_1)^2}\, \|u\|_{L^2(B(0,R_2))}\,.$$

$$(11.1.48)$$

We choose $R_2 = \frac{R + R_1}{2} \in (R_1, R)$. Then, because the polynomial p is homogeneous,

$$\frac{p(R - R_1)}{p(R - R_2)} = \frac{p(R - R_1)}{p(\frac{R - R_1}{2})}$$

is independent of R and R_1. Therefore,

$$\varepsilon = \frac{(R_2 - R_1)^2}{2c_{24}}\, \frac{p(R - R_1)}{p(R - R_2)} \sim (R - R_1)^2$$

and

$$N(\varepsilon) \sim (R - R_1)^{-\frac{2(d+1)}{\alpha}}$$

by Lemmma 11.1.2 b). Thus, when we choose

$$p(t) = t^{\frac{2(d+1)}{\alpha} + 2},$$

the coefficient of $\|u\|_{L^2(B(0,R_2))}$ in (11.1.48) is controlled.
Thus, finally

$$\|u\|_{C^{1,\alpha}(B(0,r))} \leq \frac{1}{p(R - r)} A_1$$

$$\leq c_{25} \Big(\|\Delta u\|_{C^0(B(0,R))} + \|u\|_{L^2(B(0,R))} \Big),$$

$$(11.1.49)$$

with a constant that now also depends on the radii occurring.
In the same manner, from (11.1.31) and (11.1.40), we obtain

$$\|u\|_{C^{2,\alpha}(B(0,r))} \leq c_{26} \Big(\|\Delta u\|_{C^\alpha(B(0,R))} + \|u\|_{L^2(B(0,R))} \Big)$$

$$(11.1.50)$$

for $0 < r < R$. Since $\Delta u = f$, we have thus proved (11.1.20) and (11.1.21) for $u \in C^{2,\alpha}(\Omega)$.

For $u \in W^{1,2}(\Omega)$ we consider the mollifications u_h as in Lemma A.2 of the Appendix. Let $0 < h < \mathrm{dist}(\Omega_0, \partial\Omega)$. Then

$$\int_\Omega Du_h \cdot Dv = -\int_\Omega f_h v \quad \text{for all } v \in H_0^{1,2}(\Omega),$$

and since $u_h \in C^\infty$, also

$$\Delta u_h = f_h.$$

Moreover, by Lemma A.2,

$$\|f_h - f\|_{C^0} \to 0,$$

and with an analogous proof, if $f \in C^\alpha(\Omega)$,

$$\|f_h - f\|_{C^\alpha} \to 0.$$

For $h \to 0$, the f_h therefore constitute a Cauchy sequence in $C^0(\Omega)$ or $C^\alpha(\Omega)$. Applying (11.1.20) and (11.1.21) to $u_{h_1} - u_{h_2}$, we obtain

$$\|u_{h_1} - u_{h_2}\|_{C^{1,\alpha}(\Omega_0)} \leq c_{27} \left(\|f_{h_1} - f_{h_2}\|_{C^0(\Omega)} + \|u_{h_1} - u_{h_2}\|_{L^2(\Omega)} \right)$$
$$(11.1.51)$$

or

$$\|u_{h_1} - u_{h_2}\|_{C^{2,\alpha}(\Omega_0)} \leq c_{28} \left(\|f_{h_1} - f_{h_2}\|_{C^\alpha(\Omega)} + \|u_{h_1} - u_{h_2}\|_{L^2(\Omega)} \right).$$
$$(11.1.52)$$

The limit function u thus is contained in $C^{1,\alpha}(\Omega_0)$ or $C^{2,\alpha}(\Omega_0)$, and satisfies (11.1.20) or (11.1.21). □

Part (a) of the preceding theorem can be sharpened as follows:

Theorem 11.1.3: *Let u be a weak solution of $\Delta u = f$ in Ω (Ω a bounded domain in $\mathbb{R}^d$), $f \in L^p(\Omega)$ for some $p > d$, $\Omega_0 \subset\subset \Omega$. Then $u \in C^{1,\alpha}(\Omega)$ for some α that depends on p and d, and*

$$\|u\|_{C^{1,\alpha}(\Omega_0)} \leq \text{const} \left(\|f\|_{L^p(\Omega)} + \|u\|_{L^2(\Omega)} \right).$$

Proof: Again, we consider the Newton potential

$$w(x) := \int_\Omega \Gamma(x,y)f(y)dy,$$

and

$$v^i(x) := \int_\Omega \frac{x^i - y^i}{(x-y)^d} f(y)dy.$$

Using Hölder's inequality, we obtain

$$|v^i(x)| \leq \|f\|_{L^p(\Omega)} \left(\int \frac{dy}{|x-y|^{(d-1)\frac{p}{p-1}}} \right)^{\frac{p-1}{p}},$$

and this expression is finite because of $p > d$. In this manner, one also verifies that $\frac{\partial}{\partial x^i} w = \text{const} v^i$ and obtains the Hölder estimate as in the proof of Theorem 11.1.1(a) and Theorem 11.1.2(a). □

Corollary 11.1.1: *If $u \in W^{1,2}(\Omega)$ is a weak solution of $\Delta u = f$ with $f \in C^{k,\alpha}(\Omega)$, $k \in \mathbb{N}$, $0 < \alpha < 1$, then $u \in C^{k+2,\alpha}(\Omega)$, and for $\Omega_0 \subset\subset \Omega$,*

$$\|u\|_{C^{k+2,\alpha}(\Omega_0)} \leq \text{const} \left(\|f\|_{C^{k,\alpha}(\Omega)} + \|u\|_{L^2(\Omega)} \right).$$

If $f \in C^\infty(\Omega)$, so is u.

Proof: Since $u \in C^{2,\alpha}(\Omega)$ by Theorem 11.1.2, we know that it weakly solves

$$\Delta \frac{\partial}{\partial x^i} u = \frac{\partial}{\partial x^i} f.$$

Theorem 11.1.2 then implies

$$\frac{\partial}{\partial x^i} u \in C^{2,\alpha}(\Omega) \quad (i \in \{1, \dots, d\}),$$

and thus $u \in C^{3,\alpha}(\Omega)$. The proof is concluded by induction. $\square$

11.2 The Schauder Estimates

In this section, we study differential equations of the type

$$Lu(x) := \sum_{i,j=1}^{d} a^{ij}(x) \frac{\partial^2 u(x)}{\partial x^i \partial x^j} + \sum_{i=1}^{d} b^i(x) \frac{\partial u(x)}{\partial x^i} + c(x)u(x) = f(x) \quad (11.2.1)$$

in some domain $\Omega \subset \mathbb{R}^d$. We make the following assumptions:

(A) Ellipticity: There exists $\lambda > 0$ such that for all $x \in \Omega$, $\xi \in \mathbb{R}^d$,

$$\sum_{i,j=1}^{d} a^{ij}(x)\xi_i\xi_j \geq \lambda |\xi|^2.$$

  Moreover, $a^{ij}(x) = a^{ji}(x)$ for all i, j, x.

(B) Hölder continuous coefficients: There exists $K < \infty$ such that

$$\left\| a^{ij} \right\|_{C^\alpha(\Omega)}, \left\| b^i \right\|_{C^\alpha(\Omega)}, \|c\|_{C^\alpha(\Omega)} \leq K$$

  for all i, j.

The fundamental estimates of J. Schauder are the following:

Theorem 11.2.1: *Let $f \in C^\alpha(\Omega)$, and suppose $u \in C^{2,\alpha}(\Omega)$ satisfies*

$$Lu = f \tag{11.2.2}$$

in Ω ($0 < \alpha < 1$). For any $\Omega_0 \subset\subset \Omega$, we then have

$$\|u\|_{C^{2,\alpha}(\Omega_0)} \leq c_1 \left(\|f\|_{C^\alpha(\Omega)} + \|u\|_{L^2(\Omega)} \right), \tag{11.2.3}$$

with a constant c_1 depending on $\Omega, \Omega_0, \alpha, d, \lambda, K$.

For the proof, we shall need the following lemma:

Lemma 11.2.1: *Let the symmetric matrix* $(A^{ij})_{i,j=1,...,d}$ *satisfy*

$$\lambda |\xi|^2 \leq \sum_{i,j=1}^d A^{ij} \xi_i \xi_j \leq \Lambda |\xi|^2 \quad \text{for all } \xi \in \mathbb{R}^d \tag{11.2.4}$$

with

$$0 < \lambda < \Lambda < \infty.$$

Let u satisfy

$$\sum_{i,j=1}^d A^{ij} \frac{\partial^2 u}{\partial x^i \partial x^j} = f \tag{11.2.5}$$

with $f \in C^\alpha(\Omega)$ *(0 < α < 1). For any* $\Omega_0 \subset\subset \Omega$, *we then have*

$$\|u\|_{C^{2,\alpha}(\Omega_0)} \leq c_2 \left(\|f\|_{C^\alpha(\Omega)} + \|u\|_{L^2(\Omega)} \right). \tag{11.2.6}$$

Proof: We shall employ the following notation:

$$A := (A^{ij})_{i,j=1,...,d}, \quad D^2 u := \left(\frac{\partial^2 u}{\partial x^i \partial x^j} \right)_{i,j=1,...,d}.$$

If B is a nonsingular $d \times d$–matrix, and if $y := Bx, v := u \circ B^{-1}$, i.e., $v(y) = u(x)$, we have

$$AD^2 u(x) = AB^t D^2 v(y) B,$$

and hence

$$\text{Tr}(AD^2 u(x)) = \text{Tr}(BAB^t D^2 v(y)). \tag{11.2.7}$$

Since A is symmetric, we may choose B such that $B^t A B$ is the unit matrix. In fact, B can be chosen as the product of the diagonal matrix

$$D = \begin{pmatrix} \lambda_1^{-\frac{1}{2}} & & \\ & \ddots & \\ & & \lambda_d^{-\frac{1}{2}} \end{pmatrix}$$

($\lambda_1, \ldots, \lambda_d$ being the eigenvalues of A) with some orthogonal matrix R. In this way we obtain the transformed equation

$$\Delta v(y) = f \left(B^{-1} y \right). \tag{11.2.8}$$

Theorem 11.1.2 then yields $C^{2,\alpha}$-estimates for v, and these can be transformed back into estimates for $u = v \circ B$. The resulting constants will also depend on the bounds λ, Λ for the eigenvalues of A, since these determine the eigenvalues of D and hence of B. $\qquad \square$

Proof of Theorem 11.2.1: We shall show that for every $x_0 \in \bar{\Omega}_0$ there exists some ball $B(x_0, r)$ on which the desired estimate holds. The radius r of this ball will depend only on $\mathrm{dist}(\Omega_0, \partial\Omega)$ and the Hölder norms of the coefficients a^{ij}, b^i, c. Since $\bar{\Omega}_0$ is compact, it can be covered by finitely many such balls, and this yields the estimate in Ω_0.

Thus, let $x_0 \in \bar{\Omega}_0$. We rewrite the differential equation $Lu = f$ as

$$
\sum_{i,j} a^{ij}(x_0) \frac{\partial^2 u(x)}{\partial x^i \partial x^j} = \sum_{i,j} \left(a^{ij}(x_0) - a^{ij}(x) \right) \frac{\partial^2 u(x)}{\partial x^i \partial x^j}
$$

$$
- \sum_i b^i(x) \frac{\partial u(x)}{\partial x^i} - c(x)u(x) + f(x) \tag{11.2.9}
$$

$$
=: \varphi(x).
$$

If we are able to estimate the C^α-norm of φ, putting $A^{ij} := a^{ij}(x_0)$ and applying Lemma 11.2.1 will yield the estimate of the $C^{2,\alpha}$-norm of u. The crucial term for the estimate of φ is $\sum (a^{ij}(x_0) - a^{ij}(x)) \frac{\partial^2 u}{\partial x^i \partial x^j}$. Let $B(x_0, R) \subset \Omega$. By Lemma 11.1.1

$$
\left| \sum_{i,j} \left(a^{ij}(x_0) - a^{ij}(x) \right) \frac{\partial^2 u(x)}{\partial x^i \partial x^j} \right|_{C^\alpha(B(x_0,R))}
$$

$$
\leq \sup_{i,j,x \in B(x_0,R)} \left| a^{ij}(x_0) - a^{ij}(x) \right| \left| D^2 u \right|_{C^\alpha(B(x_0,R))}
$$

$$
+ \sum_{i,j} \left| a^{ij} \right|_{C^\alpha(B(x_0,R))} \sup_{B(x_0,R)} \left| D^2 u \right|. \tag{11.2.10}
$$

Thus, also

$$
\left\| \sum \left(a^{ij}(x_0) - a^{ij}(x) \right) \frac{\partial^2 u}{\partial x^i \partial x^j} \right\|_{C^\alpha(B(x_0,R))}
$$

$$
\leq \sup \left| a^{ij}(x_0) - a^{ij}(x) \right| \|u\|_{C^{2,\alpha}(B(x_0,R))} + c_3 \|u\|_{C^2(B(x_0,R))}, \tag{11.2.11}
$$

where c_3 in particular depends on the C^α-norm of the a^{ij}.

Analogously,

$$
\left\| \sum_i b^i(x) \frac{\partial u}{\partial x^i}(x) \right\|_{C^\alpha(B(x_0,R))} \leq c_4 \|u\|_{C^{1,\alpha}(B(x_0,R))}, \tag{11.2.12}
$$

$$
\|c(x)u(x)\|_{C^\alpha(B(x_0,R))} \leq c_5 \|u\|_{C^\alpha(B(x_0,R))}. \tag{11.2.13}
$$

Altogether, we obtain

$$
\|\varphi\|_{C^\alpha(B(x_0,R))} \leq \sup_{i,j,x \in B(x_0,R)} \left| a^{ij}(x_0) - a^{ij}(x) \right| \|u\|_{C^{2,\alpha}(B(x_0,R))}
$$

$$
+ c_6 \|u\|_{C^2(B(x_0,R))} + \|f\|_{C^\alpha(B(x_0,R))}. \tag{11.2.14}
$$

By Lemma 11.2.1, from (11.2.9) and (11.2.14) for $0 < r < R$, we obtain

$$\|u\|_{C^{2,\alpha}(B(x_0,r))} \leq c_7 \sup_{i,j,x \in B(x_0,R)} |a^{ij}(x_0) - a^{ij}(x)| \|u\|_{C^{2,\alpha}(B(x_0,R))}$$
$$+ c_8 \|u\|_{C^2(B(x_0,R))} + c_9 \|f\|_{C^\alpha(B(x_0,R))} . \qquad (11.2.15)$$

Since the a^{ij} are continuous on Ω, we may choose $R > 0$ so small that

$$c_7 \sup_{i,j,x \in B(x_0,R)} |a^{ij}(x_0) - a^{ij}(x)| \leq \frac{1}{2}. \qquad (11.2.16)$$

With the same method as in the proof of Theorem 11.1.2, the corresponding term can be absorbed in the left-hand side. We then obtain from (11.2.15)

$$\|u\|_{C^{2,\alpha}(B(x_0,R))} \leq 2c_8 \|u\|_{C^2(B(x_0,R))} + 2c_9 \|f\|_{C^\alpha(B(x_0,R))} . \qquad (11.2.17)$$

By (11.1.40), for every $\varepsilon > 0$, there exists some $Q(\varepsilon)$ with

$$\|u\|_{C^2(B(x_0,R))} \leq \varepsilon \|u\|_{C^{2,\alpha}(B(x_0,R))} + Q(\varepsilon) \|u\|_{L^2(B(x_0,R))} . \qquad (11.2.18)$$

With the same method as in the proof of Theorem 11.1.2, from (11.2.18) and (11.2.17) we deduce the desired estimate

$$\|u\|_{C^{2,\alpha}(B(x_0,R))} \leq c_{10} \left(\|f\|_{C^\alpha(B(x_0,R))} + \|u\|_{L^2(B(x_0,R))} \right) . \qquad (11.2.19)$$

□

We may now state the global estimate of J. Schauder for the solution of the Dirichlet problem for L:

Theorem 11.2.2: Let $\Omega \subset \mathbb{R}^d$ be a bounded domain of class $C^{2,\alpha}$ (analogously to Definition 9.3.1, we require the same properties as there, except that (iii) is replaced by the condition that ϕ and ϕ^{-1} are of class $C^{2,\alpha}$). Let $f \in C^\alpha(\bar{\Omega})$, $g \in C^{2,\alpha}(\bar{\Omega})$ (as in Definition 9.3.2), and let $u \in C^{2,\alpha}(\bar{\Omega})$ satisfy

$$\begin{aligned} Lu(x) &= f(x) &&\text{for } x \in \Omega, \\ u(x) &= g(x) &&\text{for } x \in \partial\Omega. \end{aligned} \qquad (11.2.20)$$

Then

$$\|u\|_{C^{2,\alpha}(\Omega)} \leq c_{11} \left(\|f\|_{C^\alpha(\Omega)} + \|g\|_{C^{2,\alpha}(\Omega)} + \|u\|_{L^2(\Omega)} \right), \qquad (11.2.21)$$

with a constant c_{11} depending on $\Omega, \alpha, d, \lambda$, and K.

The *Proof* essentially is a modification of that of Theorem 11.2.1, with modifications that are similar to those employed in the proof of Theorem 9.3.3. We shall therefore provide only a sketch of the proof. We start with a simplified model situation, namely, the Poisson equation in a half-ball, from which we shall derive the general case.

As in Section 9.3, let

$$B^+(0,R) = \left\{ x = (x^1,\ldots,x^d) \in \mathbb{R}^d; |x| < R, x^d > 0 \right\}.$$

Moreover, let

$$\partial^0 B^+(0,R) := \partial B^+(0,R) \cap \{x^d = 0\},$$
$$\partial^+ B^+(0,R) := \partial B^+(0,R) \setminus \partial^0 B^+(0,R).$$

We consider $f \in C^\alpha\left(\overline{B^+(0,R)}\right)$ with

$$f = 0 \quad \text{on } \partial^+ B^+(0,R).$$

In contrast to the situation considered in Theorem 11.1.1(b), f no longer must vanish on all of the boundary of our domain $\Omega = B^+(0,R)$, but only on a certain portion of it. Again, we consider the corresponding Newton potential

$$u(x) := \int_{B^+(0,R)} \Gamma(x,y) f(y) dy. \tag{11.2.22}$$

Up to a constant factor, the first derivatives of u are given by

$$v^i(x) = \int_{B^+(0,R)} \frac{x^i - y^i}{|x-y|^d} f(y) dy \quad (i = 1,\ldots,d), \tag{11.2.23}$$

and they can be estimated as in the proof of Theorem 11.1.1(a), since there, we did not need any assumption on the boundary values.

Up to a constant factor, the second derivatives are given by

$$w^{ij}(x) = \int_{B^+(0,R)} \frac{\partial}{\partial x^j}\left(\frac{x^i - y^i}{|x-y|^d}\right) f(y) dy \quad (= w^{ji}(x)). \tag{11.2.24}$$

For $K(x-y) = \frac{\partial}{\partial x^j}\left(\frac{x^i-y^i}{|x-y|^d}\right)$, and $i \neq d$ or $j \neq d$,

$$\int_{\substack{R_1 < |y| < R_2 \\ y^d > 0}} K(y) dy = 0 \tag{11.2.25}$$

by homogeneity as in (11.1.12). Thus, for $i \neq d$ or $j \neq d$, the α-Hölder norm of the second derivative $\frac{\partial^2}{\partial x^i \partial x^j} u$ can be estimated as in the proof of Theorem 11.1.1(b). The differential equation $\Delta u = f$ implies

$$\frac{\partial^2}{(\partial x^d)^2} u = f - \sum_{i=1}^{d-1} \frac{\partial^2}{(\partial x^i)^2} u, \tag{11.2.26}$$

and so we obtain estimates for the α-Hölder norm of $\frac{\partial^2}{(\partial x^d)^2} u$ as well. We can thus estimate all second derivatives of u.

As in the proof of Theorem 11.1.2, we then obtain $C^{2,\alpha}$-estimates in $B^+(0,R)$ for solutions of

$$\Delta u = f \quad \text{in } B^+(0,R) \quad \text{with } f \in C^\alpha\left(\overline{B^+(0,R)}\right),$$

$$u = 0 \quad \text{on } \partial^0 B^+(0,R), \tag{11.2.27}$$

for $0 < r < R$:

$$\|u\|_{C^{2,\alpha}(B^+(0,r))} \leq c_{12}\left(\|f\|_{C^\alpha(B^+(0,R))} + \|u\|_{L^2(B^+(0,R))}\right). \tag{11.2.28}$$

Namely, putting

$$\varphi := \eta u$$

as in (11.1.23) with the same cutoff function as in (11.1.22), we have $\varphi = 0$ on $\partial B^+(0,R_2)$ $(0 < R_1 < R_2 < R)$, since η vanishes on $\partial^+ B^+(0,R_2)$, and u on $\partial^0 B^+(0,R_2)$. Thus, again

$$\varphi(x) = \int_{B^+(0,R)} \Gamma(x,y)\Delta\varphi(y)dy$$

is a Newton potential, and the preceding estimates can be used to deduce the same result as in Theorem 11.1.2: For $0 < r < R$,

$$\|u\|_{C^{2,\alpha}(B^+(0,r))} \leq c_{13}\left(\|f\|_{C^\alpha(B^+(0,R))} + \|u\|_{L^2(B^+(0,R))}\right). \tag{11.2.29}$$

We next consider a solution of

$$\Delta u = f \quad \text{in } B^+(0,R) \quad \text{with } f \in C^\alpha\left(\overline{B^+(0,R)}\right), \tag{11.2.30}$$

$$u = g \quad \text{on } \partial^0 B^+(0,R) \quad \text{with } g \in C^{2,\alpha}\left(\overline{B^+(0,R)}\right). \tag{11.2.31}$$

As in Section 9.3, we put $\bar{u} := u - g$. We see that $\bar{u}$ satisfies

$$\Delta \bar{u} = f - \Delta g =: \bar{f} \in C^\alpha\left(\overline{B^+(0,R)}\right) \text{ in } B^+(0,R),$$

$$\bar{u} = 0 \text{ on } \partial^0 B^+(0,R).$$

We have thus reduced our considerations to the above case (11.2.27), and so, from (11.2.29), we obtain

$$\|u\|_{C^{2,\alpha}(B^+(0,r))} \leq \|\bar{u}\|_{C^{2,\alpha}(B^+(0,r))} + \|g\|_{C^{2,\alpha}(B^+(0,r))}$$

$$\leq c_{14}\left[\|\bar{f}\|_{C^\alpha(B^+(0,R))} + \|\bar{u}\|_{L^2(B^+(0,R))} + \|g\|_{C^{2,\alpha}(B^+(0,R))}\right]$$

$$\leq c_{15}\left[\|f\|_{C^\alpha(B^+(0,R))} + \|g\|_{C^{2,\alpha}(B^+(0,r))} + \|u\|_{L^2(B^+(0,R))}\right].$$

$$\tag{11.2.32}$$

In order to finally treat the situation of Theorem 11.2.2, as in Section 9.3, we transform a neighborhood U of a boundary point $x_0 \in \partial\Omega$ with a $C^{2,\alpha}$-diffeomorphism ϕ to the ball $\mathring{B}(0, R)$, such that the portion of u that is contained in Ω is mapped to $B^+(0, R)$, and the intersection of U with $\partial\Omega$ is mapped to $\partial^0 B^+(0, R)$. Again, $\tilde{u} := u \circ \phi^{-1}$ on $B^+(0, R)$ satisfies a differential equation of the same type as $Lu = f$, $\tilde{L}\tilde{u} = \tilde{f}$, again with different constants λ, K in (A) and (B). By the preceding considerations, we obtain a $C^{2,\alpha}$-estimate for $\tilde{u}$ in $B^+(0, R/2)$. Again ϕ transforms this estimate into one for u on a subset U' of U. Since Ω is bounded, $\partial\Omega$ is compact and can thus be covered by finitely many such neighborhoods U'. The resulting estimates, together with the interior estimate of Theorem 11.2.1, applied to the complement Ω_0 of those neighborhoods in Ω, yield the claim of Theorem 11.2.2. $\square$

Corollary 11.2.1: *In addition to the assumptions of Theorem 11.2.2, suppose that $c(x) \leq 0$ in Ω. Then*

$$\|u\|_{C^{2,\alpha}(\Omega)} \leq c_{16} \left(\|f\|_{C^\alpha(\Omega)} + \|g\|_{C^{2,\alpha}(\Omega)} \right). \tag{11.2.33}$$

Proof: Because of $c \leq 0$, the maximum principle (see, e.g., Theorem 2.3.2) implies

$$\sup_\Omega |u| \leq \max_{\partial\Omega} |u| + c_{17} \sup_\Omega |f| = \max_{\partial\Omega} |g| + c_{17} \sup_\Omega |f|.$$

Therefore, the L^2-norm of u can be estimated in terms of the C^0-norms of f and g, and the claim follows from (11.2.21). $\square$

11.3 Existence Techniques IV: The Continuity Method

In this section, we wish to study the existence problem

$$Lu = f \quad \text{in } \Omega,$$
$$u = g \quad \text{on } \partial\Omega,$$

in a $C^{2,\alpha}$-region Ω with $f \in C^\alpha(\bar\Omega)$, $g \in C^{2,\alpha}(\bar\Omega)$. The starting point for our considerations will be the corresponding result for the Poisson equation, Kellogg's theorem:

Theorem 11.3.1: *Let Ω be a bounded domain of class C^∞ in $\mathbb{R}^d$, $f \in C^\alpha(\bar\Omega)$, $g \in C^{2,\alpha}(\bar\Omega)$. The Dirichlet problem*

$$\Delta u = f \quad \text{in } \Omega,$$
$$u = g \quad \text{on } \partial\Omega, \tag{11.3.1}$$

then possesses a unique solution u of class $C^{2,\alpha}(\bar\Omega)$.

Proof: Uniqueness follows from the maximum principle (see Corollary 2.1.1). For the existence proof, we first assume that f and g are of class C^∞. The variational methods of Section 8.3 yield a weak solution, which then is of class $C^\infty(\Omega)$ by Theorem 9.3.1. Moreover, by Corollary 11.2.1,

$$\|u\|_{C^{2,\alpha}(\Omega)} \le c_1 \left(\|f\|_{C^\alpha(\Omega)} + \|g\|_{C^{2,\alpha}(\Omega)} \right). \tag{11.3.2}$$

We now return to the $C^{2,\alpha}$-case. We approximate f and g by C^∞-functions f_n and g_n that are defined on Ω. Let u_n be the solution of the corresponding Dirichlet problem

$$\Delta u_n = f_n \quad \text{in } \Omega,$$
$$u_n = g_n \quad \text{on } \partial\Omega.$$

For $n \ge m$, $u_n - u_m$ then satisfies (11.3.2) on Ω, i.e.,

$$\|u_n - u_m\|_{C^{2,\alpha}(\Omega)} \le c_1 \left(\|f_n - f_m\|_{C^\alpha(\Omega)} + \|g_n - g_m\|_{C^{2,\alpha}(\Omega)} \right). \tag{11.3.3}$$

Here, the constant c_1 does not depend on the solutions; it depends only on the $C^{2,\alpha}$-geometry of the domain. We assume that f_n converges to f in $C^\alpha(\Omega)$, and g_n to g in $C^{2,\alpha}(\Omega)$, and so the u_n constitute a Cauchy sequence in $C^{2,\alpha}(\Omega)$ and therefore converge towards some $u \in C^{2,\alpha}(\Omega)$ that satisfies

$$\Delta u = f \quad \text{in } \Omega,$$
$$u = g \quad \text{on } \partial\Omega,$$

and the estimate (11.3.2). $\qquad\qquad\qquad\qquad\qquad\qquad\qquad\qquad\qquad\square$

We now state the main existence result of this chapter:

Theorem 11.3.2: *Let Ω be a bounded domain of class C^∞ in $\mathbb{R}^d$. Let the differential operator*

$$L = \sum_{i,j=1}^{d} a^{ij}(x) \frac{\partial^2}{\partial x^i \partial x^j} + \sum_{i=1}^{d} b^i(x) \frac{\partial}{\partial x^i} + c(x) \tag{11.3.4}$$

satisfy (A) and (B) from Section 11.2, and in addition,

$$c(x) \le 0 \quad \text{in } \Omega. \tag{11.3.5}$$

For any $f \in C^\alpha(\bar\Omega)$, $g \in C^{2,\alpha}(\bar\Omega)$ there then exists a unique solution $u \in C^{2,\alpha}(\bar\Omega)$ of the Dirichlet problem

$$Lu = f \quad \text{in } \Omega,$$
$$u = g \quad \text{on } \partial\Omega. \tag{11.3.6}$$

Remark: It is quite instructive to compare this result and its assumptions with Theorem 9.4.4 above.

Proof: Considering, as usual, $\bar{u} = u - g$ in place of u, we may assume $g = 0$, as our problem is equivalent to

$$L\bar{u} = \bar{f} := f - Lg \in C^{\alpha}(\Omega),$$
$$\bar{u} = 0 \quad \text{on } \partial\Omega.$$

We thus assume $g = 0$ (and write u in place of $\bar{u}$). We consider the family of equations

$$L_t u = f \quad \text{for } 0 \leq t \leq 1,$$
$$u = 0 \quad \text{on } \partial\Omega, \tag{11.3.7}$$

with

$$L_t = tL + (1-t)\Delta. \tag{11.3.8}$$

The differential operators L_t satisfy the structural conditions (A) and (B) with

$$\lambda_t = \min(1, \lambda), \quad K_t = \max(1, K). \tag{11.3.9}$$

We have $L_0 = \Delta$, $L_1 = L$. By Theorem 11.3.1, we can solve (11.3.7) for $t = 0$. We intend to show that we may then also solve this equation for all $t \in [0,1]$, in particular for $t = 1$. The latter is what is claimed in the theorem.

The operator

$$L_t : B_1 := C^{2,\alpha}(\bar{\Omega}) \cap \{u : u = 0 \quad \text{on } \partial\Omega\} \to C^{\alpha}(\bar{\Omega}) =: B_2$$

is a bounded linear operator between the Banach spaces B_1 and B_2. Let u_t be a solution of $L_t u_t = f$, $u_t = 0$ on $\partial\Omega$. By Corollary 11.2.1,

$$\|u_t\|_{C^{2,\alpha}(\Omega)} \leq c_2 \|f\|_{C^{\alpha}(\Omega)},$$

i.e.,

$$\|u\|_{B_1} \leq c_2 \|L_t u\|_{B_2}, \tag{11.3.10}$$

for all $u \in B_1$. Here, the constant c_2 does not depend on t, because by (11.3.9), the structure constants λ_t, K_t of the operators L_t can be controlled independently of t.

We want to show that for any $f \in B_2$ there exists a solution u_t of (11.3.7), i.e., of $L_t u_t = f$, in B_1. In other words, we want to show that the operators $L_t : B_1 \to B_2$ are surjective for $0 \leq t \leq 1$. This, however, follows from the general result stated as the next theorem. With that result, we then conclude the proof of Theorem 11.3.2. □

Theorem 11.3.3: *Let $L_0, L_1 : B_1 \to B_2$ be bounded linear operators between the Banach spaces B_1, B_2. We put*

$$L_t := (1-t)L_0 + tL_1 \quad for \ 0 \leq t \leq 1.$$

We assume that there exists a constant c that does not depend on t, with

$$\|u\|_{B_1} \leq c \|L_t u\|_{B_2} \quad for \ all \ u \in B_1. \tag{11.3.11}$$

If then L_0 is surjective, so is L_1.

Proof: Let L_τ be surjective for some $\tau \in [0,1]$. By (11.3.11), L_τ then is injective as well, and thus bijective. We therefore have an inverse operator

$$L_\tau^{-1} : B_2 \to B_1.$$

For $t \in [0,1]$, we rewrite the equation

$$L_t u = f \quad \text{for } f \in B_2 \tag{11.3.12}$$

as

$$L_\tau u = f + (L_\tau - L_t)u = f + (t - \tau)(L_0 u - L_1 u),$$

or

$$u = L_\tau^{-1} f + (t - \tau)L_\tau^{-1}(L_0 - L_1)u =: \Lambda u.$$

Thus, for solving (11.3.12), we need to find a fixed point of the operator $\Lambda : B_1 \to B_2$. By the Banach fixed point theorem, such a fixed point exists if we can find some $q < 1$ with

$$\|\Lambda u - \Lambda v\|_{B_1} \leq q \|u - v\|_{B_1}.$$

We have

$$\|\Lambda u - \Lambda v\| \leq \|L_\tau^{-1}\| (\|L_0\| + \|L_1\|) |t - \tau| \|u - v\|.$$

By (11.3.11), $\|L_\tau^{-1}\| \leq c$. Therefore, it suffices to choose

$$|t - \tau| \leq \frac{1}{2} (c(\|L_0\| + \|L_1\|))^{-1} =: \eta$$

for obtaining the desired fixed point. This means that if $L_\tau u = f$ is solvable, so is $L_t u = f$ for all t with $|t - \tau| \leq \eta$. Since L_0 is surjective by assumption, L_t then is surjective for $0 \leq t \leq \eta$. Repeating the preceding argument, this time for $\tau = \eta$, we obtain surjectivity for $\eta \leq t \leq 2\eta$. Iteratively, all L_t for $t \in [0,1]$, and in particular L_1, are surjective. $\qquad\square$

Basic references about Schauder's approach are [1, 9]. Our treatment of the fundamental C^α-estimate for the Poisson equation uses scaling relations in place of the usual weighted Hölder spaces and is hopefully a little simpler.

Summary

A solution of the Poisson equation

$$\Delta u = f$$

with α-Hölder continuous f is contained in the space $C^{2,\alpha}$; i.e., it possesses α-Hölder continuous second derivatives for $0 < \alpha < 1$. (This is no longer true for $\alpha = 0$ or $\alpha = 1$. For example, if f is only continuous, a solution need not be twice continuously differentiable.) By linear coordinate transformations this result can be easily extended to linear elliptic differential equations with constant coefficients. Schauder then succeeded in extending these results to solutions of elliptic equations

$$Lu(x) := \sum_{i,j} a^{ij}(x)\frac{\partial^2 u(x)}{\partial x^i \partial x^j} + \sum_i b^i(x)\frac{\partial u}{\partial x^i} + c(x)u(x) = f(x)$$

with α-Hölder continuous coefficients, by considering such an operator L as a local perturbation of an operator with constant coefficients a^{ij}, b^i, c.

The continuity method reduces the solution of

$$Lu = f$$

to that of the Poisson equation

$$\Delta u = f$$

by considering the operators

$$L_t := tL + (1-t)\Delta$$

for $0 \le t \le 1$, and showing that the set of $t \in [0,1]$ for which

$$L_t u = f$$

can be solved is open and closed (and nonempty, because the Poisson equation can be solved). The proof of closedness rests on Schauder's estimates.

Exercises

11.1 Let $K \subset \mathbb{R}^d$ be bounded, $f_n : K \to \mathbb{R}$ $(n \in \mathbb{N})$ a sequence of functions with

$$\|f_n\|_{C^\alpha(K)} \le \text{const} \quad \text{(independent of } n),$$

for some $0 < \alpha \le 1$. (Here and in the next exercise, in the case $\alpha = 1$ we consider the space $C^{0,1}$ of Lipschitz continuous functions.) Show that then $(f_n)_{n\in\mathbb{N}}$ has to contain a uniformly convergent subsequence.

11.2 Is it true that for all domains $\Omega \subset \mathbb{R}^d, 0 < \alpha < \beta \leq 1$,

$$C^\beta(\Omega) \subset C^\alpha(\Omega)?$$

11.3 Let $u \in C^{k,\alpha}(\Omega)$ satisfy

$$Lu = f$$

for some $f \in C^{k,\alpha}(\Omega)$ ($k \in \mathbb{N}, 0 < \alpha < 1$). Here, we assume that the operator L from (11.2.1) satisfies the ellipticity condition (A) as well as

$$\left\|a^{ij}\right\|_{C^{k,\alpha}(\Omega)}, \left\|b^i\right\|_{C^{k,\alpha}(\Omega)}, \left\|c\right\|_{C^{k,\alpha}(\Omega)} \leq K$$

for all i, j. Show that $u \in C^{k+2,\alpha}(\Omega_0)$ for any $\Omega_0 \subset\subset \Omega$, and

$$\|u\|_{C^{k+2,\alpha}(\Omega)} \leq c(\|f\|_{C^{k,\alpha}(\Omega)} + \|u\|_{L^2(\Omega)}),$$

with a constant c depending on K and the quantities of Theorem 11.2.1.

12. The Moser Iteration Method and the Regularity Theorem of de Giorgi and Nash

12.1 The Moser–Harnack Inequality

In this chapter, as in Chapter 9, we shall consider elliptic differential operators of divergence type. In order to concentrate on the essential aspects and not to burden the proofs with too many technical details, in this chapter we shall omit all lower-order terms and consider only solutions of the homogeneous equation. Thus, we shall investigate (weak) solutions of

$$Lu = \sum_{i,j=1}^{d} \frac{\partial}{\partial x^j} \left(a^{ij}(x) \frac{\partial}{\partial x^i} u(x) \right) = 0,$$

where the coefficients a^{ij} are (measurable and) bounded and satisfy an ellipticity condition. We thus assume that there exist constants $0 < \lambda \leq \Lambda < \infty$ with

$$\lambda |\xi|^2 \leq \sum_{i,j=1}^{d} a^{ij}(x)\xi_i\xi_j \tag{12.1.1}$$

for all x in the domain of definition Ω of u and all $\xi \in \mathbb{R}^d$, and

$$\sup_{i,j,x} \left| a^{ij}(x) \right| \leq \Lambda.$$

Definition 12.1.1: *A function $u \in W^{1,2}(\Omega)$ is called a weak subsolution of L, and we write this as $Lu \geq 0$, if for all $\varphi \in H_0^{1,2}(\Omega)$, $\varphi \geq 0$ in Ω,*

$$\int_\Omega \sum_{i,j} a^{ij}(x) D_i u D_j \varphi dx \leq 0. \tag{12.1.2}$$

Similarly, it is called a weak supersolution ($Lu \leq 0$), if we have $\geq$ in (11.1.2).

Inequalities like $\varphi \geq 0$ are assumed to hold pointwise almost everywhere, here and in the sequel. Likewise, sup and inf will denote the essential supremum and infimum, respectively. Finally, as always, $\fint$ will denote the average mean integral:

$$\fint_\Omega \varphi dx = \frac{1}{|\Omega|} \int_\Omega \varphi dx.$$

In order to familiarize ourselves with the notions of sub- and supersolutions, we shall demonstrate the following useful lemma.

Lemma 12.1.1: *(i) Let u be a subsolution, i.e. $u \in C^2(\Omega)$, $Lu \geq 0$, and let $f \in C^2(\mathbb{R})$ be convex with $f' \geq 0$. Then $f \circ u$ is a subsolution as well.*

(ii) Let u be a supersolution, $f \in C^2(\mathbb{R})$ concave with $f' \geq 0$. Then $f \circ u$ is a supersolution as well.

(iii) Let u be a solution, and $f \in C^2(\mathbb{R})$ convex. Then $f \circ u$ is a subsolution.

Proof:

$$L(f \circ u) = \sum_{i,j} \frac{\partial}{\partial x^j} \left(a^{ij} f'(u) \frac{\partial u}{\partial x^i} \right) = f''(u) \sum_{i,j} a^{ij} \frac{\partial u}{\partial x^i} \frac{\partial u}{\partial x^j} + f'(u) Lu,$$

$$(12.1.3)$$

which implies all the inequalities claimed.   □

We now wish to verify that the assertions of Lemma 12.1.1 continue to hold for *weak* (sub-, super-)solutions. We assume that $f'(u)$ and $f''(u)$ satisfy approximate integrability conditions to make the chain rules for weak derivatives

$$D_i(f \circ u) = f'(u) D_i(u)$$

and

$$D_i(f' \circ u) = f''(u) D_i u \quad \text{for } i = 1, ..., d$$

valid. (By Lemma 8.2.3 this holds if, for example,

$$\sup_{y \in \mathbb{R}} |f'(y)| + \sup_{y \in \mathbb{R}} |f''(y)| < \infty.)$$

We obtain

$$\int_\Omega \sum_{i,j} a^{ij} D_i(f \circ u) D_j \varphi = \int \sum_{i,j} a^{ij} f'(u) D_i u D_j \varphi$$

$$= \int \sum_{i,j} a^{ij} D_i u D_j (f'(u) \varphi)$$

$$- \int \sum_{i,j} a^{ij} D_i u f''(u) D_j u \varphi.$$

The last integral is nonnegative because of the ellipticity condition, if f is convex, i.e., $f''(u) \geq 0$, and $\varphi \geq 0$, and consequently yields a nonpositive contribution because of the minus sign in front of it, if u is a weak subsolution and $f'(u) \geq 0$. Therefore, under those assumptions

$$\int_\Omega \sum_{i,j} a^{ij} D_i(f \circ u) D_j \varphi \leq 0,$$

and $f \circ u$ is a weak subsolution.

In the same manner, one treats the weak versions of the other assertions of Lemma 12.1.1 to obtain the following result:

Lemma 12.1.2: *Under the corresponding assumptions, the assertions of Lemma 12.1.1 hold for weak (sub-,super-)solutions, provided that the chain rule for weak derivatives is satisfied for $f \in C^2(\mathbb{R})$.* $\qquad\square$

From Lemma 12.1.2 we derive the following result:

Lemma 12.1.3: *Let $u \in W^{1,2}(\Omega)$ be a weak subsolution of L, and $k \in \mathbb{R}$. Then*

$$v(x) := \max(u(x), k)$$

is a weak subsolution as well.

Proof: We consider the function

$$f : \mathbb{R} \to \mathbb{R},$$
$$f(y) := \max(y, k).$$

Then

$$v = f \circ u.$$

We approximate f by a sequence $(f_n)_{n \in N}$ of convex functions of class C^2 with

$$f_n(y) = f(y) \quad \text{for } y \notin \left(k - \frac{1}{n}, k + \frac{1}{n} \right)$$

and

$$|f_n'(y)| \leq 1 \quad \text{for all } y.$$

Then, as in the proofs of Lemmas 8.2.2 and 8.2.3, by an approximation argument, $f_n \circ u$ converges to $v = f \circ u$ in $W^{1,2}$. Therefore,

$$\int_\Omega \sum_{i,j} a^{ij} D_i v D_j \varphi = \lim_{n \to \infty} \int_\Omega \sum_{i,j} a^{ij} D_i(f_n \circ u) D_j \varphi$$

$$\leq 0 \quad \text{for } \varphi \in H_0^{1,2}(\Omega), \varphi \geq 0$$

by Lemma 12.1.2. $\qquad\square$

Remark: Of course, we also have a result analoguous to Lemma 12.1.3 for weak supersolutions. For $k \in \mathbb{R}$, if $u \in W^{1,2}(\Omega)$ is a weak supersolution, then so is

$$\min(u(x), k).$$

We now come to the fundamental estimates of J. Moser:

Theorem 12.1.1: *Let u be a subsolution in the ball $B(x_0, 4R) \subset \mathbb{R}^d$ ($R > 0$), and assume $p > 1$. Then*

$$\sup_{B(x_0,R)} u \le c_1 \left(\frac{p}{p-1} \right)^{\frac{2}{p}} \left(\fint_{B(x_0,2R)} (\max(u(x), 0))^p \, dx \right)^{\frac{1}{p}}, \qquad (12.1.4)$$

with a constant c_1 depending only on d and $\frac{\Lambda}{\lambda}$.

Remark: If u is positive, then obviously $\max(u, 0) = u$ in (12.1.4), and this case will constitute our main application of this result.

Theorem 12.1.2: *Let u be a positive supersolution in $B(x_0, 4R) \subset \mathbb{R}^d$. For $0 < p < \frac{d}{d-2}$, and if $d \ge 3$, then*

$$\left(\fint_{B(x_0,2R)} u^p \, dx \right)^{\frac{1}{p}} \le \frac{c_2}{\left(\frac{d}{d-2} - p \right)^2} \inf_{B(x_0,R)} u, \qquad (12.1.5)$$

with c_2 again depending on d and $\frac{\Lambda}{\lambda}$ only. If $d = 2$, this estimate holds for any $0 < p < \infty$, with a constant c_2 depending on $p, d, \frac{\Lambda}{\lambda}$ in place of $c_2 / \left(\frac{d}{d-2} - p \right)^2$.

Remark: In order to see the necessity of the condition $p < \frac{d}{d-2}$, we let L be the Laplace operator Δ and

$$u(x) = \min \left(|x|^{2-d}, k \right) \quad \text{for some } k > 0.$$

According to the remark after Lemma 12.1.3, because $|x|^{2-d}$ is harmonic on $\mathbb{R}^d \setminus \{0\}$, this is a weak supersolution on $\mathbb{R}^d$. If we then let k increase, we see that the $L^{\frac{d}{d-2}}$-norm can no longer be controlled by the infimum.

From Theorems 12.1.1 and 12.1.2 we derive Harnack-type inequalities for solutions of $Lu = 0$. These two theorems directly yield the following corollary:

Corollary 12.1.1: *Let u be a positive (weak) solution of $Lu = 0$ in the ball $B(x_0, 4R) \subset \mathbb{R}^d$ ($R > 0$). Then*

$$\sup_{B(x_0,R)} u \le c_3 \inf_{B(x_0,R)} u, \qquad (12.1.6)$$

with c_3 depending on d and $\frac{\Lambda}{\lambda}$ only.

For general domains, we have the following result:

Corollary 12.1.2: *Let u be a positive (weak) solution of $Lu = 0$ in a domain Ω of $\mathbb{R}^d$, and let $\Omega_0 \subset\subset \Omega$. Then*

$$\sup_{\Omega_0} u \leq c \inf_{\Omega_0} u, \tag{12.1.7}$$

with c depending on d, Ω, Ω_0, and $\frac{A}{\lambda}$.

Proof: This Harnack inequality on Ω_0 follows by the standard ball chain argument: Since $\bar{\Omega}_0$ is compact, it can be covered by finitely many balls $B_i := B(x_i, R)$ with $B(x_i, R) \subset \Omega$ (we choose, for example, $R < \frac{1}{4} \text{dist}(\partial\Omega, \Omega_0)$), $i = 1, \ldots, N$. Now let $y_1, y_2 \in \Omega_0$; without loss of generality $y_1 \in B_k$, $y_2 \in B_{k+m}$ for some $m \geq 1$, and the balls are enumerated in such manner that $B_j \cap B_{j+1} \neq \emptyset$ for $j = k, \ldots, k+m-1$. By applying Corollary 12.1.1 to the balls $B_k, B_{k+1}, \ldots$, we obtain

$$u(y_1) \leq \sup_{B_k} u(x) \leq c_3 \inf_{B_k} u(x)$$

$$\leq c_3 \sup_{B_{k+1}} u(x) \quad (\text{since } B_k \cap B_{k+1} \neq \emptyset)$$

$$\leq c_3^2 \inf_{B_{k+1}} u(x) \leq \ldots$$

$$\leq c_3^{m+1} \inf_{B_{k+m}} u(x) \leq c_3^{m+1} u(y_2).$$

Since y_1 and y_2 are arbitrary, and $m \leq N$, it follows that

$$\sup_{\Omega_0} u(x) \leq c_3^{N+1} \inf_{\Omega_0} u(x). \tag{12.1.8}$$

$\square$

We now start with the preparations for the proofs of Theorems 12.1.1 and 12.1.2. For positive u and a point x_0, we put

$$\phi(p, R) := \left(\fint_{B(x_0, R)} u^p dx \right)^{\frac{1}{p}}.$$

Lemma 12.1.4:

$$\lim_{p \to \infty} \phi(p, R) = \sup_{B(x_0, R)} u =: \phi(\infty, R), \tag{12.1.9}$$

$$\lim_{p \to -\infty} \phi(p, R) = \inf_{B(x_0, R)} u =: \phi(-\infty, R). \tag{12.1.10}$$

Proof: By Hölder's inequality, $\phi(p, R)$ is monotonically increasing with respect to p. Namely, for $p < p'$ and $u \in L^{p'}(\Omega)$,

$$\left(\frac{1}{|\Omega|}\int_\Omega u^p\right)^{\frac{1}{p}} \leq \frac{1}{|\Omega|^{\frac{1}{p}}}\left(\int_\Omega 1\right)^{\frac{p'-p}{pp'}}\left(\int_\Omega (u^p)^{\frac{p'}{p}}\right)^{\frac{1}{p'}} = \left(\frac{1}{|\Omega|}\int_\Omega u^{p'}\right)^{\frac{1}{p'}}.$$

Moreover,

$$\phi(p,R) \leq \left(\frac{1}{|B(x_0,R)|}\int_{B(x_0,R)}(\sup u)^p\right)^{\frac{1}{p}} = \phi(\infty,R). \qquad (12.1.11)$$

On the other hand, by the definition of the essential supremum, for any $\varepsilon > 0$ there exists some $\delta > 0$ with

$$\left|\left\{x \in B(x_0,R) : u(x) \geq \sup_{B(x_0,R)} u - \varepsilon\right\}\right| > \delta.$$

Therefore,

$$\phi(p,R) \geq \left(\frac{1}{|B(x_0,R)|}\int_{\substack{u(x)\geq\sup u-\varepsilon \\ x\in B(x_0,R)}} u^p\right)^{\frac{1}{p}} \geq \left(\frac{\delta}{|B(x_0,R)|}\right)^{\frac{1}{p}}(\sup u - \varepsilon),$$

and hence

$$\lim_{p\to\infty}\phi(p,R) \geq \sup u - \varepsilon$$

for any $\varepsilon > 0$, and thus also

$$\lim_{p\to\infty}\phi(p,R) \geq \sup u. \qquad (12.1.12)$$

Inequalities (12.1.11) and (12.1.12) imply (12.1.9), and (12.1.10) is derived similarly (or, alternatively, by applying the preceding argument to $\frac{1}{u}$). $\square$

Lemma 12.1.5:   *(i) Let u be a positive subsolution in Ω, and for $q > \frac{1}{2}$, assume*

$$v := u^q \in L^2(\Omega).$$

For any $\eta \in H_0^{1,2}(\Omega)$, we then have

$$\int_\Omega \eta^2 |Dv|^2 \leq \frac{\Lambda^2}{\lambda^2}\left(\frac{2q}{2q-1}\right)^2\int_\Omega |D\eta|^2 v^2. \qquad (12.1.13)$$

(ii) If u is a supersolution instead, this inequality holds for $q < \frac{1}{2}$.

Proof: The claim is trivial for $q = 0$. We put

$$f(u) = u^{2q} \qquad \text{for } q > 0,$$
$$f(u) = -u^{2q} \qquad \text{for } q < 0.$$

By Lemma 12.1.2, $f(u)$ then is a subsolution in case (i), and a supersolution in case (ii). The subsequent calculations are based on that fact. (In the course of the proof there will also arise integrability conditions implying the needed chain rules. For that purpose, the proof of Lemma 8.2.3 requires a slight generalization, utilizing varying Sobolev exponents, the Hölder inequality, and the Sobolev embedding theorem. We leave this as an exercise for the reader.) As a test function in (12.1.2) (or in the corresponding inequality in case (ii), we then use

$$\varphi = f'(u) \cdot \eta^2. \tag{12.1.14}$$

Then

$$\int_\Omega \sum_{ij} a^{ij}(x) D_i u D_j \varphi$$

$$= \int_\Omega \sum_{i,j} a^{ij} D_i u D_j u f''(u) \eta^2 + \int_\Omega \sum_{i,j} a^{ij} D_i u f'(u) 2\eta D_j \eta$$

$$= \int_\Omega 2|q|(2q-1) \sum_{i,j} a^{ij} D_i u D_j u \, u^{2q-2} \eta^2 + \int_\Omega 4|q| \sum_{i,j} a^{ij} D_i u \, u^{2q-1} \eta D_j \eta. \tag{12.1.15}$$

In case (i), this is ≤ 0. Applying Young's inequality to the last term, for all $\varepsilon > 0$, we obtain

$$2|q|(2q-1)\lambda \int |Du|^2 u^{2q-2} \eta^2 \leq 2|q| \Lambda \varepsilon \int |Du|^2 u^{2q-2} \eta^2$$

$$+ \frac{2|q|\Lambda}{\varepsilon} \int u^{2q} |D\eta|^2.$$

With

$$\varepsilon = \frac{2q-1}{2} \frac{\lambda}{\Lambda},$$

we thus obtain

$$\int |Du|^2 u^{2q-2} \eta^2 \leq \frac{4}{(2q-1)^2} \frac{\Lambda^2}{\lambda^2} \int u^{2q} |D\eta|^2,$$

i.e.,

$$\int |Dv|^2 \eta^2 \leq \frac{\Lambda^2}{\lambda^2} \left(\frac{2q}{2q-1}\right)^2 \int v^2 |D\eta|^2.$$

In case (ii), (12.1.15) is nonnegative, and since in that case also $2q - 1 \leq 0$, one can proceed analoguously and put

$$\varepsilon = \frac{1-2q}{2} \frac{\lambda}{\Lambda},$$

to obtain (12.1.13) in that case as well. □

We now begin the proofs of Theorems 12.1.1 and 12.1.2. Since the stated inequalities are invariant under scaling, we may assume, without loss of generality, that

$$R = 1 \quad \text{and} \quad x_0 = 0.$$

We shall employ the abbreviation

$$B_r := B(0, r).$$

Let

$$0 < r' < r \le 2r', \tag{12.1.16}$$

and let $\eta \in H_0^{1,2}(B_r)$ be a cutoff function satisfying

$$\begin{aligned}
\eta &\equiv 1 \quad \text{on } B_{r'}, \\
\eta &\equiv 0 \quad \text{on } \mathbb{R}^d \setminus B_r, \\
|D\eta| &\le \frac{2}{r - r'}.
\end{aligned} \tag{12.1.17}$$

For the proof of Theorem 12.1.1, we may assume without loss of generality that u is positive, since otherwise, by Lemma 12.1.3, we may consider the positive subsolutions

$$v_k(x) = \max(u(x), k)$$

for $k > 0$ (or the approximating subsolutions from the proof of that lemma), perform the subsequent reasoning for positive subsolutions, apply the result to the v_k, and finally let k tend to 0.

We consider once more

$$v = u^q$$

and assume that $v \in L^2(\Omega)$. By the Sobolev embedding theorem (Corollary 9.1.3), for $d \ge 3$, we obtain

$$\left(\fint_{B_{r'}} v^{\frac{2d}{d-2}} \right)^{\frac{d-2}{d}} \le c_4 \left(r'^2 \fint_{B_{r'}} |Dv|^2 + \fint_{B_{r'}} v^2 \right). \tag{12.1.18}$$

If $d = 2$ instead of $\frac{2d}{d-2}$, we may take an arbitrarily large exponent p and proceed analogously. We leave the necessary modifications for the case $d = 2$ to the reader and henceforth treat only the case $d \ge 3$. With (12.1.13) and (12.1.17), (12.1.18) yields

$$\left(\fint_{B_{r'}} v^{\frac{2d}{d-2}} \right)^{\frac{d-2}{d}} \le \bar{c} \fint_{B_r} v^2 \tag{12.1.19}$$

with

$$\bar{c} \le c_5 \left(\left(\frac{r'}{r - r'} \right)^2 \left(\frac{2q}{2q - 1} \right)^2 + 1 \right). \tag{12.1.20}$$

Thus, we get $v \in L^{\frac{2d}{d-2}}(\Omega)$. We shall iterate that step and realize that higher and higher powers of u are integrable.

We put $s = 2q$ and assume

$$|s| \ge \mu > 0,$$

choosing an appropriate value for μ later on. Because of $r \le 2r'$, then

$$\bar{c} \le c_6 \left(\frac{r'}{r - r'} \right)^2 \left(\frac{s}{s - 1} \right)^2, \tag{12.1.21}$$

with c_6 also depending on μ. Thus, by (12.1.19) and (12.1.21), since $v = u^{\frac{s}{2}}$, we get for $s \ge \mu$,

$$\phi \left(\frac{ds}{d - 2}, r' \right) = \left(\fint_{B_{r'}} v^{\frac{2d}{d-2}} \right)^{\frac{d-2}{ds}} \le c_7 \left(\frac{r'}{r - r'} \right)^{\frac{2}{s}} \left(\frac{s}{s - 1} \right)^{\frac{2}{s}} \phi(s, r) \tag{12.1.22}$$

with $c_7 = c_6^{\frac{1}{s}}$. For $s \le -\mu$, analogously,

$$\phi \left(\frac{ds}{d - 2}, r' \right) \ge \frac{1}{c_7} \left(\frac{r'}{r - r'} \right)^{-\frac{2}{|s|}} \phi(s, r) \tag{12.1.23}$$

(we may omit the term $\left(\frac{s}{s-1} \right)^{-\frac{2}{|s|}}$ here, since it is greater than or equal to 1).

We now wish to complete the proof of Theorem 12.1.1, and therefore, we return to (12.1.22). The decisive insight obtained so far is that we can control the integral of a higher power of u by that of a lower power of u. We now shall simply iterate this estimate to control even higher integral norms of u and from Lemma 12.1.4 then also the supremum of u. For that purpose, let

$$s_n = \left(\frac{d}{d - 2} \right)^n p \quad \text{for } p > 1,$$
$$r_n = 1 + 2^{-n},$$
$$r'_n = r_{n+1} > \frac{r_n}{2}.$$

Then (12.1.22) implies

$$\phi\left(s_{n+1}, r_{n+1}\right) \le c_7 \left(\frac{1 + 2^{-n-1}}{2^{-n-1}} \cdot \frac{\left(\frac{d}{d-2}\right)^n p}{\left(\frac{d}{d-2}\right)^n p - 1}\right)^{\frac{2}{p\left(\frac{d}{d-2}\right)^n}} \phi(s_n, r_n)$$

$$= c_8^{n\left(\frac{d}{d-2}\right)^{-n}} \phi(s_n, r_n),$$

and iteratively,

$$\phi\left(s_{n+1}, r_{n+1}\right) \le c_8^{\sum_{\nu=1}^n \nu\left(\frac{d}{d-2}\right)^{-\nu}} \phi(s_1, r_1) \le c_9 \left(\frac{p}{p-1}\right)^{\frac{2}{p}} \phi(p, 2). \quad (12.1.24)$$

(Since we may assume $u \in L^p(\Omega)$, therefore $\phi(s_n, r_n)$ is finite for all $n \in \mathbb{N}$, and thus any power of u is integrable.) Using Lemma 12.1.4, this yields Theorem 12.1.1.

In order to prove Theorem 12.1.2, we now assume $u > \varepsilon > 0$, in order to ensure that $\phi(\sigma, r)$ is finite for $\sigma < 0$. This does not constitute a serious restriction, because once we have proved Theorem 12.1.2 under that assumption, then for positive u, we may apply the result to $u + \varepsilon$. In the resulting inequality for $u + \varepsilon$, namely

$$\left(\fint_{B(x_0, 2R)} (u + \varepsilon)^p\right)^{\frac{1}{p}} \le \frac{c_2}{\left(\frac{d}{d-2} - p\right)^2} \inf_{B(x_0, R)} (u + \varepsilon),$$

we then simply let $\varepsilon \to 0$ to deduce the inequality for u itself.

Carrying out the above iteration analogously for $s \le -\mu$ with $r_n = 2 + 2^{-n}$, we deduce from (12.1.23) that

$$\phi(-\mu, 3) \le c_{10}\phi(-\infty, 2) \le c_{10}\phi(-\infty, 1). \quad (12.1.25)$$

By finitely many iteration steps, we also obtain

$$\phi(p, 2) \le c_{11}\phi(\mu, 3). \quad (12.1.26)$$

(The restriction $p < \frac{d}{d-2}$ in Theorem 12.1.2 arises because according to Lemma 12.1.5, in (12.1.19) we may insert $v = u^q$ only for $q < \frac{1}{2}$. The relation $p = 2q\frac{d}{d-2}$ that is needed to control the L^p-norm of u with (12.1.19), by (12.1.20) also yields the factor $\left(\frac{d}{d-2} - p\right)^{-2}$ in (12.1.15).)

The only missing step is

$$\phi(\mu, 3) \le c_{12}\phi(-\mu, 3). \quad (12.1.27)$$

Inequalities (12.1.25), (12.1.26), (12.1.27) imply Theorem 12.1.2. For the proof of (12.1.27), we shall use the theorem of John–Nirenberg (Theorem 9.1.2). For that purpose, we put

$$v = \log u, \quad \varphi = \frac{1}{u}\eta^2$$

with some cutoff function $\eta \in H_0^{1,2}(B_4)$. Then

$$\int_{B_4} \sum_{i,j} a^{ij} D_i \varphi D_j u = -\int_{B_4} \eta^2 \sum a^{ij} D_i v D_j v + \int_{B_4} 2\eta \sum a^{ij} D_i \eta D_j v.$$

Since u is a supersolution, the left-hand side is nonnegative; hence

$$\lambda \int_{B_4} \eta^2 |Dv|^2 \le \int_{B_4} \eta^2 \sum a^{ij} D_i v D_j v \le 2 \int_{B_4} \eta \sum a^{ij} D_i \eta D_j v$$

$$\le 2\Lambda \left(\int_{B_4} \eta^2 |Dv|^2 \right)^{\frac{1}{2}} \left(\int_{B_4} |D\eta|^2 \right)^{\frac{1}{2}}$$

by the Schwarz inequality, and thus

$$\int_{B_4} \eta^2 |Dv|^2 \le 4 \left(\frac{\Lambda}{\lambda} \right)^2 \int_{B_4} |D\eta|^2. \tag{12.1.28}$$

If now $B(y, R) \subset B_{3+\frac{1}{2}}$ is any ball, we choose η satisfying

$$\eta \equiv 1 \quad \text{on } B(y, R),$$
$$\eta \equiv 0 \quad \text{outside of } B(y, 2R) \cap B_4,$$
$$|D\eta| \le \frac{6}{R}.$$

With such an η, we obtain from (12.1.28)

$$\fint_{B(y,R)} |Dv|^2 \le \gamma \frac{1}{R^2} \quad \text{with some constant } \gamma.$$

Thus, by Hölder's inequality

$$\int_{B(y,R)} |Dv| \le \omega_d \sqrt{\gamma} R^{d-1}.$$

Now let α be as in Theorem 9.1.2. With $\mu = \frac{\alpha}{\omega_d \sqrt{\gamma}}$, applying that theorem to

$$w = \frac{1}{\omega_d \sqrt{\gamma}} v = \frac{1}{\omega_d \sqrt{\gamma}} \log u,$$

we obtain

$$\int_{B_3} u^\mu \int_{B_3} u^{-\mu} \le \beta^2,$$

and hence

$$\phi(\mu, 3) \le \beta^{\frac{2}{\mu}} \phi(-\mu, 3),$$

and hence (12.1.27), thus completing the proof. $\square$

A reference for this section is Moser [18].

Krylov and Safonov have shown that solutions of elliptic equations that are not of divergence type satisfy Harnack inequalities as well. In order to describe their results in the simplest case, we again omit all lower-order terms and consider solutions of

$$Mu := \sum_{i,j=1}^{d} a^{ij}(x) \frac{\partial^2}{\partial x^i \partial x^j} u(x) = 0.$$

Here the coefficients $a^{ij}(x)$ again need only be (measurable and) bounded and satisfy the structural condition (12.1.1), i.e.,

$$\lambda |\xi|^2 \le \sum_{i,j=1}^{d} a^{ij}(x)\xi_i \xi_j \quad \text{for all } x \in \Omega, \xi \in \mathbb{R}^d$$

and

$$\sup_{i,j,x} |a^{ij}(x)| \le \Lambda$$

with constants $0 < \lambda < \Lambda < \infty$.

We then have the following theorem:

Theorem 12.1.3: *Let $u \in W^{2,d}(\Omega)$ be positive and satisfy $Mu \ge 0$ almost everywhere in $B(x_0, 4R) \subset \mathbb{R}^d$. For any $p > 0$, we then have*

$$\sup_{B(x_0,R)} u \le c_1 \left(\fint_{B(x_0,2R)} u^p \, dx \right)^{1/p}$$

with a constant c_1 depending on d, $\frac{\Lambda}{\lambda}$, and p.

Theorem 12.1.4: *Let $u \in W^{2,d}(\Omega)$ be positive and satisfy $Mu \le 0$ almost everywhere in $B(x_0, 4R) \subset \mathbb{R}^d$. Then there exist $p > 0$ and some constant c_2, depending only on d and $\frac{\Lambda}{\lambda}$, such that*

$$\left(\fint_{B(x_0,R)} u^p \, dx \right)^{1/p} \le c_2 \inf_{B(x_0,R)} u.$$

As in the case of divergence-type equations (see Section 12.2 below), these results imply Harnack inequalities, maximum principles, and the Hölder continuity of solutions $u \in W^{2,d}(\Omega)$ of

$$Mu = 0 \quad \text{almost everywhere } \Omega \subset \mathbb{R}^d.$$

Proofs of the results of Krylov–Safonov can be found in Gilbarg–Trudinger [9].

12.2 Properties of Solutions of Elliptic Equations

In this section we shall apply the Moser–Harnack inequality in order to deduce the Hölder continuity of weak solutions of $Lu = 0$ under the structural condition (12.1.1). That result had originally been proved by E. de Giorgi and J. Nash independently of each other, and with different methods, before J. Moser found the proof presented here, based on the Harnack inequality.

Lemma 12.2.1: *Let $u \in W^{1,2}(\Omega)$ be a weak solution of L, i.e.,*

$$Lu = \sum_{i,j=1}^{d} \frac{\partial}{\partial x^j} \left(a^{ij}(x) \frac{\partial}{\partial x^i} u(x) \right) \geq 0 \text{ weakly,}$$

with L satisfying the conditions stated in Section 12.1. Then u is bounded from above on any $\Omega_0 \subset\subset \Omega$. Thus, if u is a weak solution of $Lu = 0$, it is bounded from above and below on any such Ω_0.

Proof: By Lemma 12.1.3, for any positive k,

$$v(x) := \max(u(x), k)$$

is a positive subsolution (by the way, in place of v, one might also employ the approximating subsolutions $f_n \circ u$ from the proof of Lemma 12.1.3). The local boundedness of v, hence of u, then follows from Theorem 12.1.1, using a ball chain argument as in the proof of Corollary 12.1.2.   □

Theorem 12.2.1: *Let $u \in W^{1,2}(\Omega)$ be a weak solution of*

$$Lu = \sum_{i,j=1}^{d} \frac{\partial}{\partial x^j} \left(a^{ij}(x) \frac{\partial}{\partial x^i} u(x) \right) = 0, \qquad (12.2.1)$$

assuming that the measurable and bounded coefficients $a^{ij}(x)$ satisfy the structural conditions

$$\lambda |\xi|^2 \leq \sum_{i,j=1}^{d} a^{ij}(x)\xi_i\xi_j, \quad |a^{ij}(x)| \leq \Lambda \qquad (12.2.2)$$

for all $x \in \Omega$, $\xi \in \mathbb{R}^d$, with constants $0 < \lambda < \Lambda < \infty$. Then u is Hölder continuous in Ω. More precisely, for any $\Omega_0 \subset\subset \Omega$, there exist some $\alpha \in (0,1)$ and a constant c with

$$|u(x) - u(y)| \leq c|x - y|^\alpha. \qquad (12.2.3)$$

for all $x, y \in \Omega_0$. α depends on d, $\frac{\Lambda}{\lambda}$, and Ω_0, c in addition on $\sup_{\Omega_0} u - \inf_{\Omega_0} u$.

Proof: Let $x \in \Omega$. For $R > 0$ and $B(x, R) \subset \Omega$, we put

$$M(R) := \sup_{B(x,R)} u, \quad m(R) := \inf_{B(x,R)} u.$$

(By Lemma 12.2.1, $-\infty < m(R) \le M(R) < \infty$.) Then

$$\omega(R) := M(R) - m(R)$$

is the oscillation of u in $B(x, R)$, and we plan to prove the inequality

$$\omega(r) \le c_0 \left(\frac{r}{R}\right)^\alpha \omega(R) \quad \text{for } 0 < r \le \frac{R}{4} \tag{12.2.4}$$

for some α to be specified. This will then imply

$$u(x) - u(y) \le \sup_{B(x,r)} u - \inf_{B(x,r)} u = \omega(r) \le c_0 \frac{\omega(R)}{R^\alpha} |x - y|^\alpha. \tag{12.2.5}$$

for all y with $|x - y| = r$. This, in turn, easily implies the claim.

We now turn to the proof of (12.2.4):

$$M(R) - u \quad \text{and} \quad u - m(R)$$

are positive solutions of $Lu = 0$ in $B(x, R)$.[1] Thus, by Corollary 12.1.1,

$$M(R) - m\left(\frac{R}{4}\right) = \sup_{B(x,\frac{R}{4})} (M(R) - u) \le c_1 \inf_{B(x,\frac{R}{4})} (M(R) - u)$$

$$= c_1 \left(M(R) - M\left(\frac{R}{4}\right)\right),$$

and analogously,

$$M\left(\frac{R}{4}\right) - m(R) = \sup_{B(x,\frac{R}{4})} (u - m(R)) \le c_1 \inf_{B(x,\frac{R}{4})} (u - m(R))$$

$$= c_1 \left(m\left(\frac{R}{4}\right) - m(R)\right).$$

(By Corollary 12.1.1, c_1 does not depend on R.) Adding these two inequalities yields

$$M\left(\frac{R}{4}\right) - m\left(\frac{R}{4}\right) \le \frac{c_1 - 1}{c_1 + 1}(M(R) - m(R)). \tag{12.2.6}$$

With $\vartheta := \frac{c_1 - 1}{c_1 + 1} < 1$, thus

---

[1] More precisely, these are nonnegative solutions, and as in the proof of Theorem 12.1.2, one adds $\varepsilon > 0$ and lets ε approach to 0.

$$\omega\left(\frac{R}{4}\right) \leq \vartheta\omega(R).$$

Iterating this inequality gives

$$\omega\left(\frac{R}{4^n}\right) \leq \vartheta^n\omega(R) \quad \text{for } n \in \mathbb{N}. \tag{12.2.7}$$

Now let

$$\frac{R}{4^{n+1}} \leq r \leq \frac{R}{4^n}. \tag{12.2.8}$$

We now choose $\alpha > 0$ such that

$$\vartheta \leq \left(\frac{1}{4}\right)^\alpha.$$

Then

$$
\begin{aligned}
\omega(r) &\leq \omega\left(\frac{R}{4^n}\right) \quad \text{since } \omega \text{ is obviously monotonically increasing} \\
&\leq \vartheta^n\omega(R) \quad \text{by (12.2.7)} \\
&\leq \left(\frac{1}{4^n}\right)^\alpha \omega(R) \\
&\leq \left(\frac{R}{4R}\right)^\alpha \omega(R) \quad \text{by (12.2.8)} \\
&= 4^{-\alpha}\left(\frac{r}{R}\right)^\alpha \omega(R),
\end{aligned}
$$

whence (12.2.4). $\qquad\square$

We now want to prove a strong maximum principle:

Theorem 12.2.2: *Let $u \in W^{1,2}(\Omega)$ satisfy $Lu \geq 0$ weakly, the coefficients a^{ij} of L again satisfying*

$$\lambda|\xi|^2 \leq \sum_{i,j} a^{ij}(x)\xi_i\xi_j, \quad |a^{ij}(x)| \leq \Lambda$$

for all $x \in \Omega$, $\xi \in \mathbb{R}^d$. If for some ball $B(y_0, R) \subset\subset \Omega$,

$$\sup_{B(y_0,R)} u = \sup_\Omega u, \tag{12.2.9}$$

then u is constant.

Proof: If (12.2.9) holds, we may find some ball $B(x_0, R_0)$ with $B(x_0, 4R_0) \subset \Omega$ and

$$\sup_{B(x_0,R_0)} u = \sup_{\Omega} u. \tag{12.2.10}$$

Without loss of generality $\sup_{\Omega} u < \infty$, because $\sup_{B(y_0,R)} u < \infty$ by Lemma 12.2.1. For

$$M > \sup_{\Omega} u$$

$M - u$ then is a positive subsolution, and we may apply Theorem 12.1.2 to it. Passing to the limit, the resulting inequalities then continue to hold for

$$M = \sup_{\Omega} u. \tag{12.2.11}$$

Thus, for $p = 1$, we get from Theorem 12.1.2

$$\fint_{B(x_0,2R_0)} (M - u) \le c \inf_{B(x_0,R_0)} (M - u) = 0$$

by (12.2.10), (12.2.11). Since by choice of M, we also have $u \le M$, it follows that

$$u \equiv M \tag{12.2.12}$$

in $B(x_0, 2R_0)$.

Now let $y \in \Omega$. We may find a chain of balls $B(x_i, R_i)$, $i = 0, \dots, m$, with $B(x_i, 4R_i) \subset \Omega$, $B(x_{i-1}, R_{i-1}) \cap B(x_i, R_i) \ne 0$ for $i = 1, \dots, m$, $y \in B(x_m, R_m)$. We already know that $u \equiv M$ on $B(x_0, 2R_0)$. Because of $B(x_0, R_0) \cap B(x_1, R_1) \ne 0$, this implies

$$\sup_{B(x_1,R_1)} u = M,$$

hence by our preceding reasoning

$$u \equiv M \quad \text{on } B(x_1, 2R_1).$$

Iteratively, we obtain

$$u \equiv M \quad \text{on } B(x_m, 2R_m),$$

and because of $y \in B(x_m, R_m)$,

$$u(y) = M.$$

Since y was arbitrary, it follows that

$$u \equiv M \quad \text{in } \Omega.$$

$\square$

As another application of the Harnack inequality, we shall now demonstrate a result of Liouville type:

Theorem 12.2.3: *Any bounded (weak) solution of $Lu = 0$ that is defined on all of $\mathbb{R}^d$, where L has measurable bounded coefficients $a^{ij}(x)$ satisfying*

$$\lambda |\xi| \leq \sum_{i,j} a^{ij}(x)\xi_i\xi_j, \quad |a^{ij}(x)| \leq \Lambda$$

for fixed constants $0 < \lambda \leq \Lambda < \infty$ and all $x \in \mathbb{R}^d$, $\xi \in \mathbb{R}^d$, is constant.

Proof: Since u is bounded, $\inf_{\mathbb{R}^d} u$ and $\sup_{\mathbb{R}^d} u$ are finite. Thus, for any

$$\mu < \inf_{\mathbb{R}^d} u,$$

$u - \mu$ is a positive solution of $Lu = 0$ on $\mathbb{R}^d$. Therefore, by Corollary 12.1.1

$$0 \leq \sup_{B(0,R)} u - \mu \leq c_3 \left(\inf_{B(0,R)} u - \mu \right)$$

for any $R > 0$ and any $\mu < \inf_{\mathbb{R}^d} u$, and passing to the limit, then this also holds for

$$\mu = \inf_{\mathbb{R}^d} u.$$

Since c_3 does not depend on R, it follows that

$$0 \leq \sup_{\mathbb{R}^d} u - \mu \leq c_3 \left(\inf_{\mathbb{R}^d} u - \mu \right) = 0,$$

and hence

$$u \equiv \text{const.}$$

□

12.3 Regularity of Minimizers of Variational Problems

The aim of this section is the proof of (a special case of) the fundamental result of de Giorgi on the regularity of minima of variational problems with elliptic Euler–Lagrange equations:

Theorem 12.3.1: *Let $F : \mathbb{R}^d \to \mathbb{R}$ be a function of class C^∞ satisfying the following conditions: For some constants $K, \Lambda < \infty$, $\lambda > 0$ and for all $p = (p_1, \ldots, p_d) \in \mathbb{R}^d$:*

(i) $\left| \frac{\partial F}{\partial p_i}(p) \right| \leq K |p| \quad (i = 1, \ldots, d)$.

(ii) $\lambda |\xi|^2 \leq \sum \frac{\partial^2 F(p)}{\partial p_i \partial p_j} \xi_i \xi_j \leq \Lambda |\xi|^2$ *for all $\xi \in \mathbb{R}^d$.*

Let $\Omega \subset \mathbb{R}^d$ be a bounded domain. Let $u \in W^{1,2}(\Omega)$ be a minimizer of the variational problem

$$I(v) := \int_\Omega F(Dv(x))dx,$$

i.e.,

$$I(u) \leq I(u + \varphi) \quad \text{for all } \varphi \in H_0^{1,2}(\Omega). \qquad (12.3.1)$$

Then $u \in C^\infty(\Omega)$.

Remark: Because of (i), there exist constants c_1, c_2 with

$$|F(p)| \leq c_1 + c_2 |p|^2. \qquad (12.3.2)$$

Since Ω is assumed to be bounded, this implies

$$I(v) = \int_\Omega F(Dv) < \infty$$

for all $v \in W^{1,2}(\Omega)$. Therefore, our variational problem, namely to minimize I in $W^{1,2}(\Omega)$, is meaningful.

We shall first derive the Euler–Lagrange equations for a minimizer of I:

Lemma 12.3.1: *Suppose that the assumptions of Theorem 12.3.1 hold. We then have for all $\varphi \in H_0^{1,2}(\Omega)$,*

$$\int_\Omega \sum_{i=1}^d F_{p_i}(Du)D_i\varphi = 0 \qquad (12.3.3)$$

(using the abbreviation $F_{p_i} = \frac{\partial F}{\partial p_i}$).

Proof: By (i),

$$\int_\Omega \sum_{i=1}^d F_{p_i}(Dv)D_i\varphi \leq dK \int_\Omega |Dv|\,|D\varphi| \leq dK \, \|Dv\|_{L^2(\Omega)} \|D\varphi\|_{L^2(\Omega)},$$

and this is finite for $\varphi, v \in W^{1,2}(\Omega)$. By a standard result of Lebesgue integration theory, on the basis of this inequality we may compute

$$\frac{d}{dt} I(u + t\varphi)$$

by differentiation under the integral sign:

$$\frac{d}{dt}I(u+t\varphi) = \int_\Omega \sum F_{p_i}(Du + tD\varphi)D_i\varphi. \tag{12.3.4}$$

In particular, $I(u+t\varphi)$ is a differentiable function of $t \in \mathbb{R}$, and since u is a minimizer,

$$\frac{d}{dt}I(u+t\varphi)|_{t=0} = 0. \tag{12.3.5}$$

Equation (12.3.4) for $t = 0$ then implies (12.3.3). □

Lemma 12.3.1 reduces Theorem 12.3.1 to the following:

Theorem 12.3.2: *Let $A^i : \mathbb{R}^d \to \mathbb{R}$, $i = 1, \ldots, d$, be C^∞-functions satisfying the following conditions: There exist constants $K, \Lambda < \infty$, $\lambda > 0$ such that for all $p \in \mathbb{R}^d$:*

(i) $|A^i(p)| \le K|p|$ $(i = 1, \ldots, d)$.
(ii) $\lambda|\xi|^2 \le \sum_{i,j=1}^d \frac{\partial A^i(p)}{\partial p_j}\xi_i\xi_j$ for all $\xi \in \mathbb{R}^d$.
(iii) $\left|\frac{\partial A^i(p)}{\partial p_j}\right| \le \Lambda$.

Let $u \in W^{1,2}(\Omega)$ be a weak solution of

$$\sum_{i=1}^d \frac{\partial}{\partial x^i}A^i(Du) = 0 \quad in\ \Omega \subset \mathbb{R}^d, \tag{12.3.6}$$

i.e., for all $\varphi \in H_0^{1,2}(\Omega)$, let

$$\int_\Omega \sum_{i=1}^d A^i(Du)D_i\varphi = 0. \tag{12.3.7}$$

Then $u \in C^\infty(\Omega)$.

The crucial step in the proof will be Theorem 12.2.1, of de Giorgi and Nash. Important steps towards Theorem 12.3.2 had been obtained earlier by S. Bernstein, L. Lichtenstein, E. Hopf, C. Morrey, and others.

We shall start with a lemma.

Lemma 12.3.2: *Under the assumptions of Theorem 12.3.2, for any $\Omega' \subset\subset \Omega$ we have $u \in W^{2,2}(\Omega')$, and moreover, $\|u\|_{W^{2,2}(\Omega')} \le c\|u\|_{W^{1,2}(\Omega)}$, where $c = c(\lambda, \Lambda, \mathrm{dist}(\Omega', \partial\Omega))$.*

Proof: We shall proceed as in the proof of Theorem 9.2.1. For

$$|h| < \mathrm{dist}(\mathrm{supp}\,\varphi, \partial\Omega),$$

$\varphi_{k,-h}(x) := \varphi(x - he_k)$ (e_k being the kth unit vector) is of class $H_0^{1,2}(\Omega)$ as well. Therefore,

$$0 = \int_{\Omega} \sum_{i=1}^{d} A^i(Du(x)) D_i \varphi_{k,-h}(x) dx$$

$$= \int_{\Omega} \sum_{i=1}^{d} A^i(Du(x)) D_i \varphi(x - he_k) dx$$

$$= \int_{\Omega} \sum_{i=1}^{d} A^i(Du(y + he_k)) D_i \varphi(y) dy$$

$$= \int_{\Omega} \sum_{i=1}^{d} A^i\left((Du)_{k,h}\right) D_i \varphi.$$

Subtracting (12.3.7), we obtain

$$\int \sum_{i} \left(A^i(Du(x + he_k)) - A^i(Du(x)) \right) D_i \varphi(x) = 0. \qquad (12.3.8)$$

For almost all $x \in \Omega$

$$A^i\left(Du(x + he_k)\right) - A^i(Du(x))$$

$$= \int_0^1 \frac{d}{dt} A^i\left(tDu(x + he_k) + (1-t)Du(x)\right) dt$$

$$= \int_0^1 \left(\sum_{j=1}^{d} A^i_{p_j}\left(tDu(x + he_k) + (1-t)Du(x)\right) D_j\left(u(x + he_k) - u(x)\right) \right) dt.$$

$$(12.3.9)$$

We thus put

$$a_h^{ij}(x) := \int_0^1 A^i_{p_j}\left(tDu(x + he_k) + (1-t)Du(x)\right) dt,$$

and using (12.3.9), we rewrite (12.3.8) as

$$\int_{\Omega} \sum_{i,j} a_h^{ij}(x) D_j \left(\frac{u(x + he_k) - u(x)}{h} \right) D_i \varphi(x) dx = 0. \qquad (12.3.10)$$

Here, because of (ii) and (iii),

$$\lambda |\xi|^2 \leq \sum_{i,j} a_h^{ij}(x) \xi_i \xi_j \leq \Lambda |\xi|^2 \quad \text{for all } \xi \in \mathbb{R}^d.$$

We may thus proceed as in Section 9.2 and put

$$\varphi = \frac{1}{h} \left(u(x + he_k) - u(x) \right) \eta^2$$

with $\eta \in C_0^1(\Omega'')$, where we choose Ω'' satisfying

$$\Omega' \subset\subset \Omega'' \subset\subset \Omega,$$

$\mathrm{dist}(\Omega'', \partial\Omega), \mathrm{dist}(\Omega', \partial\Omega'') \geq \frac{1}{4}\,\mathrm{dist}(\Omega', \partial\Omega)$, and require

$$0 \leq \eta \leq 1,$$
$$\eta(x) = 1 \quad \text{for } x \in \Omega',$$
$$|D\eta| \leq \frac{8}{\mathrm{dist}(\Omega', \partial\Omega)},$$

as well as

$$|2h| < \mathrm{dist}(\Omega'', \partial\Omega).$$

Using the notation

$$\Delta_k^h u(x) = \frac{u(x + he_k) - u(x)}{h},$$

(12.3.10) then implies

$$\lambda \int_\Omega |D\Delta_k^h u|^2 \eta^2 \leq \int_\Omega \sum_{i,j} a_h^{ij} \left(D_j \Delta_k^h u\right) \left(D_i \Delta_k^h u\right) \eta^2$$

$$= - \int_\Omega \sum_{i,j} a_h^{ij} D_j \Delta_k^h u \, 2\eta (D_i \eta) \Delta_h^k u \quad \text{by (12.3.10)}$$

$$\leq \varepsilon \Lambda \int_\Omega |D\Delta_k^h u|^2 + \frac{\Lambda}{\varepsilon} \int_\Omega |\Delta_k^h u|^2 |D\eta|^2 \quad \text{for all } \varepsilon > 0,$$

and with $\varepsilon = \frac{\lambda}{2\Lambda}$,

$$\int_\Omega |D\Delta_k^h u|^2 \eta^2 \leq c_1 \int_{\Omega''} |\Delta_k^h u|^2 \leq c_1 \int_\Omega |Du|^2$$

by Lemma 9.2.1, with c_1 independent of h. Hence

$$\left\| D\Delta_k^h u \right\|_{L^2(\Omega')} \leq c_1 \left\| Du \right\|_{L^2(\Omega)}. \tag{12.3.11}$$

Since the right hand side of (12.3.11) does not depend on h, from Lemma 9.2.2 we obtain $D^2 u \in L^2(\Omega')$ and the inequality

$$\left\| D^2 u \right\|_{L^2(\Omega')} \leq c_1 \left\| Du \right\|_{L^2(\Omega)}. \tag{12.3.12}$$

Consequently, $u \in W^{2,2}(\Omega')$. $\qquad\qquad\qquad\qquad\qquad\qquad\qquad\qquad\square$

Performing the limit $h \to 0$ in (12.3.10), with

$$a^{ij}(x) := A^i_{p_j}(Du(x)),$$
$$v := D_k u, \tag{12.3.13}$$

we also obtain

$$\int_\Omega \sum_{i,j} a^{ij}(x) D_j v D_i \varphi = 0 \quad \text{for all } \varphi \in H_0^{1,2}(\Omega).$$

By (ii), (iii), $(a^{ij}(x))_{i,j=1,\ldots,d}$ satisfies the assumptions of Theorem 12.2.1. Applying that result to $v = D_k u$ then yields the following result:

Lemma 12.3.3: *Under the assumptions of Theorem 12.2.1,*

$$Du \in C^\alpha(\Omega)$$

for some $\alpha \in (0,1)$, i.e.,

$$u \in C^{1,\alpha}(\Omega).$$

$\square$

Thus $v = D_k u$, $k = 1, \ldots, d$, is a weak solution of

$$\sum_{i,j=r}^d D_i \left(a^{ij}(x) D_j v \right) = 0. \tag{12.3.14}$$

Here, the coefficients $a^{ij}(x)$ satisfy not only the ellipticity condition

$$\lambda |\xi|^2 \le \sum_{i,j=1}^d a^{ij}(x)\xi_i\xi_j, \quad \left| a^{ij}(x) \right| \le \Lambda$$

for all $\xi \in \mathbb{R}^d$, $x \in \Omega$, $i,j = 1, \ldots, d$, but by (12.3.13), they are also Hölder continuous, since A^i is smooth and Du is Hölder continuous by Lemma 12.3.3. For the proof of Theorem 12.3.2, we thus need a regularity theory for such equations. Equation (12.3.14) is of divergence type, in contrast to those treated in Chapter 11, and therefore, we cannot apply the results of Schauder directly. However, one can develop similar methods. For the sake of variety, here, we shall present the method of Campanato as an alternative approach. As a preparation, we shall now prove some auxiliary results for equations of type (12.3.14) with constant coefficients. (Of course, these results are already essentially known from Chapter 9.)

The first result is the Caccioppoli inequality:

Lemma 12.3.4: *Let $(A^{ij})_{i,j=1,\ldots,d}$ be a matrix with $\left|A^{ij}\right| \leq \Lambda$ for all i,j, and*

$$\lambda \left|\xi\right|^2 \leq \sum_{i,j=1}^{d} A^{ij}\xi_i\xi_j \quad \text{for all } \xi \in \mathbb{R}^d$$

with $\lambda > 0$. Let $u \in W^{1,2}(\Omega)$ be a weak solution of

$$\sum_{i,j=1}^{d} D_j\left(A^{ij}D_iu\right) = 0 \quad \text{in } \Omega. \tag{12.3.15}$$

We then have for all $x_0 \in \Omega$ and $0 < r < R < \mathrm{dist}(x_0, \partial\Omega)$ and all $\mu \in \mathbb{R}$,

$$\int_{B(x_0,r)} \left|Du\right|^2 \leq \frac{c_2}{(R-r)^2} \int_{B(x_0,R)\backslash B(x_0,r)} \left|u - \mu\right|^2. \tag{12.3.16}$$

Proof: We choose $\eta \in H_0^{1,2}(B(x_0,R))$ with

$$0 \leq \eta \leq 1,$$
$$\eta \equiv 1 \quad \text{on } B(x_0,r), \text{ hence } D\eta \equiv 0 \quad \text{on } B(x_0,r),$$
$$\left|D\eta\right| \leq \frac{2}{R-r}.$$

As in Section 9.2, we employ the test function

$$\varphi = (u - \mu)\eta^2$$

and obtain

$$0 = \int \sum_{i,j} A^{ij}D_iuD_j\left((u - \mu)\eta^2\right)$$
$$= \int \sum_{i,j} A^{ij}D_iuD_ju\,\eta^2 + \int 2\sum_{i,j} A^{ij}D_iu(u - \mu)\eta D_j\eta.$$

Using the ellipticity conditions, we deduce the inequality

$$\lambda \int_{B(x_0,R)} \left|Du\right|^2\eta^2 \leq \int_{B(x_0,R)} \sum A^{ij}D_iuD_ju\,\eta^2$$
$$\leq \varepsilon\Lambda d \int_{B(x_0,R)} \left|Du\right|^2\eta^2$$
$$+ \frac{\Lambda}{\varepsilon}d \int_{B(x_0,R)\backslash B(x_0,r)} \left|D\eta\right|^2\left|u - \mu\right|^2,$$

since $D\eta = 0$ on $B(x_0,r)$. Hence, with $\varepsilon = \frac{1}{2}\frac{\lambda}{\Lambda d}$,

$$\int_{B(x_0,R)} |Du|^2 \eta^2 \leq \frac{c_2}{(R-r)^2} \int_{B(x_0,R)\setminus B(x_0,r)} |u - \mu|^2,$$

and because of

$$\int_{B(x_0,r)} |Du|^2 \leq \int_{B(x_0,R)} |Du|^2 \eta^2$$

the claim results.   □

The next lemma contains the Campanato estimates:

Lemma 12.3.5: *Under the assumptions of Lemma 12.3.4, we have*

$$\int_{B(x_0,r)} |u|^2 \leq c_3 \left(\frac{r}{R}\right)^d \int_{B(x_0,R)} |u|^2 \qquad (12.3.17)$$

as well as

$$\int_{B(x_0,r)} |u - u_{B(x_0,r)}|^2 \leq c_4 \left(\frac{r}{R}\right)^{d+2} \int_{B(x_0,R)} |u - u_{B(x_0,R)}|^2. \qquad (12.3.18)$$

Proof: Without loss of generality $r < \frac{R}{2}$. We choose $k > d$. By the Sobolev embedding theorem (Theorem 9.1.1) or an extension of this result analogous to Corollary 9.1.3,

$$W^{k,2}(B(x_0, R)) \subset C^0(B(x_0, R)).$$

By Theorem 9.3.1, now $u \in W^{k,2}\left(B\left(x_0, \frac{R}{2}\right)\right)$, with an estimate analogous to Theorem 9.2.2. Therefore,

$$\int_{B(x_0,r)} |u|^2 \leq c_5 r^d \sup_{B(x_0,r)} |u|^2 \leq c_6 \frac{r^d}{R^{d-2k}} \|u\|_{W^{k,2}\left(B\left(x_0, \frac{R}{2}\right)\right)}$$

$$\leq c_3 \frac{r^d}{R^d} \int_{B(x_0,R)} |u|^2.$$

(Concerning the dependence on the radius: The power r^d is obvious. The power R^d can easily be derived from a scaling argument, instead of carefully going through all the intermediate estimates.) This yields (12.3.17). Since we are dealing with an equation with constant coefficients, Du is a solution along with u. For $r < \frac{R}{2}$, we thus obtain

$$\int_{B(x_0,r)} |Du|^2 \leq c_7 \frac{r^d}{R^d} \int_{B\left(x_0, \frac{R}{2}\right)} |Du|^2. \qquad (12.3.19)$$

By the Poincaré inequality (Corollary 9.1.4),

$$\int_{B(x_0,r)} |u - u_{B(x_0,r)}|^2 \leq c_8 r^2 \int_{B(x_0,r)} |Du|^2. \qquad (12.3.20)$$

By the Caccioppoli inequality (Lemma 12.3.4)

$$\int_{B\left(x_0,\frac{R}{2}\right)} |Du|^2 \leq \frac{c_9}{R^2} \int_{B(x_0,R)} \left|u - u_{B(x_0,R)}\right|^2. \tag{12.3.21}$$

Then (12.3.19)–(12.3.21) imply (12.3.18).   □

We may now use Campanato's method to derive the following regularity result:

Theorem 12.3.3: *Let $a^{ij}(x)$, $i,j = 1,\ldots,d$, be functions of class C^α, $0 < \alpha < 1$, on $\Omega \subset \mathbb{R}^d$, satisfying the ellipticity condition*

$$\lambda |\xi|^2 \leq \sum_{i,j=1}^{d} a^{ij}(x)\xi_i\xi_j \quad \text{for all } \xi \in \mathbb{R}^d, x \in \Omega \tag{12.3.22}$$

and

$$\left|a^{ij}(x)\right| \leq \Lambda \quad \text{for all } x \in \Omega, i,j = 1,\ldots,d, \tag{12.3.23}$$

with fixed constants $0 < \lambda \leq \Lambda < \infty$. Then any weak solution v of

$$\sum_{i,j=1}^{d} D_j\left(a^{ij}(x)D_i v\right) = 0 \tag{12.3.24}$$

is of class $C^{1,\alpha'}(\Omega)$ for any α' with $0 < \alpha' < \alpha$.

Proof: For $x_0 \in \Omega$, we write

$$a^{ij} = a^{ij}(x_0) + \left(a^{ij}(x) - a^{ij}(x_0)\right).$$

Letting

$$A^{ij} := a^{ij}(x_0),$$

(12.3.24) becomes

$$\sum_{i,j=1}^{d} D_j\left(A^{ij}D_i v\right) = \sum_{i,j=1}^{d} D_j\left((a^{ij}(x_0) - a^{ij}(x))D_i v\right) = \sum_{j=1}^{d} D_j\left(f^j(x)\right)$$

with

$$f^j(x) := \sum_{i=1}^{d}\left((a^{ij}(x_0) - a^{ij}(x))D_i v\right). \tag{12.3.25}$$

This means that

$$\int_\Omega \sum_{i,j=1}^d A^{ij} D_i v D_j \varphi = \int_\Omega \sum_{j=1}^d f^j D_j \varphi \quad \text{for all } \varphi \in H_0^{1,2}(\Omega). \qquad (12.3.26)$$

For some ball $B(x_0, R) \subset \Omega$, let

$$w \in H^{1,2}(B(x_0, R))$$

be a weak solution of

$$\sum_{i,j=1}^d D_j \left(A^{ij} D_i w \right) = 0 \quad \text{in } B(x_0, R), \qquad (12.3.27)$$

$$w = v \quad \text{on } \partial B(x_0, R).$$

Thus w is a solution of

$$\int_{B(x_0, R)} \sum_{i,j=1}^d A^{ij} D_i w D_j \varphi = 0 \quad \text{for all } \varphi \in H_0^{1,2}(B(x_0, R)). \qquad (12.3.28)$$

Such a w exists by the Lax–Milgram theorem (see Appendix 12.3). Note that we seek $z = w - v$ with

$$B(\varphi, z) := \int \sum A^{ij} D_i z D_j \varphi$$

$$= - \int \sum A^{ij} D_i v D_j \varphi$$

$$= : F(\varphi) \quad \text{for all } \varphi \in H_0^{1,2}(B(x_0, R)).$$

Since (12.3.27) is a linear equation with constant coefficients, then if w is a solution, so is $D_k w$, $k = 1, \ldots, d$ (with different boundary conditions, of course). We may thus apply (12.3.17) from Lemma 4.3.5 to $u = D_k w$ and obtain

$$\int_{B(x_0, r)} |Dw|^2 \le c_{10} \left(\frac{r}{R} \right)^d \int_{B(x_0, R)} |Dw|^2 . \qquad (12.3.29)$$

(Here, Dw stands for the vector $(D_1 w, \ldots, D_d w)$.) Since $w = v$ on $\partial B(x_0, R)$, $\varphi = v - w$ is an admissible test function in (12.3.28), and we obtain

$$\int_{B(x_0, R)} \sum_{i,j=1}^d A^{ij} D_i w D_j w = \int_{B(x_0, R)} \sum_{i,j=1}^d A^{ij} D_i w D_j v. \qquad (12.3.30)$$

Using (12.3.27), (12.3.23) and the Cauchy–Schwarz inequality, this implies

$$\int_{B(x_0, R)} |Dw|^2 \le \left(\frac{\Lambda d}{\lambda} \right)^2 \int_{B(x_0, R)} |Dv|^2 . \qquad (12.3.31)$$

Equations (12.3.26) and (12.3.28) imply

$$\int_{B(x_0,R)} \sum_{i,j=1}^{d} A^{ij} D_i(v-w) D_j\varphi = \int_{B(x_0,R)} \sum_{i,j=1}^{d} f^j D_j\varphi$$

for all $\varphi \in H_0^{1,2}(B(x_0,R))$. We utilize once more the test function $\varphi = v-w$ to obtain

$$\int_{B(x_0,R)} |D(v-w)|^2 \leq \frac{1}{\lambda} \int_{B(x_0,R)} \sum_{i,j} A^{ij} D_i(v-w) D_j(v-w)$$

$$= \frac{1}{\lambda} \int_{B(x_0,R)} \sum_{j} f^j D_j(v-w)$$

$$\leq \frac{1}{\lambda} \left(\int_{B(x_0,R)} |D(v-w)|^2 \right)^{\frac{1}{2}} \left(\int_{B(x_0,R)} \sum_{j} |f^j|^2 \right)^{\frac{1}{2}}$$

by the Cauchy–Schwarz inequality, i.e.,

$$\int_{B(x_0,R)} |D(v-w)|^2 \leq \frac{1}{\lambda^2} \int_{B(x_0,R)} \sum_{j} |f^j|^2 . \tag{12.3.32}$$

We now put the preceding estimates together. For $0 < r \leq R$, we have

$$\int_{B(x_0,r)} |Dv|^2 \leq 2 \int_{B(x_0,r)} |Dw|^2 + 2 \int_{B(x_0,r)} |D(v-w)|^2$$

$$\leq c_{11} \left(\frac{r}{R} \right)^d \int_{B(x_0,r)} |Dv|^2 + 2 \int_{B(x_0,r)} |D(v-w)|^2$$

by (12.3.29), (12.3.31). Now

$$\int_{B(x_0,r)} |D(v-w)|^2 \leq \int_{B(x_0,R)} |D(v-w)|^2, \quad \text{since } r \leq R$$

$$\leq \frac{1}{\lambda^2} \int_{B(x_0,R)} \sum_{j} |f^j|^2 \quad \text{by (12.3.32)}$$

$$\leq \frac{1}{\lambda^2} \sup_{\substack{i,j \\ x \in B(x_0,R)}} |a^{ij}(x_0) - a^{ij}(x)|^2 \int_{B(x_0,R)} |Dv|^2$$

$$\text{by (12.3.25)}$$

$$\leq c_{12} R^{2\alpha} \int_{B(x_0,R)} |Dv|^2,$$

$$\tag{12.3.33}$$

since the a^{ij} are of class C^α. Altogether, we obtain

$$\int_{B(x_0,r)} |Dv|^2 \leq \gamma \left(\left(\frac{r}{R} \right)^d + R^{2\alpha} \right) \int_{B(x_0,R)} |Dv|^2 \qquad (12.3.34)$$

with some constant γ. If (12.3.34) did not contain the term $R^{2\alpha}$ (which is present solely for the reason that the $a^{ij}(x)$, while Hölder continuous, are not necessarily constant), we would have a useful inequality. That term, however, can be made to disappear by a simple trick. For later purposes, we formulate a somewhat more general result:

Lemma 12.3.6: Let $\sigma(r)$ be a nonnegative, monotonically increasing function satisfying

$$\sigma(r) \leq \gamma \left(\left(\frac{r}{R} \right)^{\mu} + \delta \right) \sigma(R) + \kappa R^{\nu}$$

for all $0 < r \leq R \leq R_0$, with $\mu > \nu$ and $\delta \leq \delta_0(\gamma, \mu, \nu)$. If δ_0 is sufficiently small, for $0 < r \leq R \leq R_0$, we then have

$$\sigma(r) \leq \gamma_1 \left(\frac{r}{R} \right)^{\nu} \sigma(R) + \kappa_1 r^{\nu},$$

with γ_1 depending on γ, μ, ν, and κ_1 depending in addition on κ ($\kappa_1 = 0$ if $\kappa = 0$).

Proof: Let $0 < \tau < 1$, $R < R_0$. Then by assumption

$$\sigma(\tau R) \leq \gamma \tau^{\mu} \left(1 + \delta \tau^{-\mu} \right) \sigma(R) + \kappa R^{\nu}.$$

We choose $0 < \tau < 1$ such that

$$2 \gamma \tau^{\mu} = \tau^{\lambda}$$

with $\nu < \lambda < \mu$ (without loss of generality $2\gamma > 1$), and assume that

$$\delta_0 \tau^{-\mu} \leq 1.$$

It follows that

$$\sigma(\tau R) \leq \tau^{\lambda} \sigma(R) + \kappa R^{\nu}$$

and thus iteratively for $k \in \mathbb{N}$,

$$\sigma(\tau^{k+1} R) \leq \tau^{\lambda} \sigma(\tau^k R) + \kappa \tau^{k\nu} R^{\nu}$$

$$\leq \tau^{(k+1)\lambda} \sigma(R) + \kappa \tau^{k\nu} R^{\nu} \sum_{j=0}^{k} \tau^{j(\lambda - \nu)}$$

$$\leq \gamma_0 \tau^{(k+1)\nu} \left(\sigma(R) + \kappa R^{\nu} \right)$$

(where γ_0, as well as the subsequent γ_1, contains a factor $\frac{1}{\tau}$). We now choose $k \in \mathbb{N}$ such that

$$\tau^{k+2}R < r \le \tau^{k+1}R,$$

and obtain

$$\sigma(r) \le \sigma(\tau^{k+1}R) \le \gamma_1 \left(\frac{r}{R}\right)^{\nu} \sigma(R) + \kappa_1 r^{\nu}.$$

□

Continuing with the proof of Theorem 12.3.3, applying Lemma 12.3.6 to (12.3.34), where we have to require $0 < r \le R \le R_0$ with $R_0^{2\alpha} \le \delta_0$, we obtain the inequality

$$\int_{B(x_0,r)} |Dv|^2 \le c_{13} \left(\frac{r}{R}\right)^{d-\varepsilon} \int_{B(x_0,R)} |Dv|^2 \qquad (12.3.35)$$

for each $\varepsilon > 0$, where c_{13} and R_0 depend on ε. We repeat this procedure, but this time applying (12.3.18) from Lemma 12.3.5 in place of (12.3.17). Analogously to (12.3.29), we obtain

$$\int_{B(x_0,r)} \left|Dw - (Dw)_{B(x_0,r)}\right|^2 \le c_{14} \left(\frac{r}{R}\right)^{d+2} \int_{B(x_0,R)} \left|Dw - (Dw)_{B(x_0,R)}\right|^2.$$
$$(12.3.36)$$

We also have

$$\int_{B(x_0,R)} \left|Dw - (Dw)_{B(x_0,R)}\right|^2 \le \int_{B(x_0,R)} \left|Dw - (Dv)_{B(x_0,R)}\right|^2,$$

because for any L^2-function g, the following relation holds:

$$\int_{B(x_0,R)} \left|g - g_{B(x_0,R)}\right|^2 = \inf_{\kappa \in \mathbb{R}} \int_{B(x_0,R)} |g - \kappa|^2. \qquad (12.3.37)$$

(*Proof:* For $g \in L^2(\Omega), F(\kappa) := \int_\Omega |g - \kappa|^2$ is convex and differentiable with respect to κ, and

$$F'(\kappa) = \int_\Omega 2(\kappa - g);$$

hence $F'(\kappa) = 0$ precisely for

$$\kappa = \frac{1}{|\Omega|} \int_\Omega g,$$

and since F is convex, a critical point has to be a minimizer.)

Moreover,

$$\int_{B(x_0,R)} \left| Dw - (Dv)_{B(x_0,R)} \right|^2$$

$$\leq \frac{1}{\lambda} \int_{B(x_0,R)} \sum_{i,j} A^{ij} \left(D_i w - (D_i v)_{B(x_0,R)} \right) \left(D_j w - (D_j v)_{B(x_0,R)} \right)$$

$$= \frac{1}{\lambda} \int_{B(x_0,R)} \sum_{i,j} A^{ij} \left(D_i w - (D_i v)_{B(x_0,R)} \right) \left(D_j v - (D_j v)_{B(x_0,R)} \right)$$

$$+ \frac{1}{\lambda} \int_{B(x_0,R)} \sum_{i,j} A^{ij} \left(D_i v \right)_{B(x_0,R)} \left(D_j v - D_j w \right)$$

by (12.3.30). The last integral vanishes, since $A^{ij} (D_i v)_{B(x_0,R)}$ is constant and $v - w \in H_0^{1,2}(B(x_0, R))$. Applying the Cauchy–Schwarz inequality as usual, we altogether obtain

$$\int_{B(x_0,R)} \left| Dw - (Dw)_{B(x_0,R)} \right|^2 \leq \frac{\Lambda^2}{\lambda^2} d^2 \int_{B(x_0,R)} \left| Dv - (Dv)_{B(x_0,R)} \right|^2.$$

$$(12.3.38)$$

Finally,

$$\int_{B(x_0,r)} \left| Dv - (Dv)_{B(x_0,r)} \right|^2 \leq 3 \int_{B(x_0,r)} \left| Dw - (Dw)_{B(x_0,R)} \right|^2$$

$$+ 3 \int_{B(x_0,r)} \left| Dv - Dw \right|^2$$

$$+ 3 \int_{B(x_0,r)} \left((Dv)_{B(x_0,r)} - (Dw)_{B(x_0,r)} \right)^2.$$

The last expression can be estimated by Hölder's inequality:

$$\int_{B(x_0,r)} \left(\frac{1}{|B(x_0,r)|} \int_{B(x_0,r)} (Dv - Dw) \right)^2 \leq 3 \int_{B(x_0,r)} (Dv - Dw)^2.$$

Thus

$$\int_{B(x_0,r)} \left| Dv - (Dv)_{B(x_0,r)} \right|^2$$

$$\leq 3 \int_{B(x_0,r)} \left| Dw - (Dw)_{B(x_0,r)} \right|^2 + 6 \int_{B(x_0,r)} \left| Dv - Dw \right|^2$$

$$\leq 3 \int_{B(x_0,r)} \left| Dw - (Dw)_{B(x_0,r)} \right|^2 + c_{15} R^{2\alpha} \int_{B(x_0,r)} \left| Dv \right|^2$$

$$(12.3.39)$$

by (12.3.33). From (12.3.39), (12.3.36), (12.3.38), we obtain

$$\int_{B(x_0,r)} \left|Dv - (Dv)_{B(x_0,r)}\right|^2$$

$$\leq c_{16} \left(\frac{r}{R}\right)^{d+2} \int_{B(x_0,r)} \left|Dv - (Dv)_{B(x_0,R)}\right|^2 + c_{17} R^{2\alpha} \int_{B(x_0,R)} |Dv|^2$$

$$\leq c_{16} \left(\frac{r}{R}\right)^{d+2} \int_{B(x_0,R)} \left|Dv - (Dv)_{B(x_0,R)}\right|^2 + c_{18} R^{d-\varepsilon+2\alpha},$$

$$(12.3.40)$$

applying (12.3.35) for $0 < R \leq R_0$ in place of $0 < r \leq R$. Lemma 12.3.6 implies

$$\int_{B(x_0,r)} \left|Dv - (Dv)_{B(x_0,r)}\right|^2$$

$$\leq c_{19} \left(\frac{r}{R}\right)^{d+2\alpha-\varepsilon} \int_{B(x_0,R)} \left|Dv - (Dv)_{B(x_0,R)}\right|^2 + c_{20} r^{d+2\alpha-\varepsilon}.$$

The claim now follows from Campanato's theorem (Corollary 9.1.7).   □

It is now easy to prove Theorem 12.3.2:

Proof of Theorem 12.3.2: We apply Theorem 12.3.3 to $v = Du$ and obtain $v \in C^{1,\alpha'}$, hence $u \in C^{2,\alpha'}$. We may then differentiate the equation with respect to x^k and observe that the second derivatives $D_{jk}u$, $j,k = 1,\ldots,d$, again satisfy equations of the same type. By Theorem 12.3.3, then $D^2 u \in C^{1,\alpha''}$; hence $u \in C^{3,\alpha''}$. Iteratively, we obtain $u \in C^{m,\alpha_m}$ for all $m \in \mathbb{N}$ with $0 < \alpha_m < 1$. Therefore, $u \in C^\infty$.   □

Remark: The regularity Theorem 12.3.1 of de Giorgi more generally applies to minimizers of variational problems of the form

$$I(v) := \int_\Omega F(x, v(x), Dv(x)) dx,$$

where $F \in C^\infty(\Omega \times \mathbb{R} \times \mathbb{R} \times \mathbb{R}^d)$ again satisfies conditions like (i), (ii) of Theorem 12.3.1 with respect to p, and $\frac{1}{|p|^2} F(x, v, p)$ satisfies smoothness conditions with respect to the variables x and v uniformly in p.

References for this section are Giaquinta [7],[8].

Summary

Moser's Harnack inequality says that positive weak solutions u of

$$Lu = \sum_{i,j} \frac{\partial}{\partial x^j} \left(a^{ij}(x) \frac{\partial}{\partial x^i} u(x)\right) = 0$$

satisfy an estimate of the form

$$\sup_{B(x_0,R)} u \le \text{const} \inf_{B(x_0,R)} u$$

in each ball $B(x_0, R)$ in the interior of their domain of definition Ω. Here, the coefficients a^{ij} need to satisfy only an ellipticity condition, and have to be measurable and bounded, but they need not satisfy any further conditions like continuity. Moser's inequality yields a proof of the fundamental result of de Giorgi and Nash about the Hölder continuity of weak solutions of linear elliptic differential equations of second order with measurable and bounded coefficients. These assumptions are appropriate and useful for applications to *nonlinear* elliptic equations of the type

$$\sum_{i,j} \frac{\partial}{\partial x^j} \left(A^{ij}(u(x)) \frac{\partial}{\partial x^i} u(x) \right) = 0.$$

Namely, if one does not yet know any detailed properties of the solution u, then, even if the A^{ij} themselves are smooth, one can work only with the boundedness of the coefficients

$$a^{ij}(x) := A^{ij}(u(x)).$$

Here, a nonlinear equation is treated as a linear equation with not necessarily regular coefficients.

An application is de Giorgi's theorem on the regularity of minimizers of variational problems of the form

$$\int F(Du(x))\,dx \to \min$$

under the structural conditions

(i) $|\frac{\partial F}{\partial p_i}(p)| \le K|p|$,

(ii) $\lambda|\xi|^2 \le \sum \frac{\partial^2 F(p)}{\partial p_i \partial p_j} \xi_i \xi_j \le \Lambda|\xi|^2$ for all $\xi \in \mathbb{R}^d$,

with constants $K, \Lambda < \infty, \lambda > 0$.

Exercises

12.1 Formulate conditions on the coefficients of a differential operator of the form

$$Lu = \sum_{i,j=1}^{d} \frac{\partial}{\partial x^j} \left(a^{ij}(x) \frac{\partial}{\partial x^i} u(x) \right) + \sum_{i=1}^{d} \frac{\partial}{\partial x^i} (b^i(x)u(x)) + c(x)u(x)$$

that imply a Harnack inequality of the type of Corollary 12.1.1. Carry out the detailed proof.

12.2 As in Lemma 12.1.4, let

$$\phi(p, R) = \left(\fint_{B(x_0, R)} u^p \, dx \right)^{1/p}$$

for a fixed positive $u : B(x_0, R) \to \mathbb{R}$.
Show that

$$\lim_{p \to 0} \phi(p, R) = \exp \left(\fint_{B(x_0, R)} \log u(x) \, dx \right).$$

Appendix. Banach and Hilbert Spaces. The L^p-Spaces

In the present appendix we shall first recall some basic concepts from calculus without proofs. After that, we shall prove some smoothing results for L^p-functions.

Definition A.1: *A Banach space B is a real vector space that is equipped with a norm $\|\cdot\|$ that satisfies the following properties:*

- *(i) $\|x\| > 0$ for all $x \in B$, $x \neq 0$.*
- *(ii) $\|\alpha x\| = |\alpha| \cdot \|x\|$ for all $\alpha \in \mathbb{R}$, $x \in B$.*
- *(iii) $\|x + y\| \leq \|x\| + \|y\|$ for all $x, y \in B$ (triangle inequality).*
- *(iv) B is complete with respect to $\|\cdot\|$ (i.e., every Cauchy sequence has a limit in B).*

We recall the **Banach fixed point theorem**

Theorem A.1: *Let $(B, \|\cdot\|)$ be a Banach space, $A \subset B$ a closed subset, $f : A \to B$ a map with $f(A) \subset A$ which satisfies the inequality*

$$\|f(x) - f(y)\| \leq \theta \|x - y\| \quad \text{for all } x, y \in A,$$

for some fixed θ with $0 \leq \theta < 1$.

Then f has unique fixed point in A, that is, a solution of $f(x) = x$.

For example, every Hilbert space is a Banach space. We also recall that concept:

Definition A.2: *A (real) Hilbert space H is a vector space over $\mathbb{R}$, equipped with a scalar product*

$$(\cdot, \cdot) : H \times H \to \mathbb{R}$$

that satisfies the following properties:

- *(i) $(x, y) = (y, x)$ for all $x, y \in H$.*
- *(ii) $(\lambda_1 x_2 + \lambda_2 x_2, y) = \lambda_1(x_1, y) + \lambda_2(x_2, y)$ for all $\lambda_1, \lambda_2 \in \mathbb{R}$, $x_1, x_2, y \in H$.*
- *(iii) $(x, x) > 0$ for all $x \neq 0$, $x \in H$.*
- *(iv) H is complete with respect to the norm*

$$\|x\| := (x, x)^{\frac{1}{2}}.$$

In a Hilbert space H, the following inequalities hold:

- Schwarz inequality:

$$|(x,y)| \le \|x\| \cdot \|y\|, \tag{A.1}$$

with equality precisely if x and y are linearly dependent.
- Triangle inequality:

$$\|x+y\| \le \|x\| + \|y\|. \tag{A.2}$$

Likewise without proof, we state the **Riesz representation theorem**:

Let L be a bounded linear functional on the Hilbert space H, i.e., $L : H \to \mathbb{R}$ is linear with

$$\|L\| := \sup_{x \ne 0} \frac{|Lx|}{\|x\|} < \infty.$$

Then there exists a unique $y \in H$ with $L(x) = (x,y)$ for all $x \in H$, and

$$\|L\| = \|y\|.$$

The following extension is important, too:

Theorem of Lax–Milgram: *Let B be a bilinear form on the Hilbert space H that is bounded,*

$$|B(x,y)| \le K \|x\| \|y\| \quad \text{for all } x, y \in H \text{ with } K < \infty,$$

and elliptic, or, as this property is also called in the present context, coercive,

$$|B(x,x)| \ge \lambda \|x\|^2 \quad \text{for all } x \in H \text{ with } \lambda > 0.$$

For every bounded linear functional T on H, there then exists a unique $y \in H$ with

$$B(x,y) = Tx \quad \text{for all } x \in H.$$

Proof: We consider

$$L_z(x) = B(x,z).$$

By the Riesz representation theorem, there exists $Sz \in H$ with

$$(x, Sz) = L_z x = B(x,z).$$

Since B is bilinear, Sz depends linearly on z. Moreover,

$$\|Sz\| \leq K \|z\|.$$

Thus, S is a bounded linear operator.

Because of

$$\lambda \|z\|^2 \leq B(z, z) = (z, Sz) \leq \|z\| \|Sz\|$$

we have

$$\|Sz\| \geq \lambda \|z\|.$$

So, S is injective. We shall show that S is surjective as well. In fact, there exists $x \neq 0$ with

$$(x, Sz) = 0 \quad \text{for all } z \in H.$$

With $z = x$, we get

$$(x, Sx) = 0.$$

Since we have already proved the inequality

$$(x, Sx) \geq \lambda \|x\|^2,$$

we conclude that $x = 0$. This establishes the surjectivity of S. By what has already been shown, it follows that S^{-1} likewise is a bounded linear functional on H. By Riesz's theorem, there exists $v \in H$ with

$$\begin{aligned}
Tx &= (x, v) \\
&= (x, Sz) \quad \text{for a unique } z \in H, \text{ since } S \text{ is bijective} \\
&= B(x, z) = B(x, S^{-1}v).
\end{aligned}$$

Then $y = S^{-1}v$ satisfies our claim. $\qquad \square$

The Banach spaces that are important for us here are the L^p spaces: For $1 \leq p < \infty$, we put

$$L^p(\Omega) := \Big\{ u : \Omega \to \mathbb{R} \text{ measurable,}$$
$$\text{with } \|u\|_p := \|u\|_{L^p(\Omega)} := \Big[\int_\Omega |u|^p \, dx \Big]^{\frac{1}{p}} < \infty \Big\}$$

and

$$L^\infty(\Omega) := \Big\{ u : \Omega \to \mathbb{R} \text{ measurable, } \|u\|_{L^\infty(\Omega)} := \sup |u| < \infty \Big\}.$$

Here

$$\sup |u| := \inf\{k \in \mathbb{R} : \{x \in \Omega : |u(x)| > k\} \text{ is a null set}\}$$

is the essential supremum of $|u|$.

Occasionally, we shall also need the space

$$L^p_{\text{loc}}(\Omega) := \{u : \Omega \to \mathbb{R} \text{ measurable with } u \in L^p(\Omega') \quad \text{for all } \Omega' \subset\subset \Omega\},$$

$1 \leq p \leq \infty$.

In those constructions, one always identifies functions that differ on a null set. (This is necessary in order to guarantee (i) from Definition A.1.)

We recall the following facts:

Lemma A.1: *The space $L^p(\Omega)$ is complete with respect to $\|\cdot\|_p$, and hence is a Banach space, for $1 \leq p \leq \infty$. $L^2(\Omega)$ is a Hilbert space, with scalar product*

$$(u, v)_{L^2(\Omega)} := \int_\Omega u(x)v(x)dx.$$

Any sequence that converges with respect to $\|\cdot\|_p$ contains a subsequence that converges pointwise almost everywhere. For $1 \leq p < \infty$, $C^0(\Omega)$ is dense in $L^p(\Omega)$; i.e., for $u \in L^p(\Omega)$ and $\varepsilon > 0$, there exists $w \in C^0(\Omega)$ with

$$\|u - w\|_p < \varepsilon. \tag{A.3}$$

Hölder's inequality holds: If $u \in L^p(\Omega)$, $v \in L^q(\Omega)$, $1/p + 1/q = 1$, then

$$\int_\Omega uv \leq \|u\|_{L^p(\Omega)} \cdot \|v\|_{L^q(\Omega)}. \tag{A.4}$$

Inequality (A.4) follows from Young's inequality

$$ab \leq \frac{a^p}{p} + \frac{b^q}{q}, \quad \text{if } a, b \geq 0, \quad p, q > 1, \quad \frac{1}{p} + \frac{1}{q} = 1. \tag{A.5}$$

To demonstrate this, we put

$$A := \|u\|_p, \quad B := \|v\|_p,$$

and without loss of generality $A, B \neq 0$. With $a := \frac{|u(x)|}{A}$, $b := \frac{|v(x)|}{B}$, (A.5) then implies

$$\int \frac{|u(x)v(x)|}{AB} \leq \frac{1}{p}\frac{A^p}{A^p} + \frac{1}{q}\frac{B^q}{B^q} = 1,$$

i.e., (A.4).

Inductively, (A.4) yield that if $u_1 \in L^{p_1}, \ldots, u_m \in L^{p_m}$,

$$\sum_{i=1}^{m} \frac{1}{p_i} = 1,$$

then

$$\int_{\Omega} u_1 \cdots u_m \leq \|u_1\|_{L^{p_1}} \cdots \|u_m\|_{L^{p_m}}. \tag{A.6}$$

By Lemma A.1, for $1 \leq p < \infty$, $C^0(\Omega)$ is dense in $L^p(\Omega)$ with respect to the L^p-norm. We now wish to show that even $C^\infty(\Omega)$ is dense in $L^p(\Omega)$. For that purpose, we shall use so-called mollifiers, i.e., nonnegative functions ϱ from $C_0^\infty(B(0,1))$ with

$$\int \varrho \, dx = 1.$$

Here,

$$B(0,1) := \{x \in \mathbb{R}^d : |x| \leq 1\}.$$

The typical example is

$$\varrho(x) := \begin{cases} c \exp\left(\frac{1}{|x|^2-1}\right) & \text{for } |x| < 1, \\ 0 & \text{for } |x| \geq 1, \end{cases}$$

where c is chosen such that $\int \varrho \, dx = 1$. For $u \in L^p(\Omega)$, $h > 0$, we define the mollification of u as

$$u_h(x) := \frac{1}{h^d} \int_{\mathbb{R}^d} \varrho\left(\frac{x-y}{h}\right) u(y) dy, \tag{A.7}$$

where we have put $u(y) = 0$ for $y \in \mathbb{R}^d \setminus \Omega$. (We shall always use that convention in the sequel.) The important property of the mollification is

$$u_h \in C_0^\infty\left(\mathbb{R}^d\right).$$

Lemma A.2: *For $u \in C^0(\Omega)$, as $h \to 0$, u_h converges uniformly to u on any $\Omega' \subset\subset \Omega$.*

Proof:

$$u_h(x) = \frac{1}{h^d} \int_{|x-y| \leq h} \varrho\left(\frac{x-y}{h}\right) u(y) dy$$

$$= \int_{|z| \leq 1} \varrho(z) u(x-hz) dz \quad \text{with } z = \frac{x-y}{h}. \tag{A.8}$$

Thus, if $\Omega' \subset\subset \Omega$ and $2h < \text{dist}(\Omega', \partial\Omega)$, employing

$$u(x) = \int_{|z|\leq 1} \varrho(z)u(x)dz$$

(this follows from $\int_{|z|\leq 1} \varrho(z)dz = 1$), we obtain

$$\sup_{\Omega'} |u - u_h| \leq \sup_{x\in\Omega'} \int_{|z|\leq 1} \varrho(z)\,|u(x) - u(x - hz)|\,dz,$$

$$\leq \sup_{x\in\Omega'} \sup_{|z|\leq 1} |u(x) - u(x - hz)|.$$

Since u is uniformly continuous on the compact set $\{x : \mathrm{dist}(x, \Omega') \leq h\}$, it follows that

$$\sup_{\Omega'} |u - u_h| \to 0 \quad \text{for } h \to 0.$$

$\square$

Lemma A.3: *Let $u \in L^p(\Omega)$, $1 \leq p < \infty$. For $h \to 0$, we then have*

$$\|u - u_h\|_{L^p(\Omega)} \to 0.$$

Moreover, u_h converges to u pointwise almost everywhere (again putting $u = 0$ outside of Ω).

Proof: We use Hölder's inequality, writing in (A.8)

$$\varrho(z)u(x - hz) = \varrho(z)^{\frac{1}{q}}\varrho(z)^{\frac{1}{p}}u(x - hz)$$

with $1/p + 1/q = 1$, to obtain

$$|u_h(x)|^p \leq \left(\int_{|z|\leq 1} \varrho(z)dz\right)^{\frac{p}{q}} \int_{|z|\leq 1} \varrho(z)\,|u(x - hz)|^p\,dz$$

$$= \int_{|z|\leq 1} \varrho(z)\,|u(x - hz)|^p\,dz.$$

We choose a bounded Ω' with $\Omega \subset\subset \Omega'$.
    If $2h < \mathrm{dist}(\Omega, \partial\Omega')$, it follows that

$$\int_{\Omega} |u_h(x)|^p\,dx \leq \int_{\Omega}\int_{|z|\leq 1} \varrho(z)\,|u(x - hz)|^p\,dz\,dx$$

$$= \int_{|z|\leq 1} \left(\varrho(z)\int_{\Omega} |u(x - hz)|^p\,dx\right)dz \qquad \text{(A.9)}$$

$$\leq \int_{\Omega'} |u(y)|^p\,dy$$

(with the substitution $y = x - hz$). For $\varepsilon > 0$, we now choose $w \in C^0(\Omega')$ with

$$\|u - w\|_{L^p(\Omega')} < \varepsilon$$

(compare Lemma A.1). By Lemma A.2, for sufficiently small h,

$$\|w - w_h\|_{L^p(\Omega')} < \varepsilon.$$

Applying (A.9) to $u - w$, we now obtain

$$\int_\Omega |u_h(x) - w_h(x)|^p \, dx \leq \int_{\Omega'} |u(y) - w(y)|^p \, dy$$

and hence

$$\|u - u_h\|_{L^p(\Omega)} \leq \|u - w\|_{L^p(\Omega)} + \|w - w_h\|_{L^p(\Omega)} + \|u_h - w_h\|_{L^p(\Omega)}$$
$$\leq 2\varepsilon + \|u - w\|_{L^p(\Omega')} \leq 3\varepsilon.$$

Thus u_h converges to u with respect to $\|\cdot\|_p$. By Lemma A.1, a subsequence of u_h then converges to u pointwise almost everywhere. By a more refined reasoning, in fact the entire sequence u_h converges to u for $h \to 0$.   □

Remark: Mollifying kernels were introduced into PDE theory by K.O. Friedrichs. Therefore, they are often called "Friedrichs mollifiers".

For the proofs of Lemmas A.2 and A.3, we did not need the smoothness of ρ at all. Thus, these results also hold for other kernels, and in particular for

$$\sigma(x) = \begin{cases} \frac{1}{\omega_d} & \text{for } |x| \leq 1, \\ 0 & \text{otherwise.} \end{cases}$$

The corresponding convolution is

$$u_r(x) = \frac{1}{\omega_d \, r^d} \int_\Omega \sigma\left(\frac{x - y}{r}\right) u(y) \, dy = \frac{1}{|B(x,r)|} \int_{B(x,r)} u(y) \, dy =: \fint_{B(x,r)} u,$$

i.e., the average or mean integral of u on the ball $B(x, r)$. Thus, analogously to Lemma A.3, we obtain the following result:

Lemma A.4: *Let $u \in L^p(\Omega)$, $1 \leq p < \infty$. For $r \to 0$, then*

$$\fint_{B(x,r)} u$$

converges to $u(x)$, in the space $L^p(\Omega)$ as well as pointwise almost everywhere.

For a detailed presentation of all the results that have been stated here without proof, we refer to Jost [12].

References

1. L. Bers, M. Schechter, Elliptic equations, in: L. Bers, F. John, M. Schechter: *Partial Differential Equations*, pp. 131–299, Interscience, New York, 1964

2. D. Braess, *Finite Elemente*, Springer, 1997

3. I. Chavel, *Eigenvalues in Riemannian Geometry*, Academic Press, 1984

4. R. Courant, D. Hilbert, *Methoden der Mathematischen Physik*, Vols. I and II, reprinted 1968, Springer.
   Methods of mathematical physics, Wiley-Interscience, Vol. I, 1953, Vol. II, 1962, New York (the German and English versions do not coincide, but both are highly recommended)

5. L.C. Evans, *Partial Differential Equations*, Graduate Studies in Math. 19, AMS, 1998

6. A. Friedman, *Partial Differential Equations of Parabolic Type*, Prentice Hall, 1964

7. M. Giaquinta, *Multiple Integrals in the Calculus of Variations and Nonlinear Elliptic Systems*, Princeton Univ. Press, 1983

8. M. Giaquinta, *Introduction to Regularity Theory for Nonlinear Elliptic Systems*, Birkhäuser, 1993.

9. D. Gilbarg und N. Trudinger, *Elliptic partial differential equations of second order*, Springer, 1983.

10. F. John, *Partial Differential Equations*, Springer, 1982

11. J. Jost, *Nonpositive Curvature: Geometric and Analytic Aspects*, Birkhäuser, Basel, 1997

12. J. Jost, *Postmodern Analysis*, Springer, ^{3}2005

13. J. Jost, *Dynamical Systems*, Springer, 2005

14. J. Jost, X. Li-Jost, *Calculus of Variations*, Cambridge Univ. Press, 1998

15. A. Kolmogoroff, I. Petrovsky, N. Piscounoff, Étude de l' équation de la diffusion avec croissance de la quantité de la matière et son application à un problème biologique, *Moscow Univ.Bull.Math.*1, 1937, 1-25

16. O.A. Ladyzhenskya, V.A. Solonnikov, N.N. Ural'tseva, *Linear and Quasilinear Equations of Parabolic Type*, Amer.Math.Soc., 1968

17. O.A. Ladyzhenskya, N.N. Ural'tseva, *Linear and Quasilinear Elliptic Equations*, Nauka, Moskow, 1964 (in Russian); English translation: Academic Press, New York, 1968, 2nd Russian edition 1973

18. J. Moser, On Harnack's theorem for elliptic differential equations, *Comm. Pure Appl. Math.* 14 (1961), 577–591

19. J. Murray, *Mathematical Biology*, Springer, 1989

20. G. Strang, G. Fix, *An Analysis of the Finite Element Method*, Prentice Hall, Englewood Cliffs, N.J., 1973

21. J. Smoller, *Shock Waves and Reaction-Diffusion Equations*, Springer, 1983

22. M. Taylor, *Partial Differential Equations*, Vols. I–III, Springer, 1996

23. K. Yosida, *Functional Analysis*, Springer, 1978

24. E. Zeidler, *Nonlinear Functional Analysis and its Applications*, Vols. I-IV, Springer, 1984

Index of Notation

Index

Graduate Texts in Mathematics

(continued from page ii)